AN ENGINEERING APPROACH
TO OPTIMAL CONTROL
AND ESTIMATION THEORY

AN ENGINEERING APPROACH TO OPTIMAL CONTROL AND ESTIMATION THEORY

GEORGE M. SIOURIS

Air Force Institute of Technology
Wright-Patterson AFB, Ohio

A Wiley-Interscience Publication
JOHN WILEY & SONS, INC.

New York / Chichester / Brisbane / Toronto / Singapore

Library of Congress Cataloging in Publication Data:

Siouris, George M.
    An engineering approach to optimal control and estimation theory/
George M. Siouris.
        p.   cm.
    'A Wiley-Interscience publication."
    Includes index.
    ISBN 0-471-12126-6
    1. Automatic control.   2. Control theory.   I. Title.
TJ213.S474443   1996
629.8--dc20   95-6633

To Karin

CONTENTS

PREFACE

CHAPTER 1   INTRODUCTION AND HISTORICAL PERSPECTIVE   1

CHAPTER 2   MATHEMATICAL PRELIMINARIES   6

2.1	Random Variables	6
2.2	Expectations and Moments	10
	2.2.1 Statistical Averages (Means)	11
	2.2.2 Moments	12
	2.2.3 Conditional Mean	17
2.3	The Chebychev and Schwarz Inequalities	22
2.4	Covariance	23
2.5	The Autocorrelation Function and Power Spectral Density	25
2.6	Linear Systems	31
2.7	The Classical Wiener Filter	38
2.8	White Noise	42
2.9	System Input Error Models	47
	Problems	53

**CHAPTER 3   LINEAR REGRESSION; LEAST-SQUARES
                        AND MAXIMUM-LIKELIHOOD ESTIMATION   63**

3.1	Introduction	63
3.2	Simple Linear Regression	64

vii

3.3   Least-Squares Estimation   64
      3.3.1   Recursive Least-Squares Estimator   71
3.4   The Maximum-Likelihood Estimator (MLE)   73
      3.4.1   Recursive Maximum-Likelihood Estimator   80
3.5   Bayes Estimation   82
3.6   Concluding Remarks   83
      Problems   85

CHAPTER 4   THE KALMAN FILTER   **92**

4.1   The Continuous-Time Kalman Filter   93
4.2   Interpretation of the Kalman Filter   109
4.3   The Discrete-Time Kalman Filter   111
      4.3.1   Real-World Model Errors   136
4.4   The State Transition Matrix   141
4.5   Controllability and Observability   159
      4.5.1   Observers   163
4.6   Divergence   169
4.7   The $U-D$ Covariance Algorithm in Kalman Filters   171
4.8   The Extended Kalman Filter   190
4.9   Shaping Filters and Colored Noise   201
4.10 Concluding Remarks   204
      Problems   205

CHAPTER 5   LINEAR REGULATORS   **219**

5.1   Introduction   219
5.2   The Role of the Calculus of Variations in Optimal
      Control   221
5.3   The Continuous-Time Linear–Quadratic Regulator   232
5.4   The Discrete-Time Linear–Quadratic Regulator   253
5.5   Optimal Linear-Quadratic-Gaussian Regulators   256
      5.5.1   Introduction   256
      5.5.2   Properties of the LQG Regulators   257
5.6   Pontryagin's Minimum Principle   265
5.7   Dynamic Programming and the Hamilton–Jacobi Equation   274
5.8   Concluding Remarks   291
      Problems   294

**CHAPTER 6   COVARIANCE ANALYSIS AND SUBOPTIMAL
              FILTERING**   **302**

6.1   Covariance Analysis   302
      6.1.1   Concluding Remarks   319

6.2   Suboptimal Filtering   320
     6.2.1   Concluding Remarks   324
     Problems   326

CHAPTER 7   THE $\alpha-\beta-\gamma$ TRACKING FILTERS   **329**

7.1   Introduction   329
7.2   The α-Filter   330
7.3   The $\alpha-\beta$ Tracking Filter   331
7.4   The $\alpha-\beta-\gamma$ Tracking Filter   337
7.5   Concluding Remarks   339
     Problems   340

CHAPTER 8   DECENTRALIZED KALMAN FILTERS   **347**

8.1   Introduction   347
8.2   Filter Mechanizations and Architectures   348
8.3   Parallel Kalman Filter Equations   350
8.4   Decentralized Kalman Filtering Architectures Using Reduced-Order Models   354
8.5   Probability Fusion Equations   355
8.6   Concluding Remarks   357
     Problems   358

REFERENCES   **365**

APPENDIX A   MATRIX OPERATIONS AND ANALYSIS   **370**

A.1.   Introduction   370
A.2.   Basic Concepts   370
     A.2.1   Matrix Algebra   380
     A.2.2   The Eigenvalue Problem   383
     A.2.3   Quadratic Forms   386
     A.2.4   The Matrix Inversion Lemma   387
     References   389

APPENDIX B   MATRIX LIBRARIES   **390**

     References   401

INDEX   **403**

PREFACE

Optimal control and estimation theory has grown rapidly and at the same time advanced significantly in the last three decades. During this time, many books and research papers on optimal control theory, on various levels of sophistication, have been published. Optimal control theory is playing an increasingly important role in the design of modern systems. More specifically, control problems play an important role in aerospace, as well as in other applications, where for example, temperature, pressure, and other variables must be kept at desired values regardless of disturbances. In particular, the federal government is funding a multimillion dollar effort to develop intelligent vehicle highway systems (IVHS) in the next decade.

Historically, the development of control theory in engineering emphasized stability. The design of optimal control and stabilization systems, the determination of optimal flight paths (e.g., the optimization in flight mechanics), and the calculation of orbital transfers have a common mathematical foundation in the calculus of variations. A distinction between *classical* and *modern* control theory is often made in the control community. For example, flight controllers, autopilots, and stability augmentation systems for manned aircraft are commonly designed using linearized analyses. In particular, feedback gains or adjustable parameters are manipulated in order to obtain satisfactory system transient response to control inputs and gust inputs. These linear systems, which are subject to deterministic inputs, are frequently optimized with respect to common transient response criteria such as rise time, peak overshoot, settling time, bandwidth, and so on, and which in turn depend upon the locations of the poles and zeros of the system transfer function. On the other hand, the design of a modern optimal controller requires the selection of a performance criterion.

Kalman and Bucy have investigated optimal controllers for linear systems and obtained solutions to the combined optimal control and filtering problem. In many aeronautical applications, for example, the selection of a performance criterion is based on real physical considerations such as payload, final velocity, etc. Consequently, the major approaches to optimal control are minimizing some performance index, depending on the system error and time for completely linear systems; minimizing the root-mean-square error for statistical inputs; and searching for the maximum or minimum of a function. Optimization of linear systems with bounded controls and limited control effort is important to the control engineer because the linearized versions of many physical systems can be easily forced into this general formulation.

The methods of modern control theory have been developed in many instances by pure and applied mathematicians, with the result that much of the written presentation is formal and quite inaccessible to most control engineers. Therefore, in this book I have tried to keep the theory to a minimum while placing emphasis on applications. Furthermore, even though the techniques of optimal control theory provide particularly elegant mathematical solutions to many problems, they do have shortcomings from a practical point of view. In many practical problems, for example, the optimum solution lies in a region where the performance criterion is fairly flat, rather than at a sharp minimum, so that the increased benefits in moving from a near-optimum to an optimum solution may be quite small. Nonlinearities occur in all physical systems, and an understanding of the effects that they have on various types of signals is important to engineers in many fields.

Although, as mentioned above, during the last thirty years or so a great deal of research has been done on the subject of optimal control theory, and numerous very good books have been written on the subject, I have felt that several topics of great importance to practicing engineers as well as students have not appeared in a systematic form in a book. Above all, the field is so vast that no single book or paper now available to the student or engineer can possibly give him an adequate picture of the principal results. This book is intended to fill the need especially by the practicing engineer, for a single source of information on major aspects of the subject.

My interest in optimal control theory was acquired and nurtured during my many years of research and development in industry, the government, and teaching. Therefore, my intent is to serve a broad spectrum of users, from first-year graduate-level students to experienced engineers, scientists, and engineering managers.

ORGANIZATION OF THE TEXT

The structure of the book's organization is an essential part of the presentation. The material of the book is divided into eight chapters and two appendices. Chapter 1 is an introduction, giving a historical perspective and the evolution of optimal control and estimation theory. Chapter 2 presents an overview of the basic mathematical concepts needed for an understanding of the work that follows. The topics covered in Chapter 2 include random variables, moments, covariance, the autocorrelation function and power spectral density, linear systems, the classical Wiener filter, white noise, and system input error models. Chapter 3 is concerned with linear regression, least-squares, and maximum-likelihood estimation. Among the topics covered in this chapter are simple linear regression, least-squares estimation, the recursive least-squares estimator, maximum-likelihood estimation, and the recursive maximum-likelihood estimator. The material of Chapter 4 is devoted to the development of the Kalman filter. Topics covered in this chapter include continuous-time and discrete-time Kalman filters, real-world model errors, the state transition matrix, controllability and observability, divergence, the $U-D$ covariance algorithms, and the extended Kalman filter.

Chapter 5 is devoted to linear regulators and includes a detailed discussion of the role of the calculus of variations in optimal control, the continuous-time and discrete-time linear quadratic regulator (LQR), the optimal linear quadratic Gaussian (LQG) regulator, Pontryagin's minimum principle, and dynamic programming and the Hamilton–Jacobi–Bellman equation. Chapter 6 may be considered as a natural extension of Chapter 4, and deals with covariance analysis and suboptimal filtering. Chapter 7 discusses the $\alpha-\beta-\gamma$ tracking filters. The last chapter, Chapter 8, discusses decentralized Kalman filters.

The book concludes with two appendices. Appendix A reviews matrix operations and analysis. This appendix has been added as a review for the interested reader, since modern optimal control theory leans heavily on matrix algebra. The topics covered in Appendix A are basic matrix concepts, matrix algebra, the eigenvalue problem, quadratic forms, and the matrix inversion lemma. Appendix B presents several matrix subroutines, which may be of help to the student or engineer.

The mathematical background assumed of the reader includes concepts of elementary probability theory, statistics, linear system theory, and some familiarity with classical control theory. Several illustrative examples have been included in the text that show in detail how principles discussed are applied. The examples chosen are sufficiently practical to give the reader a feeling of confidence in mastering and applying the concepts of optimal control theory. Finally, as in most textbooks, problems have been added at the end of each chapter. It is recommended that the student and/or engineer read and attempt to solve these problems. The problems have been selected with care and for the most part supplement the theory presented in the text.

Dayton, OH GEORGE M. SIOURIS
September 1995

ACKNOWLEDGMENTS

The problem of giving proper credit is a vexing one for an author. Nevertheless, preparation of this book has left me indebted to many people. I am indebted to many writers and colleagues whose work has deepened my understanding of modern control and estimation theory. Grateful acknowledgment is due to Professor William M. Brown, Head, Department of Electrical and Computer Engineering, Air Force Institute of Technology, Wright-Patterson AFB, Ohio, for his guidance and readiness to help at any time. Also I am very grateful for the advice and encouragement I received from Professor Jang Gyu Lee, Department of Control and Instrumentation Engineering, Seoul National University, Seoul, Republic of Korea, Professor Victor A. Skormin, Department of Electrical Engineering, Thomas J. Watson School of Engineering and Applied Science, Binghamton University (SUNY), Binghamton, New York, and to Dr. Guanrong Chen, Associate Professor, Department of Electrical Engineering, University of Houston, Houston, Texas. The invaluable comments and suggestions of Dr. Kuo-Chu Chang, Associate Professor, Systems Engineering Department, George Mason University, Fairfax, Virginia, and Dr. Shozo Mori of Tiburon Systems, Inc., San Jose, California, have been of considerable assistance in the preparation of the final manuscript. Dr. Chang made several corrections and improvements in portions of the manuscript. Dr. Mori read the entire manuscript, pointed out various errors, and offered constructive criticisms of the overall presentation of the text. My thanks also go to Dr. Stanley Shinners of the Unisys Corporation, Great Neck, New York, and Adjunct Professor, Department of Electrical Engineering, The Cooper Union for the Advancement of Science and Art, and to Dr. R. Craig Coulter of the Carnegie Mellon University, Robotics Institute, Pittsburgh, Pennsylvania. The enthusiastic support, and suggestions provided by both these gentlemen have materially increased its value as a text. The faults that remain are of course the responsibility of the author. Any errors that remain, I will be grateful to hear of. Finally, but perhaps most importantly, I wish to acknowledge the patience, understanding, and support of my family during the preparation of the book.

G.M.S.

CHAPTER 1

INTRODUCTION AND HISTORICAL PERSPECTIVE

Many modern complex systems may be classed as *estimation systems*, combining several sources of (often redundant) data in order to arrive at an estimate of some unknown parameters. Among such systems are terrestrial or space navigators for estimating such parameters as position, velocity, and attitude, fire-control systems for estimating impact point, and radar systems for estimating position and velocity. Estimation theory is the application of mathematical analysis to the problem of extracting information from observational data. The application contexts can be deterministic or probabilistic, and the resulting estimates are required to have some optimality and reliability properties.

Estimation is often characterized as prediction, filtering, or smoothing, depending on the intended objectives and the available observational information. Prediction usually implies the extension in some manner of the domain of validity of the information. Filtering usually refers to the extraction of the true signal from the observations. Smoothing usually implies the elimination of some noisy or useless component in the observed data. Optimal estimation always guarantees closed-loop system stability even in the event of high estimator gains. However, in classical design, for a nonminimum phase system, the closed-loop system is unstable for high controller gains.

One of the most widely used estimation algorithms is the *Kalman filter*, an algorithm which generates estimates of variables of the system being controlled by processing available sensor measurements. The Kalman filter theory, in its various forms, has become a fundamental tool for analyzing and solving a broad class of estimation problems. The Kalman filter equations, or more precisely the Kalman algorithm, save computer memory by updating the estimate of the signals between measurement times without requiring storage of all the past measurements. That is, the filter is flexible in that it can handle measurements one at a time or in batches. In short, the Kalman filter is an optimal recursive data-processing algorithm.

An important feature of the Kalman filter is its generation of a system error analysis, independent of any data inputs. Furthermore, the filter performs this error analysis in a very efficient way. One prerequisite is that the filter requires a model of the system dynamics and *a priori* noise statistics involved. Another important feature of the filter is that it includes failure detection capability. That is, a bad-data rejection technique has been developed which compares the measurement residual magnitude with its standard deviation as computed from the Kalman filter measurement update algorithm. If the residual magnitude exceeded n times the standard deviation, the measurement was rejected. The value of n used was 3, corresponding to a 3σ residual magnitude test. For example, in aided inertial navigation systems when all measuring devices (e.g., Doppler radar, the Global Positioning System, TACAN, Omega, etc.) onboard a vehicle are operating correctly, the different signals entering the optimal gain matrix [see Eqs. (4.13) and (4.20)] should be white sequences with zero mean and predictable covariance. If this condition is not met, a measuring-device failure can be detected and isolated. This is done in real time by exercising a model of the system and using the difference between the model predictions and the measurements.

At this point, it is appropriate to define and/or explain what is meant by the term *filter*. Simply stated, the algorithm is called a filter if it has the capability of ignoring, or filtering out, noise in the measured signals. Then, in conjunction with a closed-loop algorithm, the control signals are fine-tuned to bring the estimate into agreement with nominal performance, which is stored in a computer's memory. The purpose of the filter is to reconstruct the states which are not measured, and to minimize the process and measurement noise influence. Also, a filter is thought as a computer program in a central processor.

The Kalman filter theory is well developed and has found wide application in the filtering of linear or linearized systems, due primarily to its sequential nature, which is ideally suited to digital computers. Specifically, Kalman filtering techniques have seen widespread application in aerospace navigation, guidance, and control—the field where they were first used (viz., NASA's early work on the manned lunar mission, and later in the early sixties the development of the navigation systems for the Apollo and the Lockheed C-5A aircraft programs). These techniques were rapidly adapted in such diverse fields as orbit determination, radar tracking, ship motion, mobile robotics, the automobile industry (as vehicles begin to incorporate smart navigation packages), chemical process control, natural gamma-ray spectroscopy in oil- and gas-well exploration, measurement of instantaneous flow rates and estimation and prediction of unmeasurable variables in industrial processes, on-line failure detection in nuclear plant instrumentation, and power station control systems. Engineers engaged in the aforementioned areas as well as mathematicians will find Kalman filtering techniques an indispensable tool.

A major thrust of Kalman mechanizations and architectures is the use of parallel, partitioned, or decentralized versions of the standard Kalman filter. The standard Kalman filter provides the best sequential linear unbiased estimate (globally optimal estimate) when the noise processes are jointly Gaussian. Thus, the stochastic processes involved are often modeled as Gaussian ones to simplify the mathematical analysis of the corresponding estimation problems. In such a simplified case, the three best-known estimation methods—the *least-squares, maximum-likelihood,* and *Bayesian* methods—give almost identical estimates, even though the associated reliabilities may be different.

One of the prime contributing factors to the success of the present-day estimation and control theory is the ready availability of high-speed, large-memory digital computers for solving the equations.

The modern theory of estimation has its roots in the early works of A. N. Kolmogorov and N. Wiener [72]. Kolmogorov in 1941 and Wiener in 1942 independently developed and formulated the important fundamental estimation theory of *linear minimum mean-square* estimation. In particular, the filtering of continuous-time signals was characterized by the solution to the classical Wiener–Kolmogorov problem, as it came to be called later, whereas linear regression techniques based on weighted least-squares or maximum-likelihood criteria were characteristic of the treatment of discrete-time filtering problems. However, the Wiener–Kolmogorov solution, expressed as an integral equation, was only tractable for stationary processes until the early 1960s, when R. E. Kalman and later Kalman and R. Bucy revolutionized the field with their now classical papers [41, 42]. The basis of the concept was attributed by Kalman [41] to the ideas of orthogonality and *wide sense* conditional expectation discussed by J. C. Doob [27]. The results of the Kalman–Bucy filter were quickly applied to large classes of linear systems, and attempts were made at extending the results to nonlinear systems. Several authors presented a series of results to a variety of *extended Kalman filters.* These techniques were largely aimed at specific problems or classes of problems, but closed-form expressions for the error bound were not found.

Furthermore, as the filter's use gained in popularity in the scientific community, the problems of implementation on small spaceborne and airborne computers led to a *square-root* formulation to overcome the numerical difficulties associated with computer word length. The work that led to this new formulation is also discussed in this book. Square-root filtering, in one form or another, was developed in the early 1960s and can be found in the works of R. H. Battin [5], J. E. Potter [56], and S. F. Schmidt [60]. Later researches in the square-root Kalman filtering method can be found in the works of L. A. McGee and S. F. Schmidt [50, 63], G. L. Bierman [9], and N. A. Carlson [13]. Based on the work of these researchers, two different types of square-root filters have been developed. The first may be regarded as a factorization of the standard Kalman filter algorithm; it basically leads to the square-root error covariance matrix. The second involves the square root of the information matrix, which is defined as the inverse of the error covariance matrix. The desire to determine estimators that were optimal led to the more fundamental problem of determining the conditional probability. The work of Kushner [45] and Bucy [42] addressed itself to this problem for continuous-time systems. Their approach resulted in the conditional density function being expressed as either a ratio of integrals in function space or the solution to a partial differential equation similar to the *Fokker–Planck* equation. Representation of this conditional density function gave rise to an approximation problem complicated by its changing form. Thus, an approximation such as quasimoments, based on the form of the initial distribution, may become a poor choice as time progresses.

The techniques used for nonlinear problems all have limitations. Even the determination of the conditional density function becomes suboptimal due to the necessity of using approximations to describe the density function. Still other researchers formulated the problem in *Hilbert space* and generated a set of integral equations for which the kernel had to be determined. It then concentrated on the solution for this kernel by gradient techniques. Notice that from Hilbert space theory, the optimum is achieved when the error is orthogonal to the linear manifold generated by the finite polynomials of the observer. The extension to optimal control is accomplished by exploiting, as we shall see later, the duality of linear estimation and control, which is derived from duality concepts in mathematical programming.

With the above preliminaries in mind, we can now state in more precise terms the function of the Kalman filter. The Kalman filter is an optimal recursive data-processing algorithm, which generates estimates of the variables (or states) of the system being controlled by processing all available measurements. This is done in real time by utilizing a model of the system and using the difference between the model predictions and measurements. Specifically, the filter operates on the system errors and processes all available measurements, regardless of their precision to estimate the current values of the variables of interest by using the following facts: (1) knowledge of the system model and measurement-device dynamics, (2) statistical description of system noises and uncertainties, and (3) information about the initial conditions of the variables. In essence, the main function of the Kalman filter is to estimate the state vector using system sensors and measurement data corrupted by noise. The Kalman filter algorithm concerns itself with two types of estimation problems: (1) filtering (update), and (2) prediction (propagation). When the time at which an estimate of the state vector is desired coincides with the last measurement point, the estimation problem is known as filtering. Stated another way, filtering refers to estimating the state vector at the current time, based upon past measurements. When the time of interest of state-vector estimation occurs after the last measurement, the estimation problem is termed *prediction* [58].

The application of Kalman filtering theory requires the definition of a linear mathematical model of the system. With regard to the system model, a distinction is made between a *truth model*, sometimes referred to as a *real-world model*, and the *filter model*. A truth model is a description of the system dynamics and a statistical model of the errors in the system. It can represent the best available model of the true system or be hypothesized to test the sensitivity of a particular system design to modeling errors. The filter model, on the other hand, is the model from which the Kalman gains are determined. Moreover, the filter model is in general of lower order than a truth model. When the ratio of order of the truth model to that of the filter model is one, there is a perfect match between the two models. If the ratio is less than one, the filter's noise covariance matrices must be adjusted accordingly. With regard to the covariance matrix P, it is noted that if very accurate measurements are processed by the filter, the covariance matrix will become so small that additional measurements would be ignored by the filter. When this happens, only very small corrections to the estimated state will diverge from the true state. This problem is due to modeling errors, and can be corrected by using pseudonoise in the time update equations. It should be pointed out, however, that the Kalman filtering algorithms derived from complex system models can impose extremely large storage and processing requirements. Specifically, the Kalman filter, if not properly optimized, can require a large word count and excessive execution time. The implementation of a specific filter demands tradeoff between core usage and duty cycle. For example, a purely recursive loop, implementation of the matrix element computations minimizes core use, but requires more execution time because of the automatic service of zero elements. Computation of stored equations for individual matrix elements that are nonzero reduces the execution time but requires a higher word count. For this reason, the suboptimal, or simplified, filter models provide performance almost as good as the optimum filter based on the exact model. In the traditional suboptimal Kalman filter, two simulation techniques are commonly used to study the effect of uncertainties or perturbations within the system model when the system truth model is present. These two techniques are (1) covariance analysis and (2) Monte Carlo simulation (see Chapter 6). The largest sources of Kalman filter estimation error are unmodeled errors, that is, the actual system (or plant, as it is also called) differs from that being modeled by the filter.

A final note is appropriate at this point. Since there is no uniformity in the literature on optimal control and estimation theory concerning the mathematical symbols used, the reader will notice that different symbols are used to express the same concept or principle. This has been done to acquaint the reader with the various notations that will be encountered in the literature. Before we proceed with the Kalman filter and its solution, we will briefly review in Chapter 2 the mathematical concepts that are required in order to understand the work that follows.

CHAPTER 2

MATHEMATICAL PRELIMINARIES

This chapter presents certain mathematical concepts that are necessary in order to understand the work that follows. The treatment is by no means complete. For more details, the reader is referred to references [25, 27, 53, and 64] at the end of this book.

2.1 RANDOM VARIABLES

We begin this section by defining a random variable (r.v.).

Definition: A real-valued function $x(s)$ defined on a sampled space of points s will is called a *random variable* if, for every real number a, the set of points s for which $x(s) \leqslant a$ is one class of admissible sets for which a probability is defined.

From this definition, let now the real random variable $x(s)$ be such that the range of x is the real line (that is, $-\infty \leqslant x \leqslant +\infty$). Furthermore, let X be a point on the real axis. Then the function of X whose value is the probability $P(x \leqslant X)$ that the random variable x is less than or equal to X is called the *probability distribution function* (PDF) of the random variable x. The properties of the PDF can be summarized as follows:

1. The PDF exists for both discrete and continuous (see below) random variables, and it has its values between 0 and 1.

2. It is a nonnegative, continuous to the left, and nondecreasing function of the real variable x. The range is

$$P(x \leqslant -\infty) = 0 \quad \text{and} \quad P(x \leqslant +\infty) = 1.$$

3. If a and b are two real numbers such that $a < b$, and the random variable x falls in the interval $a < x \leqslant b$, then

$$P(x \leqslant b) - P(x \leqslant a) = P(a < x \leqslant b) \geqslant 0.$$

We note here that random variables are commonly defined in terms of a *Borel field*. A Borel field is a class of sets, including the empty set $\varnothing$ and space Ω, which is closed under all countable unions and intersections of its sets. A random variable is actually a function in a conventional sense. For example, a function $f(y)$ assigns values to each y according to certain rules. Similarly, a random variable x assigns numerical values (real numbers) to each sample point. Therefore, in this sense the random variable is a function of sample points and assigns a real number to each sample point according to some rule.

Random variables are of the discrete and continuous types.

Discrete Case: A random variable x is called a *discrete random variable* if x can take only a finite number of values in any finite interval. For example, the complete set of probabilities $P(x_k)$ associated with the possible values x_k of x is called the probability distribution of the discrete random variable x and can be expressed by the relation

$$P(x \leqslant X) = \sum_{x_k \leqslant X} P(x_k).$$

Continuous Case: A random variable for which the probability distribution function is everywhere continuous will be called *continuous random variable*.

The above discussion can be extended to the case of n random variables. A function which assigns a vector to each point of a sample space is called a *vector random variable*. In particular, the random variables $x_1, x_2, \ldots, x_n$ will map every element s of S (i.e., $s \in S$) in the probability space onto a point in the n-dimensional Euclidean space R^n, whose elements are vectors $\mathbf{A} = \| a_1, a_2, \ldots, a_n \|$ with real components, and in which each vector $\mathbf{A}$ has a *length* $\| \mathbf{A} \|$ given by

$$\| \mathbf{A} \| = (\mathbf{A}\mathbf{A}^T)^{1/2} = (a_1^2 + a_2^2 + \cdots + a_n^2)^{1/2}$$

(see also Section 4.4 on the Euclidean norm).

We will now define a *random process*, which is an extension of the concept of a random variable. In particular, a random process is a collection of random variables described by a set of functions whose exact values at any future instant of time are unknown. However, the statistical properties of the random variable can be described by a joint probability distribution. Therefore, at any given point in time, there is a definite probability that the random variable will lie between specified values. Next, we note that a stationary random process is one whose associated probability distribution functions do not change with time. Also, a stationary process may or may not be an ergodic process (an ergodic process is defined as one whose complete statistics can be determined from any one sample function). A random process is also called a *stochastic process*.

For a random variable whose probability distribution function P is not only continuous but also differentiable with a continuous derivative everywhere, except at a discrete set of points, we define the *probability density function $p(x)$* as the derivative of the probability distribution function:

$$p(x) = \frac{dP(x \leqslant X)}{dX}, \tag{2.1}$$

which must exist for all x. This equation can be expressed alternatively as

$$P(x \leqslant X) = \int_{-\infty}^{x} p(x)\,dx. \tag{2.2}$$

The probability density function must be a nonnegative function. That is, $p(x) \geqslant 0$. In other words, $P(x \leqslant X)$ is monotone nondecreasing. Other properties of $p(x)$ can be derived from Eq. (2.1) as follows [25, 64]:

$$P(a < x \leqslant b) = P(b) - P(a) = \int_{a}^{b} p(x)\,dx, \tag{2.3a}$$

and since $P(\infty) = 1$, it follows that

$$\int_{-\infty}^{\infty} p(x)\,dx = 1. \tag{2.3b}$$

This is also obvious from the fact that the integral in Eq. (2.3b) represents the probability of observing x in the interval $(-\infty, \infty)$, which is a certainty. Of central importance in the study of optimal control and estimation theory as well as in applications is the *Gaussian* (or *normal*) distribution. A random variable x is Gaussian, or *normally distributed*, if its probability density function $p(x)$ is of the form

$$p(x) = \frac{1}{\sqrt{2\pi}\sigma} \exp\left[-\frac{(x-m)^2}{2\sigma^2} \right], \qquad -\infty < x < \infty, \tag{2.4a}$$

where m and σ are two constant parameters with $\sigma > 0$. The corresponding probability distribution function $P(x)$ is given by [25, 53, 64]

$$P(x) = \frac{1}{\sqrt{2\pi}\sigma} \int_{-\infty}^{x} \exp\left[-\frac{(u-m)^2}{2\sigma^2}\right] du, \qquad -\infty < x < \infty. \qquad (2.4b)$$

The probability density function $p(x)$ and the corresponding distribution function $P(x)$ for $m = 0$ and $\sigma = 1$ are plotted in Figure. 2.1.

It is noted that the graph of $p(x)$ is the well-known bell-shaped curve, symmetrical about the origin. The engineering importance of Gaussian processes stems from the fact many physical processes are approximately Gaussian.

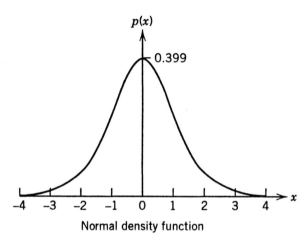

Normal density function

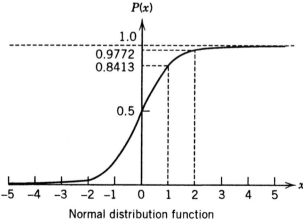

Normal distribution function

Figure 2.1 Normal (or Gaussian) probability density and distribution function of x with $m = 0$ and $\sigma = 1$.

The discussion above centered on a single random variable. For two or more variables, likewise, probability density functions become important when the random variables are continuous. The *joint probability density function* of two random variables x and y can be defined by the partial derivative

$$p(X, Y) = \frac{\partial^2}{\partial X \, \partial Y} P(x \leqslant X, \ y \leqslant Y), \tag{2.5}$$

and since $P(x, y)$ is monotone nondecreasing in both x and y, $p(x, y)$ is non-negative for all x and y, and we have from Eq. (2.5)

$$P(x \leqslant X, y \leqslant Y) = \int_{-\infty}^{Y} \int_{-\infty}^{X} p(x, y) \, dx \, dy. \tag{2.6}$$

Finally, when the region in question is the entire sample space (i.e., the entire x–y plane), we obtain, as in Eq. (2.4),

$$\int_{-\infty}^{\infty} \int_{-\infty}^{\infty} p(x, y) \, dx \, dy = 1. \tag{2.7}$$

If two or more random variables are involved, we have the so-called *multivariate normal distribution*. In particular, two random variables x and y are said to be jointly normal if their joint density function takes the form [64]

$$p(x, y) = \frac{1}{2\pi\sigma_x\sigma_y\sqrt{1-\rho^2}} \exp\left\{ -\frac{1}{2(1-\rho^2)} \right.$$
$$\left. \times \left[\left(\frac{x-m_x}{\sigma_x}\right)^2 - 2\rho\frac{(x-m_x)(y-m_y)}{\sigma_x\sigma_y} + \left(\frac{y-m_y}{\sigma_y}\right)^2 \right] \right\},$$
$$-\infty < (x, y) < \infty, \tag{2.8}$$

where ρ is the *correlation coefficient* and has the property $|\rho| \leqslant 1$ (see also Section 2.4). For the two random variables x and y, Eq. (2.8) describes a *bivariate normal distribution*. The reader will note here that the bivariate normal distribution involves five parameters, that is, m_x, m_y, $\sigma_x(>0)$, $\sigma_y(>0)$, and ρ.

2.2   EXPECTATIONS AND MOMENTS

In the preceding section we introduced the concept of random variables and their associated characteristics. In this section we will discuss the important concepts of expectations and moments, which are central in the study of optimal control and estimation theory.

2.2.1 Statistical Average (Means)

The concept of averages is of great importance in random processes. We shall now define the *mean* (or average) of a random variable. Consider a random variable x which can assume n values $x_1, x_2, \ldots, x_n$. Let an experiment, represented by x, be repeated N times ($N \rightarrow \infty$), and let $m_1, m_2, \ldots, m_n$ be the number of trials favorable to outcomes $x_1, x_2, \ldots, x_n$. Then the mean value of x, according to the traditional concept, is given by

$$\bar{x} = \frac{1}{N}(m_1 x_1 + m_2 x_2 + \cdots + m_n x_n)$$

$$= \frac{m_1}{N} x_1 + \frac{m_2}{N} x_2 + \cdots + \frac{m_n}{N} x_n \tag{2.9a}$$

In the limit as $N \rightarrow \infty$, the ratio m_i/N tends to $p_x(x_i)$ according to the relative-frequency definition of probability. Hence,

$$\bar{x} = \sum_{i=1}^{n} x_i p_x(x_i). \tag{2.9b}$$

Equation (2.9) is a meaningful definition of the mean of a random variable. The mean value is also the *expected value* (or *first moment*) of the random variable x and is denoted by the expectation operator $\mathscr{E}\{x\}$. Thus, for the discrete case,

$$\bar{x} = \mathscr{E}\{x\} = \sum_{i} x_i p_x(x_i). \tag{2.10a}$$

If the random variable is continuous, we have

$$\bar{x} = \mathscr{E}\{x\} = \int_{-\infty}^{\infty} x\, p_x(x)\, dx. \tag{2.10b}$$

If the estimate $\hat{x}$ of a parameter vector x, based on subsequent observations, satisfies

$$\mathscr{E}\{\hat{x}\} = \mathscr{E}\{x\} = \bar{x},$$

then the estimate $\hat{x}$ is called *unbiased*. An *efficient estimate* $\hat{x}$ of x is the unbiased estimate of x with minimum variance, that is,

$$\sigma_{\hat{x}}^2 = \mathscr{E}\{\|\hat{x} - x\|^2\} \leq \mathscr{E}\{\|y - x\|^2\} = \sigma_y^2$$

for all other estimates **y** of **x**. (The symbol $\|\cdot\|$ denotes the *norm* of a vector. For more details, the reader is referred to Section 4.4.)

2.2.2 Moments

The n th moment of a random variable x is defined as the expected value of the n th power of x. Thus, the n th *moment* is defined as

$$\mathscr{E}\{x^n\} = \int_{-\infty}^{\infty} x^n p_x(x)\,dx. \tag{2.11a}$$

The n th *central moment* of the variable x is its moment about its mean value m, and is given by

$$\mathscr{E}\{(x - m)^n\} = \int_{-\infty}^{\infty} (x - m)^n p_x(x)\,dx, \tag{2.11b}$$

which can also be written as

$$\mathscr{E}\{(x - \mathscr{E}\{x\})^n\} \triangleq \int_{-\infty}^{\infty} (x - \mathscr{E}\{x\})^n p(x)\,dx. \tag{2.11c}$$

The *first moment* is the mean and is easily obtained from Eq. (2.11a) with $n = 1$ as follows:

$$\mathscr{E}\{x\} = \int_{-\infty}^{\infty} x p(x)\,dx = \frac{1}{\sqrt{2\pi}\,\sigma} \int_{-\infty}^{\infty} x \exp\left[-\frac{(x - m)^2}{2\sigma^2}\right] dx$$

$$= m. \tag{2.11d}$$

The *second central moment*, which is of particular interest, is commonly denoted by the symbol σ^2, and is given the special name *variance* (or *dispersion*). Therefore, the variance is computed as follows:

$$\mathrm{var}\{x\} = \mathscr{E}\{(x - \mathscr{E}\{x\})^2\}$$

$$= \mathscr{E}\{x^2\} - \mathscr{E}^2\{x\} = \sigma^2. \tag{2.11e}$$

Equation (2.11e) says that the variance of a random variable is the mean square minus the square of its mean.

    An important theorem in connection with Gaussian distributions and central moments is the well-known *central limit theorem*, which may be stated as follows [25, 53]:

    For statistically independent samples, the probability distribution of the sample mean tends to become Gaussian as the number of statistically independent samples is increased without limit, regardless of the probability distribution of the random variable or process being sampled as long as it has a finite mean and a finite variance.

Summarized below are some examples of common continuous distributions.

1. Uniform. A continuous random variable x has a *uniform distribution* over an interval a to b $(b > a)$ if it is equally likely to take on any value in this interval. The probability density function of x is constant over the interval (a, b) and has the form (Figure 2.2)

$$p(x) = \begin{cases} \dfrac{1}{b-a}, & a \leqslant x \leqslant b, \\ 0 & \text{otherwise.} \end{cases}$$

The probability distribution function is given by

$$P(x) = \begin{cases} 0, & x < a, \\ \dfrac{x-a}{b-a}, & a \leqslant x \leqslant b, \\ 1, & x > b. \end{cases}$$

The mean and variance of x are found as follows:

$$m_x = \int_a^b x p(x)\, dx = \frac{1}{b-a} \int_a^b x\, dx = \frac{a+b}{2},$$

$$\sigma_x^2 = \frac{1}{b-a} \int_a^b \left(x - \frac{a+b}{2} \right)^2 dx = \frac{(b-a)^2}{12}.$$

2. Gaussian or Normal. The *Gaussian* or *normal* distribution was studied first in the eighteenth century, when scientists observed an astonishing degree

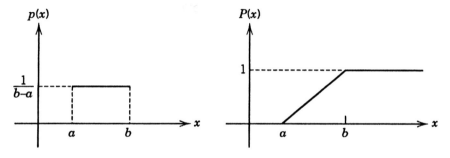

Figure 2.2 Uniform probability density and distribution function.

of regularity in errors of measurement. Specifically, they found that the patterns (distributions) they observed were closely approximated by a continuous curve which they referred to as the *normal curve of errors* and attributed to the laws of chance. As we saw earlier, the equation of the normal probability density is

$$p(x) = \frac{1}{\sqrt{2\pi}\,\sigma} \exp\left[-\frac{(x-m)^2}{2\sigma^2}\right], \qquad -\infty < x < \infty,$$

where m and σ are two parameters with $\sigma > 0$. The corresponding probability distribution function is given by

$$P(x) = \frac{1}{\sqrt{2\pi}\,\sigma} \int_{-\infty}^{x} \exp\left[-\frac{(u-m)^2}{2\sigma^2}\right] du, \qquad -\infty < x < \infty.$$

The mean and variance are respectively

$$\mathscr{E}\{x\} = m,$$

$$\mathrm{var}\{x\} = \sigma^2.$$

Frequently, use is made of the *error function* $\mathrm{erf}(\alpha)$ and its complement $\mathrm{erfc}(\alpha)$. They are commonly defined as

$$\mathrm{erf}(\alpha) \triangleq \frac{1}{\sqrt{2\pi}} \int_{-\infty}^{\alpha} e^{-x^2/2}\, dx$$

and

$$\mathrm{erfc}(\alpha) \triangleq \frac{1}{\sqrt{2\pi}} \int_{\alpha}^{\infty} e^{-x^2/2}\, dx.$$

Since these are complementary functions, it is obvious that

$$\mathrm{erf}(\alpha) + \mathrm{erfc}(\alpha) = 1.$$

A useful approximation for $\mathrm{erfc}(\alpha)$ is given by

$$\mathrm{erfc}(\alpha) \approx \frac{1}{\alpha\sqrt{2\pi}}\left(1 - \frac{1}{\alpha^2}\right)e^{-x^2/2} \qquad \text{for} \quad \alpha > 2.$$

The error in this approximation is about 10% for $\alpha = 2$ and about 1% for $\alpha = 3$. Since the normal probability density

$$p(x; m, \sigma^2) = \frac{1}{\sqrt{2\pi}\,\sigma} \exp\left[-\frac{(x-m)^2}{2\sigma^2}\right] dx, \qquad -\infty < x < \infty,$$

cannot be integrated in closed form between every pair of limits a and b, probabilities relating to the normal distribution are usually obtained from special tables. These tables pertain to the *standard normal distribution*, namely, the normal distribution which has $m = 0$ and $\sigma = 1$, and its entries are then values of

$$F(z) = \frac{1}{\sqrt{2\pi}} \int_{-\infty}^{z} e^{-\frac{1}{2}t^2} dt$$

for $z = 0.00,\ 0.01, \ldots, 3.49$. Therefore, in order to find the probability $P(a \leqslant z \leqslant b)$, where z is the value of a random variable having a standard normal distribution, we use $P(a \leqslant z \leqslant b) = F(b) - F(a)$, and if either a or b is negative, then we can use the identity

$$F(-z) = 1 - F(z).$$

In order to evaluate the above integral, the integration is simplified by substituting $t = (x - m)/\sigma$.

3. *Rayleigh.* The *Rayleigh* probability density function (Figure 2.3) is given by

$$p(x) = \begin{cases} 0 & x < 0, \\ (x/a^2)\exp[-x^2/(2a^2)], & x \geqslant 0. \end{cases}$$

where a is identified with σ (i.e., $a = \sigma$).

The mean value of x is given by

$$\bar{x} = \frac{1}{a^2} \int_{0}^{\infty} x^2 e^{-x^2/2a^2} dx$$

$$= \sqrt{\frac{\pi}{2}}\, a,$$

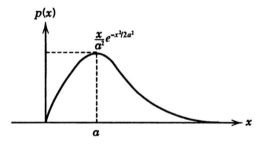

Figure 2.3 The Rayleigh probability density function.

while the variance is given by the mean-square value minus the mean value squared. The mean-square value is obtained as follows:

$$\overline{x^2} = \frac{1}{a^2} \int_0^\infty x^3 \exp\left(-\frac{x^2}{2a^2}\right) dx = 2a^2.$$

Therefore,

$$\mathrm{var}\{x\} = \sigma_x^2 = \overline{x^2} - \bar{x}^2 = 2a^2 - \frac{\pi}{2}a^2 = \left(2 - \frac{\pi}{2}\right)a^2.$$

4. Gamma Distribution. The probability density function of the gamma distributions is given by

$$p(x) = \begin{cases} \dfrac{\lambda^\eta}{\Gamma(\eta)} x^{\eta-1} e^{-\lambda x}, & x \geqslant 0, \\ 0 & \text{elsewhere,} \end{cases}$$

where $\Gamma(\eta)$ is the well-known gamma function

$$\Gamma(\eta) = \int_0^\infty u^{\eta-1} e^{-u} \, du.$$

The parameters η and λ are both positive. [Note that commonly the gamma function is tabulated in the form $\Gamma(\eta) = (\eta - 1)!$.] The distribution function of a random variable having a gamma distribution can be expressed in the form

$$P(x) = \begin{cases} \Gamma(\eta, \lambda x)/\Gamma(\eta), & x \geqslant 0, \\ 0 & \text{elsewhere.} \end{cases}$$

The mean and variance of a gamma-distributed random variable x are given by

$$m_x = \eta/\lambda, \quad \sigma_x^2 = \eta/\lambda^2.$$

5. The Chi-Square (χ^2) Distribution. The probability density function of the χ^2 distribution is give by the expression

$$p(x) = \begin{cases} \dfrac{1}{2^{n/2}\,\Gamma(n/2)} x^{(n/2)-1} e^{-x/2}, & x \geqslant 0, \\ 0 & \text{elsewhere,} \end{cases}$$

where the parameter n is referred to as its *degrees of freedom*. The mean and variance of the chi-square distribution can be expressed in the form

$$m_x = n, \qquad \sigma_x^2 = 2n.$$

6. *The Exponential Distribution.* The exponential distribution can be obtained from the gamma probability density function by letting $\eta = 1$. Thus, the exponential probability density function is expressed in the form

$$p(x) = \begin{cases} \lambda e^{-\lambda x}, & x \geqslant 0, \\ 0 & \text{elsewhere}, \end{cases}$$

where λ (> 0) is the parameter of the distribution. The distribution function is given by

$$P(x) = \begin{cases} 1 - e^{-\lambda x}, & x \geqslant 0, \\ 0 & \text{elsewhere}. \end{cases}$$

The mean and variance are given by

$$m_x = 1/\lambda, \qquad \sigma_x^2 = 1/\lambda^2.$$

Table 2.1 summarizes some of the most common continuous distributions.

2.2.3 Conditional Mean

Another important concept in the study of optimal control and estimation theory is the *conditional mean* (or *conditional expected value*). Let two random variables **x** and **y** be dependent. Given the fact that **x** has taken on a particular value, we should be able to predict the value of **y** better than if this information were not available. The conditional probability function will be defined as [58]

$$p(\mathbf{x}|\mathbf{y}) = p(\mathbf{x}, \mathbf{y})/p(\mathbf{y}) \qquad \text{for} \quad p(\mathbf{y}) \neq 0, \tag{2.12}$$

where $p(\mathbf{x}|\mathbf{y})$ is the probability of **x**, conditioned on a given value of **y**. The conditional mean (or conditional expected value) of a random variable **x** given that another random variable **y** whose value at $\mathbf{y} = y_i$ is denoted by $\mathscr{E}\{\mathbf{x}|\mathbf{y} = y_i\}$ and is defined as

$$\mathscr{E}\{\mathbf{x}|\mathbf{y} = y_i\} = \int_{-\infty}^{+\infty} x p_x(x|\mathbf{y} = y_i)\, dx \tag{2.13}$$

where x is the expected value of the random variable x (as defined earlier in this chapter, the random variable x can assume n values $x_1, x_2, \ldots, x_n$). We note that the conditional mean and covariance are not constants, but random variables, since they are functions of the conditioning random variables (or vector) **y**. Now, since $p(\mathbf{x}, \mathbf{y}) = p(\mathbf{y}|\mathbf{x})\, p(\mathbf{x})$, we have

$$p(\mathbf{x}|\mathbf{y}) = p(\mathbf{y}|\mathbf{x})p(\mathbf{x})/p(\mathbf{y}) \qquad \text{provided} \quad p(\mathbf{y}) \neq 0. \tag{2.14}$$

Table 2.1 Summary of Common Continuous Distributions

Distribution	Probability Density Function and Plot	Parameters	Mean Value (m)	Variance σ^2
Uniform	$p(x) = \begin{cases} \dfrac{1}{b-a}, & a \leqslant x \leqslant b, \\ 0 & \text{otherwise} \end{cases}$ 	$a, b > a$	$\dfrac{a+b}{2}$	$\dfrac{(b-a)^2}{12}$
Normal (Gaussian)	$p(x) = \dfrac{1}{\sqrt{2\pi}\,\sigma}\exp\left[-\dfrac{(x-m)^2}{2\sigma^2}\right],$ $-\infty < x < \infty$ 	$m, \sigma > 0$	m	σ^2

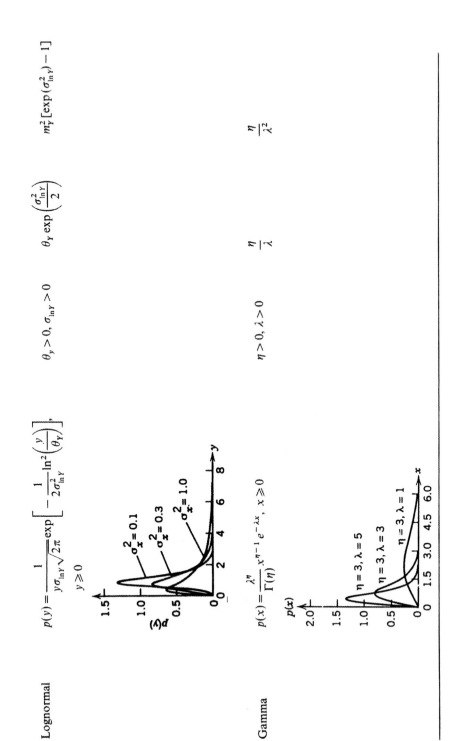

Lognormal

$$p(y) = \frac{1}{y\sigma_{\ln Y}\sqrt{2\pi}} \exp\left[-\frac{1}{2\sigma_{\ln Y}^2}\ln^2\left(\frac{y}{\theta_Y}\right)\right], \quad y \geq 0$$

$\theta_y > 0,\ \sigma_{\ln Y} > 0$

$\theta_Y \exp\left(\dfrac{\sigma_{\ln Y}^2}{2}\right)$

$m_Y^2[\exp(\sigma_{\ln Y}^2) - 1]$

Gamma

$$p(x) = \frac{\lambda^\eta}{\Gamma(\eta)} x^{\eta-1} e^{-\lambda x}, \quad x \geq 0$$

$\eta > 0,\ \lambda > 0$

$\dfrac{\eta}{\lambda}$

$\dfrac{\eta}{\lambda^2}$

Table 2.1 (*Continued*)

Distribution	Probability Density Function and Plot	Parameters	Mean Value (*m*)	Variance σ^2
Exponential	$p(x) = \lambda e^{-\lambda x}, \quad \lambda = \dfrac{1}{\theta}, \quad x \geqslant 0$	$\lambda > 0$	$\dfrac{1}{\lambda}$	$\dfrac{1}{\lambda^2}$
Chi-square	$p(x) = \dfrac{1}{2^{n/2}\Gamma(n/2)} x^{(n/2)-1} e^{-x/2}, \quad x \geqslant 0$	n = positive integer	n	$2n$

Beta

$$p(x) = \frac{\Gamma(\alpha+\beta)}{\Gamma(\alpha)\,\Gamma(\beta)}\, x^{\alpha-1}(1-x)^{\beta-1},$$

$$0 \leqslant x \leqslant 1$$

$\alpha > 0, \beta > 0$

$\dfrac{\alpha}{\alpha+\beta}$

$\dfrac{\alpha\beta}{(\alpha+\beta)^2(\alpha+\beta+1)}$

Rayleigh

$$p(x) = \begin{cases} 0, & x < 0, a > 0 \\ (x/a^2)\exp[-x^2/(2a^2)], & x \geqslant 0 \end{cases}$$

$\sqrt{\dfrac{\pi}{2}}\,a$

$\left(2-\dfrac{\pi}{2}\right)a^2$

This is known as the *Bayes formula* (or *Bayes rule*). Thus, if we regard $p(\mathbf{x})$ as the prior probability of $\mathbf{x}$ without knowledge of $\mathbf{y}$, then $p(\mathbf{x}|\mathbf{y})$ is the posterior probability of $\mathbf{x}$, given the fact that $\mathbf{y}$ has a certain value. If $\mathbf{x}$ and $\mathbf{y}$ are independent, the Bayes formula reduces to $p(\mathbf{x}|\mathbf{y}) = p(\mathbf{x})$, which implies that knowledge of $\mathbf{y}$ does not contribute to the prediction of $\mathbf{x}$. As we shall see later in Sections 4.1 and 4.2, if $\mathbf{y}$ is a measurement, $p(\mathbf{x}|\mathbf{y})$ describes $\mathbf{x}$, given the fact that the measurement has occurred.

2.3   THE CHEBYCHEV AND SCHWARZ INEQUALITIES

There are two aspects to be considered in the application of moments and expectations. The first is that of calculating moments of various orders of a random variable knowing its distribution, and the second is concerned with making statements about the behavior of the random variable when only some of its moments are available. This leads to the well-known Chebychev (also spelled Tchebysheff) inequality. Let x be a random variable with $\mathscr{E}\{x\} = m_x$, and let c be any real number. Then, if $\mathscr{E}\{(x-c)^2\}$ is finite and ε is any positive number, we have

$$P(|x-c| \geqslant \varepsilon) \leqslant \frac{1}{\varepsilon^2} \mathscr{E}\{(x-c)^2\}.$$

By choosing $c = m_x$ and $\varepsilon = k\sigma$, where $\sigma^2 = \mathrm{var}\{x\} > 0$, we obtain the *Chebychev inequality* as follows:

$$P(|x-m_x| \geqslant k\sigma_x) \leqslant 1/k^2 \tag{2.15}$$

for $k < 0$. Equation (2.15) is indicative of how the variance measures the *degree of concentration* of probability near $\mathscr{E}\{x\} = m_x$. Another way of expressing the Chebychev inequality is to note that the event $|x-m_x| < k\sigma_x$ is the complement of the event $|x-m_x| \geqslant k\sigma_x$. Hence,

$$P(|x-m_x| < k\sigma_x) \geqslant 1 - 1/k^2,$$

which says that the probability of getting a value within k standard deviations of the mean is at least $1 - 1/k^2$. The diagram in Figure 2.4 illustrates the Chebychev inequality.

The *Schwarz inequality* is also useful in the study of moments and is stated as follows:

$$\mathscr{E}^2\{xy\} = |\mathscr{E}\{xy\}|^2 \leqslant \mathscr{E}\{x^2\}\,\mathscr{E}\{y^2\}. \tag{2.16}$$

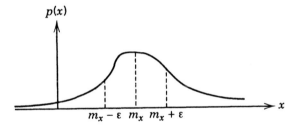

Figure 2.4   Chebychev's inequality.

2.4 COVARIANCE

The product of the mean or expectation of two independent (or *uncorrelated*) random variables x and y, $(x - m_x)(y - m_y)$, is called the *covariance* and is given by the relationship

$$P_{xy} = \mathscr{E}\{(x - m_x)(y - m_y)\}. \tag{2.17}$$

In general, for n independent random variables, the random column vector $\mathbf{x}$ will consist of n components, $x_1, x_2, \ldots, x_n$ and so will the mean $\mathbf{m}_x$. That is, if $\mathbf{x}$ is a random vector, then its *expectation* vector is the vector

$$\mathscr{E}\{\mathbf{x}\} = [\mathscr{E}\{\mathbf{x}_1\}, \mathscr{E}\{\mathbf{x}_2\}, \ldots, \mathscr{E}\{\mathbf{x}_n\}]^T,$$

where $\mathscr{E}\{\mathbf{x}\}$ is often called the mean vector of $\mathbf{x}$. Therefore, the square $(n \times n)$ covariance matrix* P can be defined in matrix form as (note that here we will drop the subscripts $\mathbf{x}$ and $\mathbf{y}$)

$$P = [\mathrm{cov}\{\mathbf{x}_i, \mathbf{x}_j\}]$$

$$= \mathscr{E}\{(\mathbf{x} - \mathscr{E}\{\mathbf{x}\})(\mathbf{x} - \mathscr{E}\{\mathbf{x}\})^T\}.$$

$$= \mathscr{E}\{(\mathbf{x} - \mathbf{m}_x)(\mathbf{x} - \mathbf{m}_x)^T)\}$$

$$= \begin{bmatrix} \mathrm{var}\{x_1\} & \mathrm{cov}\{x_1, x_2\} & \cdots & \mathrm{cov}\{x_1, x_n\} \\ \mathrm{cov}\{x_1, x_2\} & \mathrm{var}\{x_2\} & \cdots & \mathrm{cov}\{x_2, x_n\} \\ \vdots & \vdots & & \vdots \\ \mathrm{cov}\{x_1, x_n\} & \mathrm{cov}\{x_2, x_n\} & \cdots & \mathrm{var}\{x_n\} \end{bmatrix}. \tag{2.18}$$

---

*The symbol P used for the covariance matrix should not be confused with the probability symbol P in Section 2.3, nor with the probability distribution function $P(x)$.

The elements of the diagonal of the covariance matrix are the variances (σ^2) of the components, and the off-diagonal elements are the covariances. That is, the diagonal terms can be expressed as $P_{ii} \triangleq \sigma_i^2$. Note that the matrix is symmetric, so that there are only $n(n+1)/2$ distinct quantities. Also, note that the random matrix $(\mathbf{x} - \mathscr{E}\{\mathbf{x}\})(\mathbf{x} - \mathscr{E}\{\mathbf{x}\})^T$ is the outer product of the random vector $\mathbf{x} - \mathscr{E}\{\mathbf{x}\}$ with itself. As we shall see in Section 4.1, the covariance matrix is *positive definite* or *positive semidefinite*. (In addition to the definitions of positive definite and positive semidefinite given in Section 4.1, a matrix A is defined as positive definite if the quadratic form $\mathbf{x}^T A \mathbf{x} > 0$ for all nonzero $\mathbf{x}$, and positive semidefinite if and only if $\mathbf{x}^T A \mathbf{x} \geq 0$ for all nonzero $\mathbf{x}$.)

Using the vector–matrix notation, we can write the covariance matrix P of the joint density function for n *jointly normally distributed random variables* $x_1, x_2, \ldots, x_n$ as follows [53, 58]:

$$p(\mathbf{x}) = \frac{1}{(2\pi)^{n/2}|P|^{1/2}} \exp\left[-\tfrac{1}{2}(\mathbf{x} - \mathbf{m}_x)^T P^{-1}(\mathbf{x} - \mathbf{m}_x)\right], \qquad (2.19)$$

where P is a symmetric, positive definite matrix, and the determinant $|P|$ of P is nonzero. Note that $p(\mathbf{x})$ is completely characterized by specifying only the mean value of P. If P is a diagonal matrix, then $\mathbf{x} - \mathbf{m}_x$ has components that are statistically independent, since $p(\mathbf{x})$ may then be factored into a product of n scalar normal distributions. It should be pointed out that P must be positive definite (see Section 4.1 for definitions) to define $p(\mathbf{x})$. If P is only positive semidefinite, $\mathbf{x}$ can still be described by the *characteristic function*. [For a random variable x, the expectation $\mathscr{E}\{e^{jtx}\}$ is called the characteristic function of x and is expressed by the relationship $\mathscr{E}\{e^{jtx}\} = \int_{-\infty}^{\infty} e^{jtx} p(x)\, dx$.] If the random vector $\mathbf{x}$ is Gaussian with mean $\mathscr{E}\{\mathbf{x}\}$ and covariance matrix P_x, then we write

$$\mathbf{x} \sim N(\mathscr{E}\{\mathbf{x}\}, P_x).$$

Obviously, if the n random variables have zero means, Eq. (2.19) reduces to

$$p(\mathbf{x}) = \frac{1}{(2\pi)^{n/2}|P|^{1/2}} \exp\left(-\tfrac{1}{2}\mathbf{x}^T P^{-1} \mathbf{x}\right).$$

We conclude this section by saying a few words about the correlation between the random variables x and y. The amount of correlation between x and y can be conveniently expressed by a *coefficient of correlation* ρ_{xy}, which is the normalized value of P_{xy}:

$$\rho_{xy} = \frac{\mathscr{E}\{(x - \mathscr{E}\{x\})(y - \mathscr{E}\{y\})\}}{\sigma_x \sigma_y} = \frac{P_{xy}}{\sigma_x \sigma_y}, \qquad (2.20)$$

where σ_x and σ_y are the standard deviations of x and y, respectively. The magnitude of the correlation coefficient is restricted to $|\rho_{xy}| \leq 1$ ($-1 \leq \rho_{xy} \leq 1$).

2.5 THE AUTOCORRELATION FUNCTION AND POWER SPECTRAL DENSITY

In the analysis of random signals in linear systems, the parameters that are important are the correlation and the mean of the random process. If these parameters are independent of the origin, then the process satisfying these conditions is said to be *wide-sense stationary*.

The *autocorrelation function* (ACF) of a process $x(t)$ is so named because it is the correlation between two variables, both defined on the same process. It is a measure of the dependence of the amplitudes of the sample functions at $t = t_1$ on the amplitudes of the same function at $t = t_2$. The autocorrelation function is an ensemble (statistical) average of the product of two random variables. Namely, the autocorrelation function $\varphi_{xx}(\tau)$ of a signal $x(t)$ is defined as the average value of the product obtained by multiplying the signal at time t by itself at a later time $t + \tau$:

$$\varphi_{xx}(\tau) = \lim_{T \to \infty} \frac{1}{2T} \int_{-T}^{+T} x(t)x(t + \tau)\,dt$$

$$= \overline{x(t)x(t + \tau)}, \tag{2.21}$$

where the overscore denotes the averaging process. This assumes that the random process is also ergodic. For a stationary random process, the ACF is the same for each sample period and is a characteristic of the entire random process. The ACF has the following properties:

1. $\varphi_{xx}(\tau)$ is an even function:

$$\varphi_{xx}(\tau) = \varphi_{xx}(-\tau),$$

$$\overline{x(t)x(t + \tau)} = \overline{x(t)x(t - \tau)}$$

2. The following limiting cases exist:

$$\varphi_{xx}(0) = \lim_{T \to \infty} \frac{1}{2T} \int_{-T}^{+T} x(t)x(t)\,dt = \overline{x^2(t)}$$

$$\varphi_{xx}(0) \geq |\pm\varphi_{xx}(\tau)| \quad \text{(maximum value)},$$

$$\varphi_{xx}(\infty) = [\overline{x(t)}]^2.$$

In terms of the joint probability density function $p(x_1, x_2; t_1, t_2)$, the autocorrelation function can be expressed as [53]

$$\varphi_{xx}(t_1, t_2) = \overline{x_1 x_2} = \int_{-\infty}^{\infty} \int_{-\infty}^{\infty} x_1 x_2\, p(x_1, x_2; t_1, t_2)\,dx_1\,dx_2. \tag{2.22}$$

The power spectral density (PSD) of an ergodic random process is given by the Fourier transform of the autocorrelation function:

$$\Phi_{xx}(\omega) = \mathscr{F}\left[\varphi_{xx}(\tau)\right]$$

$$= \int_{-\infty}^{\infty} \varphi_{xx}(\tau)e^{-j\omega\tau}\,d\tau. \tag{2.23}$$

Through the use of an autocorrelation function, the power spectral density of the random process can be found. The relationship between the autocorrelation function and the power spectral density function is given by the Wiener–Khintchine theorem as [58]

$$\Phi_{xx}(\omega) = \int_{-\infty}^{+\infty} \varphi_{xx}(\tau)e^{-j\omega\tau}\,d\tau \tag{2.24}$$

and

$$\varphi_{xx}(\tau) = \frac{1}{2\pi}\int_{-\infty}^{+\infty} \Phi_{xx}(\omega)e^{j\omega\tau}\,d\omega. \tag{2.25}$$

The above result has been obtained for the case of the ergodic process [Eq. (2.23)]. Thus, the autocorrelation function and the power spectral density function can be expressed as a Fourier transform pair. The value of the autocorrelation function is a maximum at $\tau = 0$ and is

$$\varphi_{xx}(0) = \overline{x(t)x(t)}$$

$$= \lim_{\tau\to\infty}\frac{1}{2T}\int_{-T}^{+T} x^2(t)\,dt, \tag{2.26}$$

which is the square of the standard deviation or the variance (for zero mean). In terms of the power spectral density, we also have

$$\varphi_{xx}(0) = \mathscr{E}\{x^2\} = \frac{1}{2\pi}\int_{-\infty}^{+\infty} \Phi_{xx}(\omega)\,d\omega. \tag{2.27}$$

It should also be noted that if the random function contains a hidden periodic component, then the autocorrelation function will also contain components of the same frequency, although any phase information that the periodic component may have had is lost. Similarly, the power spectral density function computes power density spectra rather than the amplitude or phase spectra of the random signal. Analogous to the autocorrelation function, we have the cross-correlation function. The cross-correlation function $\varphi_{xy}(\tau)$ is a measure for the statistical dependence of two different processes $x(t)$ and $y(t)$. Then, for wide-sense stationary processes, in which both $x(t)$ and $y(t)$ are jointly ergodic, the cross-correlation function is defined as [53]

$$\varphi_{xy}(\tau) = \lim_{\tau \to \infty} \frac{1}{2T} \int_{-T}^{+T} x(t - \tau)\,y(t)\,dt = \overline{x(t - \tau)\,y(t)} \qquad (2.28)$$

with the following properties:

1. The cross-correlation function is an odd function:

$$\varphi_{xy}(\tau) \neq \varphi_{xy}(-\tau).$$

2. For $\tau = 0$,

$$\lim_{\tau \to \infty} \varphi_{xy}(0) = \overline{x(t)\,y(t)},$$

while for long correlation times

$$\lim_{\tau \to \infty} \varphi_{xy}(\tau) = \overline{x(t)}\ \overline{y(t)},$$

(product of the individual linear average values).

3. The cross-correlation factor is given by

$$\rho_{xy}(\tau) = \frac{\varphi_{xy}(\tau)}{\sqrt{\varphi_{xx}(0)\,\varphi_{yy}(0)}} = \frac{\overline{x(t - \tau)\,y(t)}}{\sqrt{\overline{x^2(t)y^2(t)}}}, \qquad (2.29)$$

where $-1 \leqslant \rho(\tau) \leqslant +1$.

4. The maximum value for the cross-correlation function is not $\tau = 0$, and is always smaller than the larger of $\overline{x^2(t)}$ and $\overline{y^2(t)}$:

$$\max_{\tau} |\,\varphi_{xy}(\tau)| < \max\,[\overline{x^2(t)}, \overline{y^2(t)}].$$

The characteristics of the power spectral density $\Phi_{xx}(\omega)$ can be summarized as follows:

1. The power spectral density $\Phi_{xx}(\omega)$ is a real function and is always positive.

2. The power spectrum is defined for positive and negative frequencies, and is an even function of the frequency:

$$\Phi_{xx}(\omega) = \Phi_{xx}(-\omega).$$

3. Integration over the entire frequency domain yields the mean value of the power for the process $x(t)$. Thus,

$$P = \int_{-\infty}^{\infty} \Phi_{xx}(\omega)\, d\omega \qquad (2.30a)$$

and

$$P = \int_{0}^{\infty} P_{xx}(\omega)\, d\omega \qquad (2.30b)$$

with

$$P_{xx}(\omega) = 2\Phi_{xx}(\omega) \qquad \text{for } 0 \leqslant \omega \leqslant \infty.$$

4. From Parseval's theorem, we have

$$\int_{-\infty}^{+\infty} x^2(t)\, dt = \frac{1}{2\pi} \int_{-\infty}^{+\infty} |\hat{X}(j\omega)|^2\, d\omega, \qquad (2.31)$$

where $\hat{X}(j\omega)$ is the amplitude density in the interval $-T \leqslant t \leqslant +T$. In terms of the amplitude density, the PSD is

$$\Phi_{xx}(\omega) = \frac{1}{2\pi} \lim_{T \to \infty} \frac{|\hat{X}_T(j\omega)|^2}{2T}. \qquad (2.32)$$

The mean power is

$$\overline{x^2(t)} = \lim_{T \to \infty} \frac{1}{2T} \int_{-T}^{+T} x^2(t)\, dt$$

$$= \int_{-\infty}^{+x} \lim_{T \to \infty} \frac{1}{2\pi} \frac{|\hat{X}(j\omega)|^2}{2T}\, d\omega$$

$$= \int_{-\infty}^{+\infty} \Phi_{xx}(\omega)\, d\omega$$

$$= P. \qquad (2.33)$$

5. Since the mean power of a real, physical signal is always finite, as $\omega \to \infty$ the PSD $\Phi_{xx}(\omega) \to 0$. For this reason, as we shall see in Section 2.8, a white-noise process with PSD $P_{xx}(\omega) = P_0 = \text{constant}$ is not physically realizable. The power spectrum for white noise is illustrated in Figure 2.5.

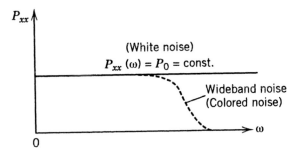

Figure 2.5 Power-spectrum representation for white and colored noise.

The mean power of stationary process $x(t)$ in the frequency interval $[\omega_1, \omega_2]$ is

$$\overline{\Delta x^2(t)} = \int_{\omega_1}^{\omega_2} P_{xx}(\omega)\, d\omega \approx P_{xx}(\omega_m)\Delta\omega \tag{2.34}$$

so that the power spectrum is

$$P_{xx}(\omega) = \frac{\overline{\Delta x^2(t)}}{\Delta\omega}. \tag{2.35}$$

The calculation of the mean power discussed in item 3 above can be illustrated by means of an example. Consider the measured power spectral density

$$P_{xx}(\omega) = P_0 \frac{1}{1 + (\omega/\omega_0)^2} \tag{2.36}$$

of the signal $x(t)$, where the measurement device's corresponding limiting frequency is ω_1. Therefore, the mean power can be calculated as follows:

$$P = \int_0^{\omega_1} P_{xx}(\omega)\, d\omega = P_0 \int_0^{\omega_1} \frac{d\omega}{1 + (\omega/\omega_0)^2}$$

$$= P_0 \omega_0 \arctan \frac{\omega_1}{\omega_0}. \tag{2.37}$$

The normalized mean power, $P = f(\omega_1)$, is illustrated in Figure 2.6.

For a sinusoidal signal $x(t) = \hat{X} \sin \omega t$, the mean value can be calculated from the expression [22]

$$\overline{x(t)} = \lim_{T \to \infty} \frac{1}{2T} \int_{-T}^{+T} x(t)\, dt$$

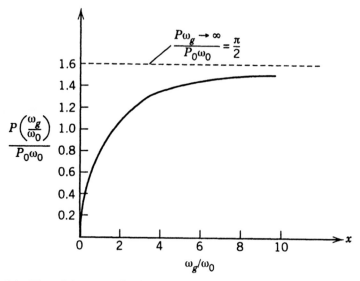

Figure 2.6 Plot of the normalized mean power with limiting frequency dependence.

$$= \frac{1}{2T} \int_{-T}^{+T} \hat{X} \sin \omega t \, dt = 0 \tag{2.38}$$

where the quadratic mean value is

$$\overline{x^2(t)} = \frac{1}{2T} \int_{-T}^{+T} \hat{X}^2 \sin^2 \omega t \, dt = \frac{\hat{X}^2}{2}. \tag{2.39}$$

Finally, the effective value is given by

$$X_{\text{eff}} = \tilde{X} = \frac{\hat{X}}{\sqrt{2}}. \tag{2.40}$$

The autocorrelation function for a periodic, sinusoidal time function $x(t) = a \sin \omega_0 t$ can be calculated from the general equation for the quadratic mean value [Eq. (2.21)]:

$$\varphi_{xx}(\tau) = \lim_{T \to \infty} \frac{1}{2T} \int_{-T}^{+T} x(t)x(t + \tau) \, dt$$

$$= \overline{x(t) \times (t + \tau)}. \tag{2.41}$$

Since here we are concerned with a periodic function, the limit as $T \to \infty$ is not needed. Thus,

$$\varphi_{xx}(\tau) = \frac{a^2}{2T} \int_{-T}^{+T} \sin \omega_0 t \, \sin \left[\omega_0(t - \tau) \right] dt.$$

$$= \frac{a^2}{2} \cos \omega_0 \tau. \tag{2.42a}$$

The maximum value yields the quadratic mean value $\overline{x^2(t)}$. Therefore,

$$\varphi_{xx}(0) = \frac{a^2}{2} = \overline{x^2(t)}. \tag{2.42b}$$

The same procedure can be used for a signal of the form $x(t) = a \cos \omega_0 t$. For $\tau \geqslant 0$,

$$\varphi_{xx}(\tau) = \frac{2a^2}{T} \int_{-T/4+\tau}^{T/4} \cos \omega_0 t \cos \left[\omega_0(t - \tau) \right] dt$$

$$\varphi_{xx}(\tau) = \frac{2a^2}{T} \left[\cos \omega_0 \tau \int_{-T/4+\tau}^{T/4} \cos^2 \omega_0 t \, dt + \sin \omega_0 \tau \right.$$

$$\left. \int_{-T/4+\tau}^{T/4} \sin \omega_0 t \cos \omega_0 t \, dt \right]$$

and

$$\varphi_{xx}(\tau) = \frac{a^2}{2\pi} \left[(\pi - \omega_0 \tau) \cos \omega_0 \tau + \sin \omega_0 \tau \right]. \tag{2.43a}$$

For $\tau = 0$,

$$\varphi_{xx}(0) = \frac{a^2}{2}. \tag{2.43b}$$

2.6 LINEAR SYSTEMS

A linear system may be characterized by its response to a unit impulse function. Consider now the transmission of signals through linear systems. First of all, the sampling property of the impulse function allows us to represent a signal $f(t)$ by the convolution integral as [22]

$$f(t) = f(t) * \delta(t) = \int_{-\infty}^{\infty} f(\tau) \delta(t - \tau) \, d\tau. \tag{2.44}$$

This representation can be viewed as an expression of $f(t)$ in terms of impulse components, where $f(t)$ is expressed as a continuous sum (integral) of the impulse function. The convolution integral mathematically relates the input and output in terms of the unit impulse response:

$$f_0(t) = \int_{-\infty}^{\infty} h(\tau) f_i(t - \tau) \, d\tau \tag{2.45}$$

where $f_0(t)$ is the system output, $h(\tau)$ is the impulse response, and $f_i(t - \tau)$ is the system input delayed τ seconds. The transfer function relating the system output to the system input is therefore [26]

$$G(j\omega) = \frac{F_0(j\omega)}{F_i(j\omega)} \tag{2.46}$$

where $F_i(j\omega)$ and $F_0(j\omega)$ are the Fourier transforms of the input and output, respectively. Therefore, the impulse response and the transfer function are Fourier transform pairs of each other, and may be expressed as

$$H(t) = \frac{1}{2\pi} \int_{-\infty}^{\infty} G(j\omega) e^{j\omega t} \, d\omega \tag{2.47}$$

and

$$G(j\omega) = \int_{-\infty}^{\infty} h(t) e^{-j\omega t} \, dt. \tag{2.48}$$

Figure 2.7 represents a typical linear filter.

In general, if a linear time-invariant system has a unit impulse response $h(t)$, then it follows that the response $r(t)$ to the signal $f(t)$ will be given by [72]

$$r(t) = \int_{-\infty}^{\infty} f(\tau) h(t - \tau) \, d\tau. \tag{2.49}$$

This follows from Eq. (2.44) and the principle of superposition. Thus,

$$r(t) = f(t) * h(t) \tag{2.50}$$

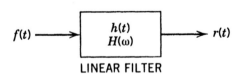

LINEAR FILTER

Figure 2.7   Signals into and out of a linear filter.

and

$$R(\omega) = F(\omega) H(\omega), \qquad (2.51)$$

where

$$r(t) \leftrightarrow R(\omega)$$

and

$$h(t) \leftrightarrow H(\omega). \qquad (2.52)$$

The double arrow represents the correspondence between the time of frequency domains. That is, $f(t) \leftrightarrow F(\omega)$ denotes that $F(\omega)$ is the direct Fourier transform of $f(t)$ and that $f(t)$ is the inverse Fourier transform of $F(\omega)$.

As noted in Eqs. (2.24) and (2.25), this Fourier transform pair is given by

$$F(\omega) = \int_{-\infty}^{\infty} f(t) e^{-j\omega t} \, dt, \qquad (2.53)$$

$$f(t) = \frac{1}{2\pi} \int_{-\infty}^{\infty} F(\omega) e^{j\omega t} \, d\omega. \qquad (2.54)$$

It is important to point out here that there is a certain amount of symmetry in the equations defining the two domains. For example, we expect that the effect on the frequency domain due to differentiation in the time domain should be similar to the effect on the time domain due to differentiation in the frequency domain.

The unit-impulse response $h(t)$ is the response of the system to a unit impulse applied at $t = 0$. Obviously, for any physical system, the response cannot appear before $t = 0$. Hence, $h(t) = 0$ for $t < 0$. Such signals are commonly called *causal* signals. For all physical realizable systems, $h(t)$ is causal. One can postulate systems for which $h(t)$ is not causal, but such systems are physically unrealizable. The causality of $h(t)$ is the criterion of physical realizability in the time domain. In the frequency domain, this criterion implies that a necessary and sufficient condition for a magnitude function $|H(\omega)|$ to be physically realizable is that [72]

$$\int_{-\infty}^{\infty} \frac{|\ln|H(\omega)||}{1 + \omega^2} \, d\omega < \infty. \qquad (2.55)$$

This is the so-called Wiener–Paley criterion of physical realizability. Note that the magnitude of the function $|H(\omega)|$ must be square-integrable in order for the Wiener–Paley criterion to be valid. A system whose magnitude function

$H(\omega)$ violates the Wiener–Paley criterion has noncausal impulse response, and is therefore physically unrealizable.

In summarizing the results of this section, we note that a general relationship exists between the random signals into and out of a linear system. Letting $x(t)$ denote the input signal and $y(t)$ the output, the power spectral densities of these signals are related by

$$\Phi_{yy}(j\omega) = G(j\omega)\,G(-j\omega)\,\Phi_{xx}(j\omega), \tag{2.56}$$

where $G(s)$ is the system transfer function (with $s = j\omega$), and $G(j\omega)$, and $G(-j\omega)$ are complex conjugates of each other. Note that since the power spectral densities are always real functions, they are always even functions of $j\omega$. In particular, suppose that the power spectral density of the input is constant over the frequency range of interest (white noise — see Section 2.8):

$$\Phi_{xx}(j\omega) = N,$$

and that the filter has a single time lag τ_c:

$$G(s) = \frac{1}{\tau_c s + 1} = \frac{\beta}{s + \beta},$$

$$\beta = \frac{1}{\tau_c}.$$

Then the output spectral density is, from Eq. (2.56),

$$\Phi_{yy}(j\omega) = \frac{N\beta^2}{\omega^2 + \beta^2}. \tag{2.57}$$

From Eq. (2.56), the following power spectral density relationship exists for the system input and output:

$$\Phi_{yy}(\omega) = |G(j\omega)|^2\,\Phi_{xx}(j\omega),$$

or

$$|G(j\omega)|^2 = \frac{\Phi_{yy}(\omega)}{\Phi_{xx}(\omega)} = G(j\omega)G^*(j\omega). \tag{2.58}$$

where, as before, $G(j\omega)$ and $G*(j\omega)$ are complex conjugates of each other with poles and zeros in the s half plane. Diagrammatically, the transfer function is illustrated in Figure 2.8. The output power P_0 corresponding to the quadratic mean value $\overline{y^2(t)}$ is

$$P_0 = \overline{y^2(t)} = \frac{1}{2\pi} \int_{-\infty}^{+\infty} \Phi_{yy}(\omega)\, d\omega = \int_0^\infty P_{yy}(\omega)\, d\omega,$$

$$P_0 = \overline{y^2(t)} = \frac{1}{2\pi} \int_{-\infty}^{+\infty} |G(j\omega)|^2 \Phi_{xx}(\omega)\, d\omega,$$

(2.59)

where the transfer function is plotted in Figure 2.9.

Frequently, random signals or noise $y(t)$ are assumed to have power spectra of this form. The parameter β is called the *bandwidth*, and $\tau_c = 1/\beta$ is called the *correlation time*. The total power in the random process $y(t)$ can be determined from the expression.

$$\text{Power} = \sigma_x^2 = \frac{1}{2\pi} \int_{-\infty}^{\infty} \Phi_{xx}(j\omega)\, d\omega.$$

(2.60)

Therefore, using Eq. (2.60) we have

$$\sigma_y^2 = \frac{N\beta}{2}.$$

Equation (2.60) clearly illustrates the reason for calling $\Phi(j\omega)$ the "power

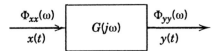

Figure 2.8 Input–output representation for the transfer function $G(j\omega)$.

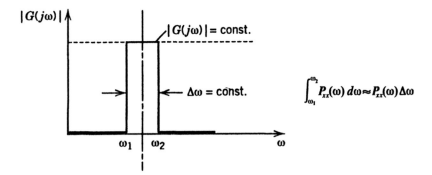

Figure 2.9 Bandpass characteristic for the power spectrum $P_{xx}(\omega)$.

spectral density." Using the relationship $\sigma_y^2 = N\beta/2$, Eq. (2.57) can be written in the form

$$\Phi_{yy}(j\omega) = \frac{2\sigma_y^2 \beta}{\omega^2 + \beta^2}, \tag{2.61}$$

in which the total power σ_y^2 appears explicitly. Using Eq. (2.61), the autocorrelation function of $y(t)$ is

$$\varphi_{yy}(\tau) = \sigma_y^2 e^{-\tau/\tau_c} = \sigma_y^2 e^{-\beta|\tau|}, \tag{2.62}$$

which shows that τ_c is the time constant of the autocorrelation function. Hence the name "correlation" time. Also, Eq. (2.62) shows why noise of this form is known as "exponentially correlated noise." Let us consider as an example the following first-order Markov process*:

$$\frac{d\varepsilon(t)}{dt} = -\beta\varepsilon(t) + w(t), \qquad \beta > 0, \tag{2.63}$$

where $w(t)$ is a white-noise driving process with output spectral density Φ_0, and β is a correlation parameter. This first-order differential equation appears often in the modeling of sensors and/or physical processes. Equation (2.63) can be solved using standard Laplace transform techniques. Taking the Laplace transform of the equation, we have

$$s\varepsilon(s) + \beta\varepsilon(s) = \omega(s),$$

or

$$\varepsilon(s)(s + \beta) = \omega(s),$$

so that

$$G(s) = \frac{\varepsilon(s)}{\omega(s)} = \frac{1}{s + \beta}.$$

Since

$$\Phi_0(\omega) = G(j\omega)G(-j\omega)\Phi_i(\omega) = \Phi_i(\omega)|G(j\omega)|^2,$$

where $\Phi_i(\omega)$ is the input power spectral density, we have

$$\Phi_0(\omega) = \frac{1}{\beta - j\omega}\frac{1}{\beta + j\omega}\Phi_i = \frac{1}{\beta^2 + \omega^2}\Phi_i,$$

---

*A first-order Markov process is generated by the output of a first-order lag function driven by white noise (see Section 2.9 for a more detailed description).

and since

$$\overline{f^2(t)} = \frac{1}{\pi} \int_0^\infty \Phi(\omega)\, d\omega,$$

the mean quadratic value is therefore

$$\overline{\varepsilon^2(t)} = \frac{1}{\pi} \int_\infty^\infty \frac{1}{\beta^2 + \omega^2} \Phi_i\, d\omega$$

$$= \frac{1}{\pi} \Phi_i \frac{1}{\beta} \text{ are tan}\left(\frac{\omega}{\beta}\right)\Big|_0^\infty = \frac{\Phi_0}{\pi\beta}\frac{\pi}{2} = \frac{\Phi_i}{2\beta}. \qquad (2.64)$$

Next, let us calculate the autocorrelation function. From Eq. (2.25),

$$\varphi_{xx}(\tau) = \frac{1}{2\pi} \int_{-\infty}^\infty \Phi_{xx}(\omega)\, e^{j\omega t}\, d\omega.$$

This integral can be evaluated by means of the Cauchy integral theorem as follows:

$$\varphi_{xx} = \frac{1}{2\pi} \int_{-\infty}^\infty \frac{\Phi_i}{(\beta - j\omega)(\beta + j\omega)} e^{j\omega\tau}\, d\omega$$

$$= \frac{1}{2\pi} \int_{-\infty}^\infty \frac{\Phi_i}{(\omega - j\beta)(\omega - j\beta)} e^{j\omega\tau}\, d\omega$$

$$= \begin{cases} \dfrac{1}{2\pi} \Phi_i\, 2\pi j \left[\dfrac{e^{j\omega\tau}}{\omega + j\beta}\right]_{\omega = j\beta} = \dfrac{1}{2\pi} \Phi_i\, 2\pi j \dfrac{e^{-\beta\tau}}{2j\beta}, & \tau > 0, \\[2ex] -\dfrac{1}{2\pi} \Phi_i\, 2\pi j \left[\dfrac{e^{j\omega\tau}}{\omega - j\beta}\right]_{\omega = -j\beta} = \dfrac{1}{2\pi} \Phi_i\, 2\pi j \dfrac{e^{\beta\tau}}{2j\beta}, & \tau > 0 \end{cases}$$

$$= \Phi_i \frac{e^{-\beta|\tau|}}{2\beta}. \qquad (2.65)$$

Note that the Wiener–Khintchine equations, Eqs. (2.24) and (2.25), are also written in the form

$$\Phi_{xx}(w) = \int_{-\infty}^\infty \varphi_{xx}(\tau) e^{-j\omega\tau}\, d\tau, \qquad (2.66a)$$

$$\varphi_{xx}(\tau) = \frac{1}{2\pi} \int_{-\infty}^\infty \Phi_{xx}(\omega) e^{j\omega\tau}\, d\omega. \qquad (2.66b)$$

Some authors prefer this convention. For the autocorrelation example just solved, we selected the previous one.

As a final example, the determination of the autocorrelation function $\varphi_{yy}(\tau)$ for a simple case will now be illustrated. For this example, let

$$\varphi_{xx}(\tau) = \delta(\tau),$$

$$h(t) = \begin{cases} 1, & 0 < t < T, \\ 0 & \text{otherwise.} \end{cases}$$

From Eq. (2.56) we have

$$G(s) = \frac{1}{s} - \frac{e^{-sT}}{s},$$

$$\Phi_{yy}(s) = G(-s)\,G(s)\,\Phi_{xx}(s)$$

$$= \left(\frac{1}{-s} - \frac{e^{sT}}{-s}\right)\left(\frac{1}{s} - \frac{e^{-sT}}{s}\right)$$

$$= \frac{1}{s^2}(e^{sT} + e^{-sT} - 2),$$

so that

$$\varphi_{yy}(\tau) = -2\tau u(\tau) + (\tau - T)u(\tau - T) + (\tau + T)u(\tau + T).$$

2.7 THE CLASSICAL WIENER FILTER

Given a statistical description of a stationary, scalar random signal (or message) $x(t)$ and stationary random noise $n(t)$, the problem is to find a linear, time-invariant filter that will operate on the observable signal $z(t) = x(t) + n(t)$ and minimize [72]

$$\mathscr{E}\{(\hat{x}(t) - x(t))^2\}, \tag{2.67}$$

where $\hat{x}(t)$ is the output of the filter. Specifically, the Wiener process was devised by N. Wiener as a simple model for Brownian motion. As we saw in the previous section, the solution to this problem is expressed in terms of the filter impulse response $h(t)$ (see Figure. 2.10) [72]

$$\hat{x}(t) = \int_{-\infty}^{\infty} h(\tau)\,z(t - \tau)\,d\tau. \tag{2.68}$$

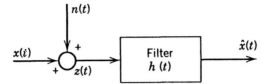

Figure 2.10 Representation of the classical Wiener filter.

The impulse response of the optimum filter, $h_0(\tau)$, must satisfy the Wiener–Hopf equation

$$\int_{-\infty}^{\infty} h_0(\tau)\,\varphi_{zz}(t-\tau)\,d\tau = \varphi_{zx}(t), \qquad t \geqslant 0, \tag{2.69}$$

where φ_{zz} and φ_{zx} are the autocorrelation and cross-correlation functions, respectively [22]. The solution to this integral equation is usually accomplished by using Fourier transforms and solving for the transfer function of the optimum filter in the frequency domain.

When the signal and noise are uncorrelated, this transfer function is given by

$$H_0(s) = \frac{1}{G(s)}\left[\frac{\Phi_{xx}(s)}{G^*(-s)}\right] \tag{2.70}$$

where

$$\Phi_{xx}(s) + \Phi_{vv}(s) = G(s)G^*(-s).$$

Here $\Phi_{xx}(s)$ and $\Phi_{vv}(s)$ are the spectral densities of the signal and noise, respectively; $G(s)$ has all of its poles and zeros in the left half plane, and $G^*(-s)$ has all of its poles and zeros in the right half plane. The fraction in the square brackets indicates the partial fraction expansion at the left-half-plane poles. Consequently, the solution to the Wiener problem amounts to specifying the optimum filter by its impulse response, which is obtained from the solution of an integral equation. However, this approach of filtering has several disadvantages. These are:

1. It is difficult to measure the statistical characteristics of random processes (that is, the autocorrelation and cross-correlation functions), which is the starting point of the Wiener problem. Furthermore, the autocorrelation and cross-correlation functions that appear in the Wiener–Hopf equation must be computed from previous knowledge and stored on line.

2. The problem ends with the optimum filter defined as an impulse response. However, in general, there is no simple method for synthesizing a filter with a prescribed impulse response. The numerical determination of the optimal impulse response is not suited for machine computation.

3. The classical Wiener filter provides the solution only to the one-dimensional problem. However, Wiener's results have been extended to include the case of the n-dimensional signal and noise for a certain class of problems. Therefore, the order of the filter, n, must be fixed before computation begins. As more data become available, that is, as n increases, then all computations must be repeated.

4. The classical Wiener filter is valid for stationary processes only. Extending the treatment to time-varying problems is very difficult.

5. The solution of the Wiener–Hopf equation to find the filter parameters requires the inversion of the autocorrelation matrix, which is a nontrivial computational problem when n is large.

The classical Wiener filter can also be represented diagrammatically as shown in Figure 2.11.

Our objective here is to find the realizable impulse response $h(t)$ to minimize the cost function [72]

$$\sigma_c^2 = \mathscr{E}\{e(t)^2\}, \tag{2.71}$$

where $e(t) = y_i(t) - \hat{y}(t)$,
$\quad y_i(t) = $ ideal response,
$\quad \hat{y}(t) = $ ideal impulse response.

The above expression is known as the mean-square error.

The fundamental Wiener–Hopf integral equation discussed earlier must be solved in order to obtain the transfer function $H(s)$. For simplicity, it will be assumed that the Wiener filter is a linear, time-invariant filter, so that the response $\mathbf{y}(t)$ is related to the directly available measurements $\mathbf{x}(t) + \mathbf{n}(t)$ via a standard superposition integral of the form

$$\hat{y}(t) = \int_{-\infty}^{\infty} h(t - \tau)[\mathbf{x}(\tau) + \mathbf{n}(\tau)]\, d\tau$$

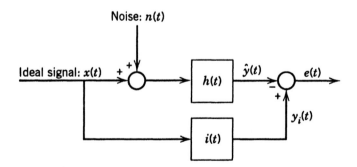

Figure 2.11   Wiener filter formulation with ideal impulse response.

$$= \int_{-\infty}^{\infty} h(\tau)[\mathbf{x}(t-\tau) + \mathbf{n}(t-\tau)] \, d\tau. \tag{2.72}$$

From the Hilbert space projection theorem, the measurements must be orthogonal (i.e., uncorrelated) to the error estimation $e(t)$. Thus, assuming that the signal $\mathbf{x}(t)$ and the noise $\mathbf{n}(t)$ are not correlated, zero expectation is encountered as

$$0 = \mathscr{E}\{e(t)[\mathbf{x}(w) + \mathbf{n}(w)]^T\} \qquad \text{for} \quad 0 \leqslant w \leqslant t$$

$$= \mathscr{E}\{[y_i(t) - \hat{y}(t)][\mathbf{x}^T(w) + \mathbf{n}^T(w)]\}$$

$$= \mathscr{E}\left\{\left(y_i(t) - \int_{-\infty}^{\infty} h(\tau)[\mathbf{x}(t-\tau) + \mathbf{n}(t-\tau)] \, d\tau\right)[\mathbf{x}^T(w) + \mathbf{n}^T(w)]\right\}.$$

(Note: The signal and noise are assumed to be vectors).

From the above discussion, the Wiener–Hopf equation can be written in the form [72]

$$\int_{-\infty}^{\infty} i(\tau)R_{xx}(t-\tau) \, d\tau = \int_{-\infty}^{\infty} h(\tau)[R_{xx}(t-\tau) + R_{nn}(t-\tau)] \, d\tau, \tag{2.73}$$

where $R_{xx}(\tau)$ and $R_{nn}(\tau)$ are known autocorrelation functions. The unknown function $h(\tau)$ can be solved for using frequency-domain methods, such as Fourier or bilateral Laplace transforms. Therefore, solving for $H(s)$ results in

$$I(s)\Phi_{xx}(s) = H(s)[\Phi_{xx}(s) + \Phi_{nn}(s)],$$

or

$$H(s) = I(s) \, \Phi_{xx}(s) \, [\Phi_{xx}(s) + \Phi_{nn}(s)]^{-1}, \tag{2.74}$$

which in general is not realizable, since the right-hand side of the first equation has poles in both the right and left half s-planes. A realizable solution can be obtained using the matrix spectral factorization method.

Most of the difficulties associated with Wiener filtering depend on the fixed integer n, which is the order of the filter. The recursive algorithm is designed to allow for updating the estimate as more data are obtained, which saves on-line storage space and processing time for the onboard digital computer used to implement the algorithm. Finally, the Wiener filter deals with stationary signals or the constant parameters of prescribed signal forms, whereas the Kalman filter deals with time-varying signals expressed as random processes. Furthermore, as we shall see in the next chapter, the Kalman filter approach has the virtue of exhibiting immediately the form of the optimal filter and makes use of either finite or infinite data with equal facility. The filtering objective may be to improve the signal-to-noise ratio (SNR) for the measured data prior to information processing. This is the standard *"real time"* filtering operation, in which measurements up until the present time are used (recursively) to provide the *"best"* estimate of the signal. The prediction problem, on the other hand, uses past measurements to estimate the future value of the signal, while the smoothing problem uses the complete data set to *"better"* estimate past signal values.

2.8   WHITE NOISE

The term *white noise* is applied to random processes whose power-density spectrum is uniform over the entire frequency range, *"white"* implying that it contains all *"colors"* of frequencies in equal strength. Such a process is physically unrealizable, since it implies infinite average power. However, white noise is a useful concept to the engineer in modeling a physical process (or system), because many random processes occurring in nature are approximately Gaussian and stationary, and have a power spectrum which is flat (i.e., constant) up to frequencies far higher than the maximum frequency at which a system is capable of significant response. Specifically, white noise can be viewed as a limiting form of exponentially correlated noise. The limit is the situation where the correlation time approaches zero, while the power spectral density is held constant and equal to $\Phi_{xx}(0)$.

Such infinite-power models are used to generate error models. For example, white noise is used to drive a shaping filter which produces a finite power level over a particular frequency range or system bandpass. Thus, within the bandpass of the shaping filter, the output of the filter looks identical to the power-limited noise of the system being modeled. The output of the shaping filter is then used as an input to a linear system model. Furthermore, the linear model of the physical system can be driven by white Gaussian noise in order to represent wideband noises with essentially constant power over the system bandpass.

From the discussion of Section 2.5, a continuous Gaussian process $w(t)$ is a zero-mean white-noise process if [37]

$$\mathscr{E}\{\mathbf{w}(t)\} = 0, \qquad \mathrm{cov}(\mathbf{w}(t),\mathbf{w}(\tau)) = \boldsymbol{\Psi}_w(t)\delta(t-\tau),$$

where $\boldsymbol{\Psi}_w(t)$ is a constant matrix. As noted earlier, the exponentially correlated noise is a zero-mean stationary process having the following autocorrelation function and power spectral density:

$$\varphi_x(\tau) = \sigma_x^2\, e^{-\beta|T|} = \sigma_x^{2\,-|T|/\tau_c}, \tag{2.75}$$

$$\Phi_x(\omega) = \frac{2\sigma_x^2\,\beta}{\omega^2 + \beta^2}, \tag{2.76}$$

where σ_x^2 = variance of $x(t)$,

$\quad \beta = 1/$correlation time, τ_c.

Therefore, as the correlation time approaches zero,

$$\varphi_x(\tau) = \sigma_x^2 \delta(\tau),$$

$$\Phi_x(\omega) = 2\sigma_x^2 \beta, \tag{2.77}$$

where $\delta(\tau)$ is the Dirac delta function.

By way of an example, let us now determine the power spectral density for white noise whose autocorrelation function $\varphi_{xx}(\tau) = 2\pi S_0 \delta(\tau)$. Substituting this autocorrelation function in Eq. (2.24), we have

$$\Phi_{xx}(\omega) = \int_{-\infty}^{\infty} \varphi_{xx}(\tau) e^{-j\omega\tau} \, d\tau$$

$$= \int_{-\infty}^{+\infty} 2\pi S_0 \delta(\tau) e^{-j\omega\tau} \, d\tau. \tag{2.78}$$

Since for a delta function

$$\int_{-\infty}^{+\infty} \delta(\tau) e^{-j\omega\tau} \, d\tau = e^{-j\omega\tau} \Big|_{\tau=0} = 1,$$

then $\Phi_{xx}(\omega) = S_0$. Similarly, for this power spectral density, the autocorrelation function is

$$\varphi_{xx}(\tau) = S_0 \int_{-\infty}^{\infty} e^{j\omega\tau} \, d\omega,$$

$$\int_{-\infty}^{+\infty} e^{j\omega\tau} \, d\omega = 2\pi \delta(\tau),$$

$$\varphi_{xx}(\tau) = 2\pi S_0 \delta(\tau) = \pi P_0 \delta(\tau). \tag{2.79}$$

where P_0 is the constant power. The autocorrelation function and power spectrum for white noise are illustrated in Figure 2.12. It indicates that the white noise has no correlation time, even for points arbitrarily close together, yet is has infinite mean-square value.

Since, as stated earlier, the autocorrelation function is an even function, $[\varphi_{xx}(\tau) = \varphi_{xx}(-\tau)]$, we have [1, 48]

$$\Phi_{xx}(\omega) = \frac{1}{\pi} \int_0^{\infty} \varphi_{xx}(\tau) \cos \omega\tau \, d\tau \tag{2.80a}$$

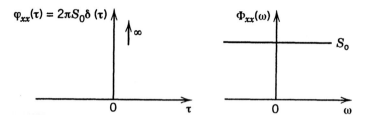

Figure 2.12   White-noise autocorrelation function and power spectrum.

and

$$\varphi_{xx}(\tau) = 2 \int_0^\infty \Phi_{xx}(\omega) \cos \omega\tau \, d\omega. \tag{2.80b}$$

In the limiting case,

$$P = \int_0^\infty \Phi_{xx}(\omega) \, d\omega$$

with

$$P_{xx}(\omega) = 2\Phi_{xx}(\omega) \qquad \text{for} \quad 0 \leqslant \omega \leqslant \infty,$$

$$\varphi_{xx}(0) = 2 \int_0^\infty \Phi_{xx}(\omega) \, d\omega = \int_0^\infty P_{xx}(\omega) \, d\omega.$$

As a second example, we will calculate the autocorrelation function and the quadratic mean value of the power spectral density given by

$$\Phi_{xx}(\omega) = \frac{S_0}{1 + \omega^2 T^2}. \tag{2.81}$$

Using Eq. (2.25), we have

$$\varphi_{xx}(\tau) = \int_{-\infty}^\infty \frac{S_0 e^{j\omega t}}{1 + \omega^2 T^2} \, d\omega.$$

This integral can be evaluated by using the residue theorem:

$$f(t) = \frac{1}{2\pi j} \int_{c-j\infty}^{c+j\infty} F(s) e^{st} \, ds = \mathscr{L}^{-1}\{F(s)\}. \tag{2.82}$$

Thus,

$$\varphi_{xx}(\tau) = \frac{S_0 \pi}{T} e^{-|\tau|/T}. \tag{2.83}$$

The quadratic mean value is obtained by setting $\tau = 0$. Therefore,

$$\varphi_{xx}(0) = \overline{x^2(t)} = \frac{S_0 \pi}{T}. \tag{2.84}$$

On the other hand, if we are given the autocorrelation function of a certain process as

$$\varphi_{xx}(\tau) = A_0 \, e^{-\beta |\tau|}$$

then the related power spectral density is obtained form the Wiener–Khintchine relation [Eq. (2.24)] as

$$\Phi_{xx}(\omega) = \frac{1}{2\pi} \int_{-\infty}^{\infty} A_0 \, e^{-\beta |\tau|} \, e^{-j\omega t} \, d\tau$$

$$= \frac{A_0}{\pi} \frac{\beta}{\beta^2 + \omega^2}. \tag{2.85}$$

From the above discussion, it can be said that many stochastic processes appearing in engineering possess a power spectral density of the general form [22]

$$\Phi(\omega) = a^2 \left(\frac{\omega_n^2}{\omega_n^2 + \omega^2} \right). \tag{2.86}$$

Example We shall illustrate the derivation of Eq. (2.86) with an example. Consider the case of tracking a moving target. In some target tracking problems, the target is assumed to have a constant velocity. Instead, in this example we will consider a target having an acceleration described by a discrete Poisson distribution, and with constant power spectral density (i.e., a white-noise process). We begin by letting the stochastic process $X(t)$ possess the following characteristics:

1. $X(t)$ can accept both values $+\beta$ and $-\beta$ (see Figure 2.13).
2. The probability that $X(t)$ changes sign in the time interval $(t, t + \Delta t)$ is equal to $\mu \Delta t$ as long as $\mu \Delta t \ll 1$.
3. The probabilities for the numbers of sign changes in the time interval $(t, t + T)$ does not depend outside this interval.

The probability for n changes of sign within the time interval $(0, T)$ can be described by means of the Poisson distribution [64]

$$p(n, T) = \frac{(\mu T)^n}{n!} e^{-\mu T}, \qquad n = 0, 1, 2, \ldots \, . \tag{2.87}$$

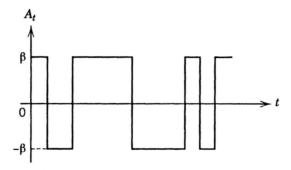

Figure 2.13 Poisson square wave for target acceleration.

If the interval $(0, T)$ is subdivided into λ subintervals Δt with $\Delta t \ll 1$, then the probability that no sign changes appear during the first subinterval is equal to $1 - \mu \Delta t$. Correspondingly the probability that no sign change appears in any of the λ subintervals is equal to $(1 - \mu \Delta t)^{\lambda}$. Replacing now Δt by T/λ, the following results:

$$\lim_{\lambda \to \infty} \left(1 - \frac{\mu T}{\lambda} \right) = e^{-\mu T}. \tag{2.88}$$

Since $p(0, T) = e^{-\mu T}$, we have proved Eq. (2.87) for $n = 0$. We are now ready to compute the autocorrelation function $\varphi_{xx}(\omega)$ of the process $X(t)$:

$$\varphi_{xx}(\tau) = \mathscr{E}\{X(t + \tau)X(t)\}, \tag{2.89}$$

where $\mathscr{E}\{\ \}$ is the expected-value operator. Introduce now a new process

$$Y(t, \tau) = X(t + \tau)X(t). \tag{2.90}$$

$Y(t, \tau)$ can take the values $y_1 = c^2$ and $y_2 = -c^2$ with the probabilities

$$P\{Y(t, \tau) = y_1\} = p(y_1), \tag{2.91}$$

$$P\{Y(t, \tau) = y_2\} = p(y_2), \tag{2.92}$$

where $p(y_1)$ [$p(y_2)$] is equal to the probability that the process $X(t)$ in the interval $(t, t + T)$ with $T = |\tau|$ possesses an even [odd] number of sign changes:

$$p(y_1) = \sum_{\lambda = 0}^{\infty} p(2\lambda, T) = e^{-\mu T} \sum_{\lambda = 0}^{\infty} \frac{(\mu T)^2}{(2\lambda)!} \tag{2.93}$$

$$p(y_2) = \sum_{\lambda = 0}^{\infty} p(2\lambda + 1, T) = e^{-\mu T} \sum_{\lambda = 0}^{\infty} \frac{(\mu T)^{2\lambda + 1}}{(2\lambda + 1)!}. \tag{2.94}$$

Thus, the autocorrelation function can be computed as follows:

$$\varphi_{xx}(\tau) = \mathscr{E}\{Y(t,\tau)\} = \sum_{i=1}^{2} y_i p(y_i)$$

$$= y_1\,p(y_1) + y_2\,p(y_2)$$

$$= c^2\,e^{-\mu T} \sum_{\lambda=0}^{\infty} \left[\frac{(\mu T)^{2\lambda}}{(2\lambda)!} - \frac{(\mu T)^{2\lambda+1}}{(2\lambda+1)!} \right]$$

$$= c^2\,e^{-\mu T} \sum_{\lambda=0}^{\infty} (-1) \frac{(\mu T)^{\lambda}}{\lambda!}$$

$$= c^2\,e^{-2\mu T}.$$

$$= c^2\,e^{-2\mu|\tau|}. \tag{2.95}$$

The power spectral density (PSD) follows from this autocorrelation function:

$$\Phi_{xx}(\omega) = c^2 \int_{-\infty}^{\infty} e^{-2\mu|\tau|}\,e^{-i\omega\tau}\,d\tau = c^2\,\frac{4\mu}{4\mu^2 + \omega^2}, \tag{2.96}$$

and with $4\mu^2 = \omega_x^2$,

$$\Phi_{xx}(\omega) = \left(\frac{2c^2}{\omega_x}\right) \frac{\omega_x^2}{\omega_x^2 + \omega^2} = \left(\frac{2c^2}{\omega_x}\right) \frac{1}{1 + (\omega/\omega_x)^2}. \tag{2.97}$$

In the limit as $\omega_x \to \infty$, $c \to \infty$, and $2c^2/\omega_x = a^2 = $ constant, Eq. (2.92) becomes

$$\lim_{\omega_x \to \infty} \Phi_{xx}(\omega) = a^2 = \text{constant}. \tag{2.98}$$

This is a white-noise process, which cannot be physically realized. Within the bandpass of the system of interest, however, this white noise can be made to look identical to the real wideband noise.

2.9 SYSTEM INPUT ERROR MODELS

Real-time error estimation requires the use of valid noise models if optimal estimation is to be efficiently applied. If an optimal filter is used, in which the error sources are included in the state vector to obtain an augmented state vector, it is then convenient to use filtered or integrated white-noise sources for the models. More specifically, when the original system inputs, say $\delta u(t)$, are not white noises or wideband noises which can be reasonably approximated as white, then a model of a system which can generate the error component $\delta u(t)$ from a white noise must be introduced. Therefore, the original system becomes a state variable, which is included in the state vector $\delta\mathbf{x}(t)$. This is done for all correlated inputs, so that the final augmented system has only white-noise inputs. Such models have been effectively applied to gyroscopes and other instruments. Other errors and their equations are accelerometers, gravity uncertainties, baroaltimeter errors, radar altimeter errors, Doppler radar errors, terrain correlation errors, and satellite positioning system receiver errors. Components of these errors are modeled as random constants (or bias), random walks, and first- and second-order Markov processes. We will now describe in some detail these statistical error models [37, 48].

White Noise: White noise is uncorrelated in time (i.e., it has zero correlation time). The characteristics of white noise $w(t)$ are

$$\text{a)} \ \mathscr{E}\{w(t)\} = 0 \quad \text{and} \quad \text{b)} \ \psi_{xx}(\tau) = \sigma^2 \delta(\tau).$$

Specifically, this process can be viewed as a limiting form of exponentially correlated noise. The limit is the situation where the correlation time approaches zero while the magnitude of the PSD is held constant and equal to $\Phi_{xx}(0)$. This results in an autocorrelation–PSD pair given as

$$\psi_{xx}(\tau) = \sigma^2 \delta(\tau),$$

$$\Phi_{xx}(\omega) = 2\sigma^2 \beta, \qquad \beta = 1/\tau_c,$$

where σ^2 is the variance and τ_c is the correlation time.

Random Constant (or Bias): A random constant is modeled as the output of an integrator with zero input and a (Gaussian) random initial condition which has mean zero and variance P_0. This type of model is suitable for describing an instrument bias that changes each time the instrument is turned on, but remains constant while it is on. Therefore, the random constant is a nondynamic quantity whose amplitude is fixed. Bias can be introduced as a state variable. For example, if we wish to estimate the gyroscope bias, it will be necessary to have the gyroscope bias as a state variable. The equation for the random constant is

$$\dot{x}(t) = 0.$$

Random Walk: A ramdon walk (also referred to by other names, such as integrated white noise, Wiener process, and Brownian motion) is the output of an integrator driven by zero-mean, white, Gaussian noise. The defining equations are given below:

$$\frac{dx(t)}{dt} = w(t), \qquad x(t_0) \triangleq 0,$$

$$\mathscr{E}\{w(t)\} = 0, \qquad\qquad\qquad\qquad\qquad (2.99a)$$

$$\mathscr{E}\{w(t)w(t+\tau)\} = Q\,\delta(t-\tau)$$

$$\mathscr{E}\{x^2(t)\} = Q(t-t_0) = 2\sigma^2(t-t_0),$$

where Q is the strength of the white Gaussian noise $w(t)$ and $\delta(\tau)$ is the Dirac delta function. The random walk is suitable for describing errors that grow without bound or are slowly varying (a pseudonoise driver addition to a random constant integrator model). Random walk effectively models a noise source that is equivalent to integrated broadband noise. Direct measurement of the random-walk content can be made by examining the time history of the variance of the noise. A single integration of white noise produces a linear growth in variance, so that the slope, if one exists, gives a measure of the strength of the random walk. This in turn can be converted to a coefficient for use in the optimal estimator.

Analyses of gyroscope drift rate have shown that a random walk, or possibly a random walk plus a first-order Markov process, is a reasonable approximation for the statistical characteristics of the drift rate data. However, it should be noted that to include both a bias state and a random walk would be indistinguishable from the merely altering initial condition of the random walk. For many purposes, it is recommended that gyroscope drift rate be modeled just as a random walk with an appropriately chosen power density for the noise source.

In addition, random walk produces a distinct signature in a power spectral density plot. The integrated white noise will then appear as a first-order slope corresponding to -20 db/decade in amplitude [48]. The PSD of the derivative of the variable being considered for the random walk will, of course, be flat. The gain associated with the slope gives a measure of the strength of the random walk. Furthermore, the derivative, that is, the broadband function that is being integrated to produce the random walk, can also be evaluated directly. The rms level of the derivative integrated with respect to time gives an estimate of the standard deviation of the random walk process in this case. The amplitude of a random walk process is obtained from its PSD or by measuring its mean square-growth rate. Specifically, the mean-square growth rate and PSD are given by

$$\mathscr{E}\{x^2(t)\} = \sigma^2 t, \tag{2.99b}$$

$$\Phi_{xx}(\omega) = 2\sigma^2/\omega^2, \tag{2.99c}$$

where the PSD is given in (distance units)^2/Hz.

From Eqs. (2.75) and (2.99b), the signatures of the white noise and random walk processes can be plotted in the σ^2–t plane as shown in the sketch in Figure 2.14.

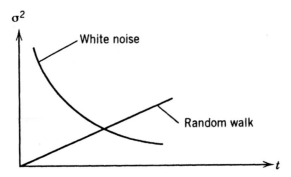

Figure 2.14 White-noise and random-walk "signatures."

First-Order Markov: A first-order Markov* process is the output of a first-order lag driven by a zero-mean, white, Gaussian noise of strength Q. The model is described by the following first-order differential equation [71]:

$$\frac{dx(t)}{dt} = -\beta x(t) + w(t), \qquad \beta = 1/\tau_c,$$

$$\mathscr{E}\{x(t)^2\} = Q\tau_0/2,$$

(2.100)

where

$$Q = 2\sigma^2/\tau_c = 2\sigma^2 \beta, \qquad \text{or} \quad \sigma^2 = Q/2\beta,$$

$$\sigma = [\mathscr{E}\{x(t)^2\}]^{1/2},$$

$$\tau_c = \text{correlation time.}$$

The discrete-time equivalent of Eq. (2.100) is given by

$$x(t_{k+1}) = e^{-\beta T} x(t_k) + w(t_k),$$

where T is sample time interval. A solution of the differential equation is

$$x(t) = \int_0^1 e^{-\beta(t-\tau)} w(\tau)\, d\tau.$$

A first-order Markov model is used to represent exponentially time-correlated noises. That is, a first-order Markov process produces an exponential

---

*Also called Gauss–Markov process.

correlation. Thus, the autocorrelation kernel function for this process is

$$\psi_{xx}(\tau) = \mathscr{E}\{x(t)x(t+\tau)\} = \sigma^2 e^{-|\tau|/\tau_c}, \tag{2.101}$$

where τ is the time interval. From the above expression, it can be seen that the autocorrelation function is an even, declining exponential function. As we saw earlier, the sum of a component of this form plus a random walk is sometimes found to fit gyroscope drift-rate data reasonably well over a limited bandwidth. However, the correlation time τ_c for a gyroscope drift rate is rather difficult to define. Correlation times reported in the literature have ranged from seconds to hours; these values are correlated primarily with the sampling period used in the recording the data.

Second-Order Markov: A second-order Markov process is one whose autocorrelation function has the general form [1, 58]

$$\psi_{xx}(\tau) = \mathscr{E}\{x(t)x(t+\tau)\}$$

$$= \frac{\sigma^2}{\cos\eta} e^{-\zeta\omega_n|\tau|} \cos\left(\sqrt{1-\zeta^2}\,\omega_n|\tau| - \eta\right). \tag{2.102}$$

With properly chosen parameters, this process produces an exponentially damped simusoidal correlation. Furthermore, the above autocorrelation function can be generated by passing a white, Gaussian noise $w(t)$ of strength $Q = 1$ through a second-order system. The state differential equation for this process is

$$\dot{x}_1(t) = x_2(t) + aw(t),$$

$$\dot{x}_2(t) = -\omega_n^2 x_1(t) - 2\zeta\omega_n x_2(t) + cw(t),$$

or in state-space representation

$$\begin{bmatrix} \dot{x}_1(t) \\ \dot{x}_2(t) \end{bmatrix} = \begin{bmatrix} 0 & 1 \\ -\omega_n^2 & -2\zeta\omega_n \end{bmatrix} \begin{bmatrix} x_1(t) \\ x_2(t) \end{bmatrix} + \begin{bmatrix} a \\ c \end{bmatrix} w(t),$$

where $x_1(t)$ is the output, and the parameters a, b, and c are

$$a = [(2\sigma^2/\cos\eta)\omega_n \sin(\alpha - \eta)]^{1/2},$$

$$b = [(2\sigma^2/\cos\eta)\omega_n^3 \sin(\alpha + \eta)]^{1/2},$$

$$c = b - 2a\zeta\omega_n$$

with

$$\alpha = \tan^{-1}\left(\zeta/\sqrt{1-\zeta^2}\right)$$

and are chosen so as to fit empirical data. A simpler form of a second-order Markov process can be described by the equation

$$\ddot{x} + 2\zeta\omega_n\dot{x} + \omega_n^2 x = \omega_n^2 w(t),$$

where $w(t)$ is a white-noise forcing function, with statistics

$$\mathscr{E}\{w(t)\} = 0,$$

$$\mathscr{E}\{w(t)w(\tau)\} = Q\,\delta(t - \tau).$$

In state-space notation this equation can be written as

$$\begin{bmatrix} \dot{x}_1(t) \\ \dot{x}_2(t) \end{bmatrix} = \begin{bmatrix} 0 & 1 \\ -\omega_n^2 & -2\zeta\omega_n \end{bmatrix} \begin{bmatrix} x_1(t) \\ x_2(t) \end{bmatrix} + \begin{bmatrix} 0 \\ c \end{bmatrix} w(t),$$

with $a = 0$ and $c = \omega_n^2$. Note that η and ζ ($\zeta \in (0.707, 1]$) are also determined from empirical data.

All modeled errors are assumed to be independent with initial covariances described by

$$P_{ij}(0) = 0, \qquad i \neq j,$$

$$P_{ii}(0) = x_i^2,$$

where x_i is the initial condition on the standard deviation of the ith truth state at time t_0. As mentioned earlier in this section, the models given above are the ones most commonly used to model other error sources as well. For instance, vertical deflections of gravity over many regions exhibit an autocorrelation function, with distance as the shift variable (see Chapter 4). Vertical deflections can be approximated by an exponential function or an exponentially damped sinusoidal function. However, for a spatially referenced process such as vertical deflections, wind gusts, etc., the correlation properties are defined in terms of distance. That is, exponential correlation will be characterized by a correlation length. As before, the autocorrelation function can be represented by the equation

$$\psi_{xx}(d) = \mathscr{E}\{x(t)x(t + d)\}, \tag{2.103a}$$

$$\psi_{xx}(d) = \sigma_x^2 e^{-|d|/l_c}, \tag{2.103b}$$

where x and d are length measures, and l_c is the correlation distance.

The spatial process just described can be converted into a temporal process. A *temporal process* experienced by a moving vehicle is one for which the scale factor relating time to distance is velocity. Therefore, the correlation time is the time required to move through the correlation distance. The autocorrelation

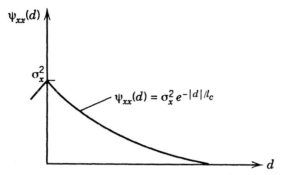

Figure 2.15   Autocorrelation function for vertical deflection of gravity.

function is as follows:

$$\psi_{xx}(d) = \sigma_x^2 \, e^{-|d|/l_c},$$

and is plotted in Figure 2.15. Both spatial and temporal processes have the same autocorrelation function, except that the former is related to the correlated distance while the latter is related to the velocity (i.e., $\tau_c = l_c/v$, where v is the velocity). Since all the error sources are modeled as the solutions to differential equations driven by white noise, these error-source states are added to the original state vector to produce an augmented state vector, which satisfies the equation

$$\delta\dot{\mathbf{x}} = F\,\delta\mathbf{x} + G\mathbf{w}, \qquad \delta\mathbf{x}(0) \text{ given},$$

where $\delta\mathbf{x}$ includes errors in position, velocity, attitude, gyroscope, accelerometer, etc. It should be noted that the solution $\delta\mathbf{x}(t)$ is a random process having first and second-order statistics. Table 2.2 summarizes the various models which find extensive use in the modeling of dynamic systems [48, 61].

PROBLEMS

2.1   In the figure given below, the probability density function $p(x) = c_1 =$ constant in the region $\xi \geqslant a$, $\xi < a + b$, and $p(x) = 0$ outside this region.
(a) Determine the constant c_1.
(b) Using Eq. (2.3a), determine the probability in the region $a \leqslant \xi < a + (b/3)$.

2.2   Calculate the constant c for the probability density

$$p(x) = c/(e^x + e^{-x})$$

and the probability that $x \leqslant 1$. [Hint: Use Eqs. (2.3b) and (2.2).]

Table 2.2 Stochastic Processes for Modeling of Dynamic Systems

Process	State Differential Equation	Block Diagram	Autocorrelation Function	Power Spectral Density
White noise	$w(t) = x(t)$		$\psi_{xx}(\tau) = S_0\,\delta(\tau)$	$\Phi_{xx}(\omega) = S_0 = \text{(constant)}$
Random constant (or bias)	$\dfrac{dx(t)}{dt} = 0$		$\psi_{xx}(\tau) = S_0 + m_0^2$	$\Phi_{xx}(\omega) = 2\pi(S_0 + m_0^2)\delta\omega$
Random walk (or Brownian motion)	$\dfrac{dx(t)}{dt} = w(t)$		$\mathscr{E}\{x_1(t), x_2(t)\} = \sigma^2 \min\{t_1, t_2\}$ $[\text{if } t_1 = t_2,\ \mathscr{E}\{x^2(t)\} = \sigma^2 t]$	$\Phi_{xx}(\omega) = 2\sigma^2/\omega^2$

First-order Markov (or exponentially time-correlated)

$$\frac{dx(t)}{dt} = \frac{1}{\tau_c}x(t) + w(t)$$

$$\psi_{xx}(\tau) = \sigma^2\, e^{-|\tau|/\tau_c}$$

$$\Phi_{xx}(\omega) = \frac{2\sigma^2/\tau_c}{\omega^2 + (1/\tau_c)^2}$$

Second-order Markov

$$\dot{x}_1(t) = x_2(t) + aw(t),$$

$$\dot{x}_2(t) = -\omega_n^2 x_1(t) - 2\zeta\omega_n x_2(t) + cw(t)$$

$$\psi_{xx}(\tau) = \frac{\sigma^2}{\cos\eta}\, e^{-\zeta\omega_n|\tau|}$$
$$\cos\left(\sqrt{1+\zeta^2}\,\omega_n|\tau| - \eta\right)$$

$$\Phi_{xx}(\omega) = \frac{a^2\omega^2 + b^2}{\omega^4 + 2\omega_n^2(2\zeta^2 - 1)\omega^2 + \omega_n^4}$$

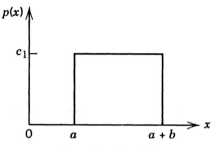

Figure Problem 2.1

2.3 For the triangular probability density shown in the diagram below, the densities are given by

$$p_1(x) = \frac{x}{2a^2 r} - \frac{x_0}{2a^2 r}, \qquad x_0 \leqslant x \leqslant x_0 + 2ar,$$

$$p_2(x) = \frac{-x}{2a^2(1-r)} + \frac{x_0 + 2a}{2a^2(1-r)}, \qquad x_0 + 2ar \leqslant x \leqslant x_0 + 2ar,$$

where $2a$ is the length of the base, r is the factor of symmetry, and x_0 is the initial value. Determine the probability distribution for the following limits:

$$x = x_0,$$

$$x = x_0 + 2ar,$$

$$x = x_0 + 2a.$$

2.4 Determine the first and second moments for the probability density function of Problem 2.1, using the following relations:

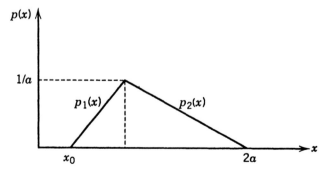

Figure Problem 2.3

1st moment: $\mathscr{E}\{x\} = \bar{x} = \displaystyle\int_{-\infty}^{+\infty} xp(x)dx;$

2nd moment: $\mathscr{E}\{x^2\} = \displaystyle\int_{-\infty}^{+\infty} x^2 p(x)\,dx.$

2.5 Let x and y be two statistically independent random variables. A new random variable z can be defined as the product of x and y:

$$z = xy.$$

Derive an expression for the probability function $p(z)$ of the new random variable z in terms of the probability density functions $p(x)$ and $p(y)$ of the original random variables x and y. [Hint: Use Eq. (2.6).]

2.6 Given two statistically independent random variables with probability density functions

$$p(x) = \begin{cases} 1/\pi\sqrt{1-x^2} & \text{for} \quad |x| < 1, \\ 0 & \text{for} \quad |x| > 1 \end{cases}$$

and

$$p(y) = \begin{cases} y \, \exp(-y^2/2) & \text{for} \quad y \geqslant 0, \\ 0 & \text{for} \quad y \geqslant 0 \end{cases}$$

Show that their product has a Gaussian probability density function.

2.7 Prove *Chebychev's inequality*,

$$\Pr\{|X - m_x| > \varepsilon\} \leqslant \sigma_x^2/\varepsilon^2$$

where X is a random variable with mean m_x, and variance σ_x^2, and ε is any positive number. (*Hint:* Use Figure 2.4 in Section 2.3.)

2.8 Let the random variable y be defined as the sum of N random variables x_n as follows:

$$y = \sum_{n=1}^{N} x_n.$$

Derive an expression for the variance of y when

(a) the random variables x_n are uncorrelated;
(b) they are correlated.

2.9 The random variable x may assume only nonnegative integral values, and the probability that it assumes the specific value m is given by the

Poisson probability distribution

$$P(x = m) = \lambda^m e^{-\lambda}/m!.$$

Determine the mean and variance of x. (Note that the Poisson probability density function can also be written in the form $p(\lambda t) = (\lambda t)^k e^{-\lambda t}/k!$, $k = 0, 1, 2, \ldots$; see the example in Section 2.8.)

2.10 Two discrete random variables x and y each have a Poisson distribution (as in Problem 2.9). If x and y are statistically independent random variables, determine the probability distribution of their sum $z = x + y$.

2.11 The *error function*, erf(α), and its complement, erfc(α), were given as

$$\text{erf}(\alpha) \triangleq \frac{1}{\sqrt{2\pi}} \int_{-\infty}^{\alpha} e^{-x^2/2} dx$$

and

$$\text{erfc}(\alpha) \triangleq \frac{1}{\sqrt{2\pi}} \int_{\alpha}^{\infty} e^{-x^2/2} dx.$$

Integrate by parts the complementary error function erfc(α) to establish the bound

$$\frac{1}{\sqrt{2\pi}\alpha}\left(1 - \frac{1}{\alpha^2}\right)\exp\left(-\frac{\alpha^2}{2}\right) < \text{erfc}(\alpha) < \frac{1}{\sqrt{2\pi}\alpha}\exp\left(-\frac{\alpha^2}{2}\right), \qquad \alpha > 0.$$

2.12 Let x be a Gaussian random variable with probability density function of the form

$$p(x) = \frac{\exp[-(x-a)^2/2b^2]}{b\sqrt{2\pi}}$$

(a = mean; b = standard deviation). This equation is also known as the *normal* probability density function.

(a) Derive an expressions for the characteristic function of x.

(b) Using the results from part (a), derive an expression for the mean and standard deviation of x, and express $p(x)$ in terms of these quantities and x.

2.13 Let the observation $z(t)$ be given as the sum of the signal $s(t)$ and a wideband noise $r(t)$ in the form $z(t) = s(t) + r(t)$. Determine the autocorrelation function of the total signal.

2.14 The autocorrelation function of a stationary random process is given by

$$\psi_{xx}(\tau) = A^2\, e^{-k|\tau|}.$$

Determine the power spectral density $S_{xx}(\omega)$ of this process. [Use the expression $S(\omega) = (1/2\pi)\int_{-\infty}^{+\infty} \psi(\tau)\, e^{-j\omega\tau}\, d\tau$]

2.15 As an extension of Problem 2.13, consider the three stationary random processes having the sample functions $x(t)$, $y(t)$, and $z(t)$, respectively. Derive an expression for the autocorrelation function of the *sum* of these three processes under the assumptions:

(a) that the processes are correlated;

(b) that the processes are all uncorrelated;

(c) that the processes are uncorrelated and all have zero means.

2.16 Consider the random process defined by the sample functions $x(t) = y^2(t)$, where the sample function $y(t)$ is given by $y(t) = a\cos(t + \phi)$.

(a) Show that if a is not a constant with probability one, then

$$\mathscr{E}\{x\} \neq \langle x(t)\rangle.$$

Here $\langle x(t)\rangle$ is defined as the *time average* of a sample function and is given by

$$\langle x(t)\rangle = \lim_{T\to\infty} \frac{1}{2T} \int_{-T}^{+T} x(t)\, dt$$

if the limit exists. [Note that this result is independent of T, then $\mathscr{E}\{\langle x(t)\rangle\} = m_x$.]

(b) Show that the integrability condition stated by the equation

$$\int_{-\infty}^{+\infty} |\psi_{xx}(\tau) - m_x^2|\, dt < \infty$$

is not statisfied by this random process.

2.17 Given two periodic functions of time

$$x(t) = \sum_{\nu = -\infty}^{\infty} a_\nu \exp(j\nu\omega_x t),$$

$$y(t) = \sum_{\mu = -\infty}^{\infty} b_\mu \exp(j\mu\omega_x t),$$

with incommensurable fundamental periods and zero average values. Show that the cross-correlation function is zero.

2.18 Determine the arithmetic mean and the quadratic mean of the sinusoidal wave $x(t) = A \sin \omega t$, where A is the amplitude of the signal.

2.19 For the cosine wave $x(t) = a \cos \omega_0 t$ shown below, calculate the auto-correlation function $\psi_{xx}(\tau)$.

2.20 The diagram below illustrates two time functions $x(t)$ and $y(t)$ with

$$y(t) = a,$$

$$x(t - \tau) = (a/T)(t - \tau).$$

Determine the cross-correlation functions $\psi_{xx}(\tau)$ and $\psi_{xy}(-\tau)$. [Note that since the cross-correlation function is not an even function, $\psi_{xy}(\tau) \neq \psi_{xy}$; however, $\psi_{xx}(\tau) = \psi_{yx}(-\tau)$.]

2.21 For *white noise* with autocorrelation function $\psi_{xx}(\tau) = 2\pi S_0 \delta(\tau)$, where S_0 is the white-noise power spectral density, calculate the power spectral density $S_{xx}(\omega)$.

2.22 For the diagram below, calculate the autocorrelation function of a wideband noise with bandwidth ω_g. [Note that the mean power is given by $\psi_{xx}(0) = \overline{x^2(t)} = 2S_0\omega_g$.]

2.23 Calculate the autocorrelation function for the approximate valid power spectrum of the bell-shaped impulse given by

$$S_{xx}(\omega) = \frac{a^2}{12\beta} \exp\left(\frac{-\omega^2}{2\beta^2}\right), \qquad x(t) = a e^{-\beta^2 t^2}$$

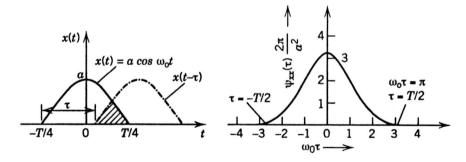

Figure Problem 2.19

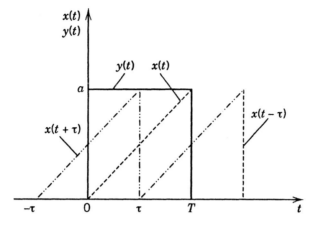

Figure Problem 2.20

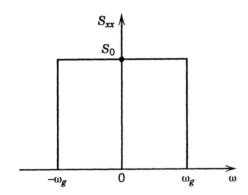

Figure Problem 2.22

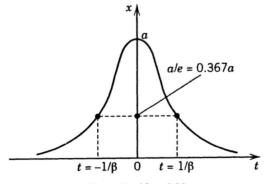

Figure Problem 2.23

2.24   Calculate the power spectrum of a random, stationary process with autocorrelation function $\psi_{xx}(\tau) = A^2\, e^{-k|\tau|} \cos \alpha\tau$.

2.25   The autocorrelation function of a random, stationary process is given by $\psi_{xx}(\tau) = A^2\, e^{-k|\tau|}$.

(a) Calculate the power spectrum $S_{xx}(\omega)$ of the process.

(b) How large is the mean power which is generated across a 1-Ω resistor when amplitude A is given by $A = 0.5\,V$.

CHAPTER 3

LINEAR REGRESSION; LEAST-SQUARES AND MAXIMUM-LIKELIHOOD ESTIMATION

3.1 INTRODUCTION

The beginnings of classical estimation theory can be attributed to R. A. Fisher, who introduced the maximum-likelihood estimation method, which from a theoretical point of view has become the most important general method of estimation [29, 30]. This chapter discusses briefly the different methods used in parameter estimation. Specifically, one often is interested in the *identification of system parameters*, that is, the evaluation of coefficients in a known mathematical model from measured input–output data. For example, one may be interested in evaluating the aerodynamic derivatives of an aircraft from flight measurements. These derivatives are contained in the coefficients of linear differential equations relating the input and output variables of the aircraft in flight. This problem can therefore be reduced to a parameter estimation which uses the tools of estimation theory. As will be discussed below, there are several methods for the solution. In essence, their differences are due to the variety of assumptions regarding the prior probability and an optimality criterion. Therefore, in this chapter we will discuss the simple linear regression, the "least-squares" estimator, and the maximum-likelihood estimator. In the first two methods, the optimality criterion is a scalar quantity, termed the cost function. Its minimization requires the minimization of the sum of the squares of deviations between data points and corresponding points derived from the solution. The maximum-likelihood estimation method is based on the maximization of probability density functions. In the present discussion we shall use the symbol θ to represent the unknown parameter vector to be estimated. Furthermore, we shall assume that θ is constant.

3.2 SIMPLE LINEAR REGRESSION

Our objective here is to develop a model which will allow us to estimate the response of a certain process and be able to place prediction intervals about these estimates. Simple linear regression will assist us in accomplishing this objective. Consider a random quantity (i.e., a random variable) Y which is a function of one or more independent (deterministic*) variables $x_1, x_2, \ldots, x_m$. More specifically, given a sample of Y-values with their associated values of x_i ($i = 1, 2, \ldots, m$), we are interested in estimating on the basis of this sample the relationship between Y and the independent variables $x_1, x_2, \ldots, x_m$. In the present discussion we will assume that the random variable Y is a function of only one independent variable, and that their relationship is linear. The term linear implies that the mean of Y, $\mathscr{E}\{Y\}$, is known to be a linear function of x. Mathematically, this is expressed as [64].

$$\mathscr{E}\{Y\} = \alpha + \beta x, \tag{3.1}$$

where the two unknown constants α and β represent the intercept and slope, respectively. In physical problems, the constant α may be a bias term and the β-term may represent a time ramp. These unknown constants (or population parameters) are to be estimated from a sample of Y-values with their associated values of x. (Data used in experimentation and process control are almost always sample data; a subset of population data.) Thus, the true regression line is not known. Furthermore, since all observations do not lie exactly on the line, there is obviously some error in our straight-line estimate. In order to incorporate this error in the equation for predicting Y, given any x-value, we use the relation

$$Y = \alpha + \beta x + e, \tag{3.2}$$

where e represents the error term, which is assumed to be distributed normally about 0 and have a variance σ^2; the e's have equal variability of all levels of x, and the error values are assumed to be independent. Note that the variance of e is identical with the variance of Y.

The model (3.2) is applicable for the population data (i.e., the set of all possible x and Y values). The random variable e is defined by

$$e = Y - (\alpha + \beta x). \tag{3.3}$$

3.3 LEAST-SQUARES ESTIMATION

The above discussion considered only a simple sample $Y(Y = Y_1, Y_2, \ldots, Y_n)$. In the discussion that follows, we will consider the sample to consist of the

---

* A deterministic system is one whose process noise is zero (i.e., noise-free).

pairs (x_j, Y_j), $j = 1, 2, \ldots, n$. One mathematical method for estimating the regression parameters α and β is the classical *method of least squares*. The method of deterministic least squares was first used by the German mathematician and astronomer Karl Friedrich Gauss (1777–1855) in his well-known paper "Theoria Motus" in 1809, dealing with the problem of orbit determination.

One application of this method to *least-squares curve fitting*, whereby one wishes to obtain the polynomial of a given order that best fits a set of data points. The name is derived from the property of minimizing the sum of squared deviations from the line, that is, $\sum e_i^2$. That is, the estimates of $\hat{\alpha}$ and $\hat{\beta}$ are chosen so that the sum of the squared differences between the observed sample values y_i and the estimated expected value of Y, $\hat{y}_i = \hat{\alpha} + \hat{\beta}x_i$, is minimized. Thus,

$$e_i = y_1 - \hat{y}_1 = y_i - (\hat{\alpha} + \hat{\beta}x_i). \tag{3.4}$$

Therefore, the least-squares estimates $\hat{\alpha}$ and $\hat{\beta}$ of α and β, respectively, are found by minimizing

$$J = \sum_{i=1}^{n} e_i^2 = \sum_{i=1}^{n} [y_i - (\hat{\alpha} + \hat{\beta}x_i)]^2. \tag{3.5}$$

The sample-value pairs are (x_1, y_1), $(x_2, y_2), \ldots, (x_n, y_n)$ while the e's are the *residuals*. Therefore, the least-squares estimates of α and β can be found by taking the partial derivatives of Eq. (3.5) and setting them equal to zero, resulting in

$$\hat{\beta} = \frac{\sum_{i=1}^{n}(x_i - \bar{x})(y_i - \bar{y})}{\sum_{i=1}^{n}(x_i - \bar{x})^2}, \tag{3.6}$$

$$\hat{\alpha} = \bar{y} - \hat{\beta}\bar{x}, \tag{3.7}$$

where

$$\bar{x} = \frac{1}{n}\sum_{i=1}^{n} x_i \quad \text{and} \quad \bar{y} = \frac{1}{n}\sum_{i=1}^{n} y_i.$$

In certain physical problems, such as data reduction, where one has to deal with a large number of test data, it is instructive to apply vector–matrix notation to the above discussion. Assume, as before, that we have the observed sample values (x_1, y_1), $(x_2, y_2), \ldots, (x_n, y_n)$. The system of observed regression equations can be written in the form of Eq. (3.2) as

$$y_i = \alpha + \beta x_i + e_i, \qquad i = 1, 2, \ldots, n. \tag{3.8}$$

Assuming a linear observation model of the form

$$\mathbf{z} = H\mathbf{x} + \mathbf{v}$$

where $\mathbf{v}$ is an $\mathbf{m}$-vector of measurement noise, and letting

$$C = \begin{bmatrix} 1 & x_1 \\ 1 & x_2 \\ \vdots & \vdots \\ 1 & x_n \end{bmatrix}, \quad \mathbf{y} = \begin{bmatrix} y_1 \\ y_2 \\ \vdots \\ y_n \end{bmatrix}, \quad \mathbf{e} = \begin{bmatrix} e_1 \\ e_2 \\ \vdots \\ e_n \end{bmatrix},$$

$$\boldsymbol{\theta} = \begin{bmatrix} \alpha \\ \beta \end{bmatrix},$$

then Eq. (3.8) can be written in a compact vector–matrix form as

$$\mathbf{y} = C\boldsymbol{\theta} + \mathbf{e}. \tag{3.9}$$

This is a linear relation in which a set of measurements $\mathbf{y}$ are linearly related to the unknown quantities $\boldsymbol{\theta}$. (Note that the vector $\mathbf{y}$ may represent a vector of test data.) From Eq. (3.5), the sum of the squared residuals takes the form

$$J(\hat{\boldsymbol{\theta}}) = \mathbf{e}^T \mathbf{e} = (\mathbf{y} - C\hat{\boldsymbol{\theta}})^T (\mathbf{y} - C\hat{\boldsymbol{\theta}}). \tag{3.10a}$$

Often, the cost function of $\hat{\boldsymbol{\theta}}$ is written in the form

$$j(\hat{\boldsymbol{\theta}}) = \tfrac{1}{2}(\mathbf{y} - C\hat{\boldsymbol{\theta}})^T (\mathbf{y} - C\hat{\boldsymbol{\theta}}). \tag{3.10b}$$

It should be noted that the quadratic cost function can be evaluated without prior knowledge of $\boldsymbol{\theta}$ or $\mathbf{e}$ ($\mathbf{e} = \mathbf{y} - C\boldsymbol{\theta}$).

The least-squares estimate of $\boldsymbol{\theta}$, $\hat{\boldsymbol{\theta}}$, can be found by minimizing $J(\hat{\boldsymbol{\theta}})$. That is, a necessary condition for the least-squares estimator is

$$\left. \frac{\partial J(\hat{\boldsymbol{\theta}})}{\partial \hat{\boldsymbol{\theta}}} \right|_{\theta = \hat{\theta}_{LS}} = 0. \tag{3.11}$$

Before we proceed in determining the lest-squares estimate $\hat{\boldsymbol{\theta}}$, the following lemma is required:

Lemma: For any square matrix $\mathbf{A}$, one has the following derivatives:

$$\frac{\partial}{\partial \mathbf{x}}(\mathbf{x}^T \mathbf{A} \mathbf{z}) = \mathbf{A}\mathbf{z},$$

$$\frac{\partial}{\partial \mathbf{x}}(\mathbf{z}^T \mathbf{A} \mathbf{z}) = \mathbf{A}^T \mathbf{z},$$

$$\frac{\partial}{\partial \mathbf{x}}(\mathbf{x}^T \mathbf{A} \mathbf{x}) = (\mathbf{A} + \mathbf{A}^T)\mathbf{x}.$$

Using the above lemma, and carrying out the operation indicated by Eq. (3.11) on Eq. (3.10), the solution for $\hat{\theta}$ is obtained as

$$\left.\frac{\partial J(\hat{\theta})}{\partial \hat{\theta}}\right|_{\theta=\hat{\theta}} = C^T(y - C\hat{\theta}) = 0$$

or

$$\hat{\theta} = [C^T C]^{-1} Cy = C^T y \tag{3.12}$$

where $C^+ = [C^T C]^{-1} C^T$ is the pseudoinverse of C, provided that C has more rows than columns. The solution is unique if C is of full rank, that is, the columns of C are linearly independent (note that $C^+ C = I$).

Equation (3.12) represents the least-squares estimate of $\hat{\theta}$. Also, this equation implies that all available measurements y are processed together at one time, in the so-called "batch processing" mode. Note that the inverse of the matrix $C^T C$ exists (this implies nonzero determinant) if there are at least two distinct values of x_i present in the sample. It can be shown that Eq. (3.12) is identical to Eqs. (3.6) and (3.7). In order to verify this fact, we note the following operations [64]:

$$C^T C = \begin{bmatrix} 1 & 1 & \cdots & 1 \\ x_1 & x_2 & \cdots & x_n \end{bmatrix} \begin{bmatrix} 1 & x_1 \\ 1 & x_2 \\ \vdots & \vdots \\ 1 & x_n \end{bmatrix} = \begin{bmatrix} n & n\bar{x} \\ n\bar{x} & \sum_{i=1}^{n} x_i^2 \end{bmatrix},$$

$$C^T y = \begin{bmatrix} 1 & 1 & \cdots & 1 \\ x_1 & x_2 & \cdots & x_n \end{bmatrix} \begin{bmatrix} y_1 \\ y_2 \\ \vdots \\ y_n \end{bmatrix} = \begin{bmatrix} n\bar{y} \\ \sum_{i=1}^{n} x_i y_i \end{bmatrix}.$$

Therefore, $\hat{\theta}$ is given by

$$\hat{\theta} = [C^T C]^{-1} C^T y = \begin{bmatrix} n & n\bar{x} \\ n\bar{x} & \sum_{i=1}^{n} x_i^2 \end{bmatrix}^{-1} \begin{bmatrix} n\bar{y} \\ \sum_{i=1}^{n} x_i y_i \end{bmatrix}$$

$$= \begin{bmatrix} \bar{y} - \hat{\beta}\bar{x} \\ \dfrac{\sum_{i=1}^{n}(x_i - \bar{x})(y_i - \bar{y})}{\sum_{i=1}^{n}(x_i - \bar{x})^2} \end{bmatrix}.$$

A special case of the preceding problem occurs when $n = 2$. This case is called *linear regression* analysis, and is an attempt to find the best (in the least-squares sense) straight line to fit a given set of data.

Finally, Eq. (3.12) can be modified by the inclusion of a weighting matrix W. As a result, Eq. (3.12) assumes the modified form

$$\hat{\boldsymbol{\theta}} = [C^T W C]^{-1} C^T W \mathbf{y}, \tag{3.13}$$

where W is assumed to be positive definite and symmetric $(m \times m)$. This weighting matrix may be used to assign different costs to each of the errors $(\mathbf{y} - C\hat{\boldsymbol{\theta}})_i$. For equal weighting of each $\mathbf{y}$, the weighting matrix W becomes an identity, that is, $W = I$. This method is often referred to as *weighted least squares*. In it, if one desires to match certain data points more closely than others, a weighting coefficient can be assigned to each term in the sum to be minimized. Note that some authors determinine the weighted least-squares estimate of by minimizing the quadradic cost function [48, 58]

$$J(\hat{\boldsymbol{\theta}}) = \tfrac{1}{2}(\mathbf{y} - C\hat{\boldsymbol{\theta}})^T W (\mathbf{y} - C\hat{\boldsymbol{\theta}}). \tag{3.14}$$

Before we leave this section, it is of interest to determine the variance of the least-squares estimation error. To this end, consider the measurement model, Eq. (3.9). The estimation error* is given by

$$\tilde{\boldsymbol{\theta}}_{LS} \triangleq \boldsymbol{\theta} - \hat{\boldsymbol{\theta}}_{LS} = \boldsymbol{\theta} - [C^T W C]^{-1} C^T W \mathbf{y} = -[C^T W C]^{-1} C^T W \mathbf{e}, \tag{3.15}$$

and the variance of the least-squares estimation error is given by

$$\text{var}\{\tilde{\boldsymbol{\theta}}_{LS}\} = \mathscr{E}\{\tilde{\boldsymbol{\theta}}_{LS}\} = -[C^T W C]^{-1} C^T W \mathscr{E}\{\mathbf{e}\}. \tag{3.16}$$

Here we require that the expected value of the estimator $\tilde{\boldsymbol{\theta}}_{LS}$ be equal to the expected value of the parameter $\boldsymbol{\theta}$. When an estimator satisfies this property, the estimator is said to be *unbiased*. Specifically, an unbiased estimator is defined as one whose expected value is the same as that of the quantity being estimated (i.e., $\mathscr{E}\{\hat{\boldsymbol{\theta}}\} = \mathscr{E}\{\boldsymbol{\theta}\}$ for all values of $\boldsymbol{\theta}$; see also Chapter 4, Section 4.3 for a definition).

From the above discussion we note that if the expected value of $\tilde{\boldsymbol{\theta}}_{LS}$ is

$$\mathscr{E}\{\tilde{\boldsymbol{\theta}}_{LS}\} = \mathscr{E}\{\mathbf{e}\} = 0, \tag{3.17}$$

then

$$\text{var}\{\boldsymbol{\theta}_{LS}\} = \text{var}\{-(C^T W C)^{-1} C^T W \mathbf{e}\}$$

$$= (C^T W C)^{-1} C^T W V_e W C (C^T W C)^{-1}, \tag{3.18}$$

---

* The estimation error, $\tilde{\boldsymbol{\theta}}$, is a defined as $\tilde{\boldsymbol{\theta}} = \boldsymbol{\theta} - \hat{\boldsymbol{\theta}}$; in words, this means the error equals the true values minus the estimate.

where V_e is the variance of e (i.e., $\mathscr{E}\{ee^T\}$). If the measurement model is given by $z = Hx + v$, then Eq. (3.18) may be written in the form

$$\text{var}\{\tilde{\boldsymbol{\theta}}_{LS}\} = \mathscr{E}\{\tilde{\boldsymbol{\theta}}_{LS}\tilde{\boldsymbol{\theta}}_{LS}^T\} = (H^T W H)^{-1} H^{-T} W V_v WH(H^T WH)^{-1} \quad (3.19)$$

where again $\mathscr{E}\{vv^T\} = V_v$. An extension of the above discussion leads to the concept of *linear minimum-variance* (LMV) estimation. For a linear minimum-variance estimator with no *a priori* information about $\boldsymbol{\theta}$, with $V^{-1} = 0$, the estimator is given by [58]

$$\tilde{\boldsymbol{\theta}}_{LMV} = (H^T V^{-1} H)^{-1} H^T V^{-1} z, \quad (3.20)$$

and the variance of the estimation error by

$$\text{var}\{\tilde{\boldsymbol{\theta}}_{LMV}\} = \mathscr{E}\{\tilde{\boldsymbol{\theta}}_{LMV}\tilde{\boldsymbol{\theta}}_{LMV}^T\} = (H^T V^{-1} H)^{-1}. \quad (3.21)$$

Note that the solution of Eq. (3.20) requires the inversion of an $N \times N$ matrix $H^T V^{-1} H$. Furthermore, we note that if $H = I$, then the error variance is equal to V, the noise variance. The error variance is independent of the measurement z, so that the variance can be computed off line to determine if the accuracy is acceptable, before any measurements are taken. Comparing Eqs. (3.13) and (3.20), we note that if $W = V^{-1}$, then the least-squares estimator will be a linear minimum-variance estimator. Consequently, since the least-squares estimator is linear and the linear minimum-variance estimator has minimum variance, we conclude that

$$\text{var}\{\tilde{\boldsymbol{\theta}}_{LMV}\} \leqslant \text{var}\{\tilde{\boldsymbol{\theta}}\} \quad (3.22)$$

for all W.

As an example, consider estimating a scalar constant. In general, in order to estimate an N-dimensional vector constant x, we make linear measurements, which are degraded or corrupted by zero-mean Gaussian noise of the form

$$z_k = H_k x + n_k$$

where z_k is the M-dimensional vector of observations, and H_k and n_k are $M \times N$ and $M \times 1$ matrices, respectively. Thus,

$$z_k^T = [z_1, z_2, ..., z_M],$$

$$H_k^T = [H_1, H_2, ..., H_M],$$

$$n_k^T = [n_1, n_2, ..., n_M].$$

As before, the estimate of $\mathbf{x}$, designated $\hat{\mathbf{x}}_k$, can be obtained by minimizing the cost function $J(\hat{\mathbf{x}}_k)$ given by

$$J(\hat{\mathbf{x}}_k) = \tfrac{1}{2}(\mathbf{z}_k - H_k\hat{\mathbf{x}}_k)^T R_k^{-1}(\mathbf{z}_k - H_k\hat{\mathbf{x}}_k)$$

where R_k^{-1} is an $M \times M$ weighting matrix. Taking the gradient of $J(\hat{\mathbf{x}}_k)$,

$$\frac{\partial J(\hat{\mathbf{x}}_k)}{\partial \hat{\mathbf{x}}_k}\bigg|_{\hat{\mathbf{x}}_k = \hat{\mathbf{x}}_{LS}} = H_k^T R_k^{-1}(\mathbf{z}_k - H_k\hat{\mathbf{x}}_{kLS}) = 0$$

or

$$\hat{\mathbf{x}}_{k_{LS}} = [H_k^T R_k^{-1} H_k]^{-1} H_k^T R_k^{-1}\mathbf{z}_k.$$

By definition, the factor in the square brackets is the $N \times N$ covariance matrix. That is,

$$P_k \triangleq [H_k^T R_k^{-1} H_k]^{-1},$$

or

$$P_k^{-1} = H_k^T R_k^{-1} H_k.$$

If we make additional measurements of $\mathbf{x}$, then

$$\mathbf{z}_{k+1} = H_{k+1}\mathbf{x} + \mathbf{v}_{k+1},$$

so that

$$\hat{\mathbf{x}}_{k+1_{LS}} = (H_{k+1}^T R_{k+1}^{-1} H_{k+1})^{-1} H_{k+1}^T R_{k+1}^{-1}\mathbf{z}_{k+1}$$

Since

$$P_{k+1}^{-1} = P_k^{-1} + H_{k+1}^T R_{k+1}^{-1} H_{k+1},$$

so that

$$P_{k+1} = (P_k^{-1} + H_{k+1}^T R_{k+1}^{-1} H_{k+1})^{-1},$$

we have, using the well-known matrix inversion lemma [58],

$$P_{k+1} = P_k - P_k H_{k+1}^T (R_{k+1} + H_{k+1} P_k H_{k+1}^T)^{-1} H_{k+1} P_k.$$

Assume now that the unknown x is a scalar and that R_k^{-1} is the identity matrix. Then, for a series of k scalar observations,

$$z_i = x + n_i,$$

where $H = [1, 1, \ldots, 1]^T$. Therefore, from

$$\hat{\mathbf{x}}_{k_{LS}} = [\mathbf{H}^T \mathbf{H}]^{-1} \mathbf{H}^T z_k$$

$$= \frac{1}{k} \sum_{i=1}^{k} z_i.$$

The product $\mathbf{H}^T z$ sums the measurements, which are divided by k to produce the estimate. Similarly, if another measurement is made, then,

$$\hat{\mathbf{x}}_{k+1} = \frac{1}{k+1} \sum_{i=1}^{k+1} z_i.$$

This leads to the recursive least-squares estimator treated in the following section.

3.3.1 Recursive Least-Squares Estimator

In batch-processing algorithms (see Section 4.3) all the measurements are processed together to provide the estimate of a constant vector. In recursive estimation, the prior estimate can be used as the starting point for a sequential estimation algorithm that assigns proper relative weighting to the old and new data. Again, consider the linear observation equation

$$\mathbf{z} = H\mathbf{x} + \mathbf{n},$$

where $\mathbf{x}$ is the constant state vector of dimension n, $\mathbf{z}$ is an m-dimensional measurement (or output) vector, H is an $m \times m$ observation matrix, and $\mathbf{n}$ is an m-dimensional error vector. Our objective here is to compute an estimate of the state, denoted by $\hat{\mathbf{x}}$, from the measurement $\mathbf{z}$. Furthermore, we can write the measurement residual as

$$\mathbf{n} = \mathbf{z} - H\mathbf{x}.$$

From Chapter 4 [Eq. (4.14)–(4.20)], we can write, without proof, the equations for the "recursive mean-value estimator" in the form [48]

$$\hat{\mathbf{x}}_k = \hat{\mathbf{x}}_{k-1} + K_k(\mathbf{z}_k - H_k \hat{\mathbf{x}}_{k-1}), \tag{3.23}$$

$$K_k = P_{k-1} H_k^T (H_k P_{k-1} H_k^T + R_k)^{-1}, \tag{3.24}$$

$$P_k = (P_{k-1}^{-1} + H_k^T R_k^{-1} H_k)^{-1} \tag{3.25}$$

where the subscript k is a *time index*, K_k is an $m \times n$ gain matrix, and P_k is an $n \times n$ matrix that represents the estimation error at the kth sampling instant.

Example We will now illustrate the recursive least-squares method with a simple example. Let us estimate the angle $\hat{\theta}$ as shown in Figure 3.1. In aircraft navigation, the angle $\hat{\theta}$ can be identified as relative bearing to destination of a TACAN beacon (TACAN measures the range and bearing to the beacon).

In this problem, we will take the observation to be in the form

$$\mathbf{y} = C\hat{\theta} + \varepsilon. \tag{1}$$

Here we want to find the estimate of the vector $\hat{\theta}$ that minimizes the cost function

$$J(\hat{\theta}) = \tfrac{1}{2}(\mathbf{y} - C\hat{\theta})^T (\mathbf{y} - C\hat{\theta}). \tag{2}$$

The cost function can be written in the standard least-squares form as follows:

$$J(\hat{\theta}) = \tfrac{1}{2} \sum_{i=1}^{m} (\mathbf{y} - C\hat{\theta})_i^2 \Rightarrow \text{minimum}. \tag{3}$$

The residual or error in the observation is $\varepsilon(t_i) = \mathbf{y}(t_i) - C\hat{\theta}$. Minimization is accomplished by taking the gradient of $J(\hat{\theta})$ and setting the result equal to

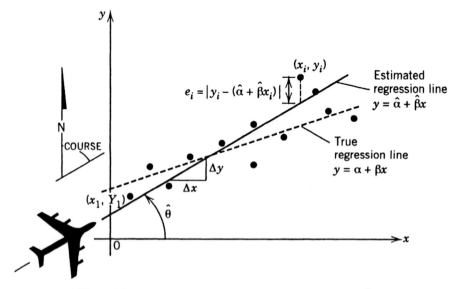

Figure 3.1   Least-square method of estimating the angle $\hat{\theta}$.

zero. Thus,

$$\left. \frac{\partial J(\hat{\theta})}{\partial \theta} \right|_{\theta = \hat{\theta}} = 0. \tag{4}$$

After n-measurements, the estimate of the parameter $\hat{\theta}$ can be obtained from

$$\hat{\theta} = \left(\sum_{i=1}^{n} C_i^T C_i \right)^{-1} \sum_{i=1}^{n} C_i y_i. \tag{5}$$

This equation involves the inversion of an $n \times n$ matrix. Now, if the covariance of the estimation error is given by

$$P_n^* = \sigma^2 (C^T C)^{-1} = \sigma^2 \left(\sum_{i=1}^{n} C_i^T C_i \right)^{-1} = \sigma^2 P_n \tag{6}$$

then from Eqs. (3.23)–(3.25) the recursive least-squares estimation of $\hat{\theta}$ becomes

$$\hat{\theta}(t_n) = \hat{\theta}(t_{n-1}) + K(t_n) [y(t_n) - C^T(t_n)\hat{\theta}(t_{n-1})], \tag{7}$$

$$K(t_n) = P_{n-1}^* C(t_n) [\sigma^2 + C^T(t_n) P_{n-1} C(t_n)]^{-1}, \tag{8}$$

$$P_n^* = P_{n-1}^* - P_{n-1}^* C(t_n) [\sigma^2 + C^T(t_n) P_{n-1}(t_n)]^{-1} C^T(t_n) P_{n-1}^*. \tag{9}$$

In order to solve these recursive equations, the initial values of $\hat{\theta}(0)$ and $P^*(0)$ must be known.

3.4 THE MAXIMUM-LIKELIHOOD ESTIMATOR (MLE)

The maximum-likelihood estimate is the expected value of the quantity being estimated, given the measurement data. Like the method of least squares, the maximum-likelihood estimation method can be traced to K. F. Gauss, and it was used extensively by R. A. Fisher in 1922 [29]. The maximum-likelihood estimator is one of the most useful estimators discussed in this chapter because of its simplicity, minimal amount of required statistical information, and provision of unbiased estimates. In particular, one of the important properties of the maximum-likelihood estimator is its invariance under invertible transformation.

Again, assume that an observation sequence z is a linear function of θ:

$$z = H\theta + n \tag{3.26}$$

where $\mathbf{z} = (z_1, z_2, \ldots, z_l)$ is an $l \times 1$ observation sequence, $\mathbf{H}$ is an $l \times n$ matrix, $\boldsymbol{\theta}$ is an $n \times 1$ unknown parameter vector, and $\mathbf{n}$ is an $l \times 1$ noise vector. One method of solving the maximum-likelihood problem is to consider the *Bayesian* estimation. The conditional probability density function* $p(\mathbf{z}|\boldsymbol{\theta})$ of the observation $\mathbf{z}$, given the unknown parameter $\boldsymbol{\theta}$, is assumed to be known. Furthermore, one is interested in determining an *a posteriori* conditional density function of the form $p(\boldsymbol{\theta}|\mathbf{z})$. Using the Bayes rule, we can write the conditional density $p(\boldsymbol{\theta}|\mathbf{z})$ in the form (see also Section 2.2.3)

$$p(\boldsymbol{\theta}|\mathbf{z}) = \frac{p(\mathbf{z}|\boldsymbol{\theta})\, p(\boldsymbol{\theta})}{p(\mathbf{z})}, \tag{3.27}$$

where $p(\boldsymbol{\theta})$ is the *a priori* probability density function of $\boldsymbol{\theta}$, and $p(\mathbf{z})$ is the probability density function of the observations.

Before we proceed with the development of the maximum-likelihood estimator, we need to define the *maximum a posteriori* (MAP) estimator $\hat{\boldsymbol{\theta}}_{MAP}$. It should be pointed out here that the deterministic least-squares approach is equivalent to the MAP approach when the appropriate assumption are introduced. Since $p(\mathbf{z})$ does not depend on $\boldsymbol{\theta}$, an equivalent definition of MAP estimation is [58]

$$p(\hat{\boldsymbol{\theta}}_{MAP})\, p(\mathbf{z}|\hat{\boldsymbol{\theta}}_{MAP}) \geqslant p(\hat{\boldsymbol{\theta}})\, p(\mathbf{z}|\hat{\boldsymbol{\theta}}) \qquad \text{for all} \quad \hat{\boldsymbol{\theta}} \neq \hat{\boldsymbol{\theta}}_{MAP}. \tag{3.28}$$

In many cases of interest, $\hat{\boldsymbol{\theta}}_{MAP}$ occurs at a stationary point of $p(\boldsymbol{\theta}|\mathbf{z})$, expressed as

$$\frac{\partial p(\boldsymbol{\theta}|\mathbf{z})}{\partial \boldsymbol{\theta}}\bigg|_{\boldsymbol{\theta} = \boldsymbol{\theta}_{MAP}} = 0 \tag{3.29a}$$

or its equivalent

$$\frac{\partial p(\mathbf{z}|\boldsymbol{\theta})\, p(\boldsymbol{\theta})}{\partial \boldsymbol{\theta}}\bigg|_{\boldsymbol{\theta} = \boldsymbol{\theta}_{MAP}} = 0. \tag{3.29b}$$

In Gaussian problems, one normally finds the stationary points of $\ln p(\boldsymbol{\theta}|\mathbf{z})$ rather than $p(\boldsymbol{\theta}|\mathbf{z})$, since $\ln(\cdot)$ is a monotonically increasing function, so that $\hat{\boldsymbol{\theta}}_{MAP}$ is given by

$$\frac{\partial \ln p(\boldsymbol{\theta}|\mathbf{z})}{\partial \boldsymbol{\theta}}\bigg|_{\boldsymbol{\theta} = \boldsymbol{\theta}_{MAP}} = \frac{\partial \ln p(\mathbf{z}|\boldsymbol{\theta})}{\partial \boldsymbol{\theta}} + \frac{\partial \ln p(\boldsymbol{\theta})}{\partial \boldsymbol{\theta}}\bigg|_{\boldsymbol{\theta} = \boldsymbol{\theta}_{MAP}} = 0. \tag{3.30}$$

---

* Sometimes the notation $p_{z|\theta}(\mathbf{z}|\boldsymbol{\theta})$ is encountered in the literature. This notation means: *the probability density of the random variable* $\mathbf{z}$ *at the value* $\mathbf{z}$, *given that the variable* $\boldsymbol{\theta}$ *has the value* $\boldsymbol{\theta}$. The MLE chooses the value of $\boldsymbol{\theta}$ which yields the largest probability of the measurement actually observed; that is, it is the parameter value for which the observed measurement was most likely.

For this reason, one frequently works with the logarithm, $\ln p(\theta|z)$, which is called the *log likelihood function*.

Now, the maximum-likelihood estimate is commonly defined as follows: for an observation z, $\hat{\theta}_{ML}(z)$ is the maximum-likelihood estimate of the parameter θ if

$$p(z|\hat{\theta}_{ML}) \geqslant p(z|\hat{\theta}) \tag{3.31}$$

for any estimate $\hat{\theta} \neq \hat{\theta}_{ML}$. That is, $\hat{\theta}_{ML}(z)$ maximizes the likelihood function $p(z|\theta)$ for a given z. From the above discussion, the maximum-likelihood estimates are obtained by maximizing the log likelihood function over the unknown parameters following substitution of the observation in place of the z variable. Therefore, from Eq. (3.27), we can write

$$\frac{\partial \ln p(z|\theta)}{\partial \theta}\bigg|_{\theta = \theta_{MAP}} = 0. \tag{3.32}$$

Similarly, we have

$$\frac{\partial \ln p(z|\theta)}{\partial \theta}\bigg|_{\theta = \theta_{ML}} = 0. \tag{3.33}$$

indicating that $\hat{\theta}_{MAP} = \hat{\theta}_{ML}$. Therefore, if we have no *a priori* knowledge of θ, then the maximum-likelihood and MAP estimates are equal. In other words, the maximum-likelihood estimate corresponds to the limiting case of a MAP estimate in which the *a priori* knowledge approaches zero.

Another method of solving the maximum-likelihood estimation problem is to consider the conditional probability function for z, conditioned on a given value of θ. In the present discussion, n is taken as a zero-mean, Gaussian-distributed observation with covariance matrix R. The conditional probability density can thus be written as

$$p(z|\theta) = \frac{1}{(2\pi)^{n/2}|R|^{1/2}} \exp\left[-\tfrac{1}{2}(z - H\theta)^T R^{-1}(z - H\theta) \right], \tag{3.34}$$

where the covariance matrix R^{-1} is of dimension $n \times n$ and corresponds to the weighting matrix W in Eq. (3.13). Consequently, in order to maximize $p(z|\theta)$, one must minimize the exponent in the brackets. The log (note that here we use log and ln interchangeably) likelihood function of Eq. (3.34) is

$$\log p(z|\theta) = -\frac{n}{2}\log(2\pi) - \tfrac{1}{2}\log \det R - \tfrac{1}{2}(z - H\theta)^T R^{-1}(z - H\theta). \tag{3.35}$$

If R is known, the maximum-likelihood estimate $\hat{\theta}_{ML}$ coincides with the least-squares estimate, corresponding to the weighting R^{-1}. Thus,

$$\hat{\theta}_{ML} = (H^T R^{-1} H)^{-1} H^T R^{-1} z \tag{3.36}$$

[compare with Eq. (3.20)]. Note that to obtain the maximum-likelihood estimate in Eq. (3.36), the observation (or measurement) matrix H must have full rank. In order to see this, we need merely to note that $\hat{\theta}$, which minimizes Eq. (3.35) for a given R and observation z, minimizes the least-squares criterion

$$\hat{\theta} \rightarrow \tfrac{1}{2}(z - H\theta)^T R^{-1} (z - H\theta).$$

At this point, the maximum-likelihood estimation method will be illustrated by means of an example.

Example Consider the general normal frequency function

$$P(x|x_0, \sigma) = \frac{1}{\sigma\sqrt{2\pi}} \exp\left[-\tfrac{1}{2}(x - x_0/\sigma)^2\right],$$

where x_0 is the arithmetic mean of the random variable x. This normal frequency function contains two adjustable parameters x_0 and σ (the standard deviation). Furthermore, this frequency function attains its maximum value $1/(\sigma\sqrt{2\pi})$ at the point x_0, and decreases asymptotically to zero as $x \rightarrow \pm\infty$ (note that the distribution is symmetrical with respect to x_0, which represents the mean) [53]. When an observed set of statistical data is to be represented by a normal frequency function, the parameters of the latter are generally so chosen that the mean and the standard deviation of the theoretical function are the same as those of the observational data. That is,

$$x_0 = \bar{x}, \qquad \sigma^2 = s^2 = \frac{1}{N}\sum_i (x_i - \bar{x})^2.$$

Let us now assume that we have observed a sample of N independent values $x_1, x_2, \ldots, x_N$ of the random variable x. In the parent population the values of x are known to be distributed normally according to the equation

$$p(x|x_0, \sigma) = \frac{1}{\sigma\sqrt{2\pi}} \exp\left[-\tfrac{1}{2}(x - x_0/\sigma)^2\right],$$

We now wish to estimate x_0 and σ. From the principle of maximum likelihood we know that our estimates of x_0 and σ should be made so that the probability of drawing a sample like the observed set becomes a maximum. Specifically, this probability should be larger when the parent population is defined by the maximum likelihood estimates of x_0 and σ. The probability density at the sample point is

$$p(x_1, x_2, \ldots, x_N | x_0, \sigma) = \frac{1}{(\sigma \sqrt{2\pi})^N} \exp \left[-\tfrac{1}{2}\sigma^2 \Sigma_i^N (x_i - x_0)^2 \right].$$

From the above discussion, the maximizing conditions are

$$\frac{\partial p}{\partial x_0} = p(x_1, x_2, \ldots, x_N | x_0, \sigma) \left[\frac{1}{\sigma^2} \sum_{i=1}^{N} (x_1 - x_0) \right] = 0,$$

$$\frac{\partial p}{\partial \sigma} = \frac{1}{\sigma^3} \left[\sum_{i=1}^{N} (x_1 - x_0)^2 - N\sigma^2 \right] p(x_1, x_2, \ldots, x_N | x_0, \sigma) = 0,$$

and the solutions for x_0 and σ which correspond to a maximum are

$$\hat{x}_0 = \frac{1}{N} \sum_{i=1}^{N} x_i \equiv \bar{x},$$

$$\hat{\sigma}^2 = \frac{1}{N} \sum_{i=1}^{N} (x_i - \bar{x})^2 \equiv s^2.$$

The properties of the maximum-likelihood estimate which are valid when the error is small are generally referred to as asymptotic. Under reasonably general conditions, the following can be proved [1, 58]:

1. The solution of the likelihood equation

$$\frac{\partial \ln p(z|\theta)}{\partial \theta} \bigg|_{\theta = \hat{\theta}_{ML}} = 0$$

   converges in probability to the correct value of θ as $N \to \infty$. Any estimate with this property is called *consistent*.
2. The maximum-likelihood estimate is asymptotically Gaussian.
3. The maximum-likelihood estimate is asymptotically efficient. That is,

$$\lim_{N \to \infty} \frac{\mathrm{var}\{\hat{\theta}_{ML} - \theta\}}{\left(-\mathcal{E}\left\{ \dfrac{\partial^2 \ln p(z|\theta)}{\partial \theta^2} \right\} \right)^{-1}} = 1. \tag{3.37}$$

Finally, we note from the *Cramér–Rao inequality* that for any *unbiased* estimator in which the error variance is conditioned,

$$\mathrm{var}\{\hat{\theta}_{ML} - \theta\} \geqslant \frac{1}{\mathcal{E}\left\{ \left[\dfrac{\partial \ln p(z|\theta)}{\partial \theta} \right]^2 \right\}} \tag{3.38a}$$

or alternatively,

$$\text{var}\{\hat{\boldsymbol{\theta}}_{ML} - \boldsymbol{\theta}\} \geqslant \left[-\mathscr{E}\left\{\frac{\partial^2}{\partial\boldsymbol{\theta}^2}\ln p(\mathbf{z}|\boldsymbol{\theta})\right\} \right]^{-1}. \tag{3.38b}$$

The Cramér–Rao bound lets us define and find an efficient estimate. Note that the Cramér–Rao bound applies only to estimates of scalars. There is a form of the bound for estimating vectors, but it is more complex. Also, for any unbiased estimate $\hat{\boldsymbol{\theta}}$ of a scalar, the conditional variance is bound by

$$\text{var}\{\hat{\boldsymbol{\theta}}|\boldsymbol{\theta}\} \geqslant \left[-\mathscr{E}\left\{\frac{\partial^2}{\partial\boldsymbol{\theta}^2}\ln p(\mathbf{z}|\boldsymbol{\theta})\right\} \right]^{-1}. \tag{3.38c}$$

As an extension of the Cramér–Rao bound, and assuming vector functions, we note that the diagonal elements of the conditional variance matrix $\text{var}\{\hat{\boldsymbol{\theta}}|\boldsymbol{\theta}\}$ are each greater than or equal to the corresponding diagonal elements of J^{-1}, where J is a square matrix given by

$$\begin{aligned} J &= \mathscr{E}\left[\frac{\partial\ln p(\mathbf{z}|\boldsymbol{\theta})}{\partial\boldsymbol{\theta}} \left(\frac{\partial\ln p(\mathbf{z}|\boldsymbol{\theta})}{\partial\boldsymbol{\theta}}\right)^T \right] \\ &= -\mathscr{E}\left[\frac{\partial}{\partial\boldsymbol{\theta}}\left(\frac{\partial}{\partial\boldsymbol{\theta}}\ln p(\mathbf{z}|\boldsymbol{\theta})\right)^T \right] \end{aligned} \tag{3.38d}$$

The matrix J is commonly called the *Fisher information matrix*. Note that in general, we can also write Eq. (3.38b) as

$$\mathscr{E}\{(\hat{\boldsymbol{\theta}} - \boldsymbol{\theta})(\hat{\boldsymbol{\theta}} - \boldsymbol{\theta})^T\} \geqslant J^{-1}$$

if J^{-1} exists.

The maximum-likelihood estimation method will now be illustrated with another example.

Example Consider a system of two transmitted signals s_1 and s_2:

$$\mathbf{S} = \begin{bmatrix} s_1 \\ s_2 \end{bmatrix} = \begin{bmatrix} x_{11} & x_{12} \\ x_{21} & x_{22} \end{bmatrix} \begin{bmatrix} A_1 \\ A_2 \end{bmatrix} = \mathbf{x}\mathbf{A},$$

and the received signal $\mathbf{R}$ given by

$$\mathbf{R} = \begin{bmatrix} r_1 \\ r_2 \end{bmatrix} = \mathbf{S} + \mathbf{n} = \begin{bmatrix} s_1 \\ s_2 \end{bmatrix} + \begin{bmatrix} n_1 \\ n_2 \end{bmatrix},$$

where the A's are two nonrandom parameters of interest and $\mathbf{n}$ is a zero-mean, Gaussian additive noise (note that n_1 and n_2 are independent). After observing

R, we want to estimate the value of **A**. The estimate will be denoted by $\hat{\mathbf{A}}$. Here we will determine whether the maximum-likelihood estimates of $\mathbf{A}_1$ and $\mathbf{A}_2$ are unbiased, compute the variance of the ML estimates $\mathbf{A}_1$ and $\mathbf{A}_2$ and test if the ML estimates are efficient. Since the noises are independent,

$$p(\mathbf{R}|\mathbf{A}) = \frac{1}{(2\pi)^{n/2}|P|^{1/2}} \exp\left[-\tfrac{1}{2}(\mathbf{R} - \mathbf{S})^T P^{-1} (\mathbf{R} - \mathbf{S}) \right], \tag{1}$$

where

$$P = \sigma^2 I.$$

Taking the logarithm of the density function yields [note that $n = 2$ in (1)]

$$\ln p(\mathbf{R}|\mathbf{A}) = k - \tfrac{1}{2}[(\mathbf{R}^T - \mathbf{S}^T) P^{-1} (\mathbf{R} - \mathbf{S})]. \tag{2}$$

In order to obtain $\hat{\mathbf{A}}_{\mathrm{ML}}$, differentiate Eq. (2) and equate the result to zero. Thus,

$$\frac{\partial \ln p(\mathbf{R}|\mathbf{A})}{\partial A_1} = \begin{bmatrix} 1 \\ 0 \end{bmatrix}^T \mathbf{x}^T P^{-1} [\mathbf{R} - \mathbf{x}\mathbf{A}] \Big|_{A = \hat{A}}, \tag{3}$$

$$\frac{\partial \ln p(\mathbf{R}|\mathbf{A})}{\partial A_2} = \begin{bmatrix} 0 \\ 1 \end{bmatrix}^T \mathbf{x}^T P^{-1} [\mathbf{R} - \mathbf{x}\mathbf{A}] \Big|_{A = \hat{A}}. \tag{4}$$

Combining Eq. (3) and (4),

$$\mathbf{x}^T P^{-1} [\mathbf{R} - \mathbf{x}\hat{\mathbf{A}}]. \tag{5}$$

Consequently, if the inverse of x exists, then

$$\hat{\mathbf{A}} = \mathbf{x}^{-1} \mathbf{R} \tag{6}$$

and

$$\mathcal{E}\{\hat{\mathbf{A}}\} = \mathcal{E}\{\mathbf{x}^{-1}[\mathbf{x}\mathbf{A} + \mathbf{n}]\} = \mathbf{A}, \tag{7}$$

implying that the ML estimates are unbiased. To compute the variances, we can use the expression

$$\hat{\mathbf{A}}(\mathbf{R}) - \mathbf{A}_i = \sum_{i=1}^{k} k_{ij}(\mathbf{A}) \frac{\partial \ln p(\mathbf{R}|\mathbf{A})}{\partial A_i}$$

$$= \mathbf{A} - \mathbf{K}(\mathbf{A})\nabla_{\mathbf{A}} [\ln p(\mathbf{R}|\mathbf{A})].$$

Substituting Eqs. (5) and (6) into Eq. (8) results in

$$\mathbf{x}^{-1} \mathbf{R} - \mathbf{A} = \mathbf{K}(\mathbf{A})[\mathbf{x}^T P^{-1}] [\mathbf{R} - \mathbf{x}\mathbf{A}]$$

or

$$\mathbf{K}^{-1}(\mathbf{A}) = \mathbf{x}^T \mathbf{P}^{-1} \mathbf{x}, \tag{9}$$

so that the estimate is efficient.

3.4.1 Recursive Maximum-Likelihood Estimator

In this section we will develop a sequential form of the maximum-likelihood estimator, such that significant computational savings can be achieved. In essence, we will examine the sequential estimation of a constant vector $\mathbf{x}$. Before we proceed with the development, we must recall that the maximum-likelihood estimation method does not assume any prior knowledge of the statistics of $\mathbf{x}$. We start the development by assuming that we can make linear measurements which are corrupted by zero-mean noise of the form

$$\mathbf{z}_p = \mathbf{H}_p \mathbf{x} + \mathbf{v}_p \tag{3.39a}$$

or in component form

$$\begin{bmatrix} z_1 \\ z_2 \\ \vdots \\ z_p \end{bmatrix} = \begin{bmatrix} h_1^T \\ h_2^T \\ \vdots \\ h_p^T \end{bmatrix} \mathbf{x} + \begin{bmatrix} v_1 \\ v_2 \\ \vdots \\ v_p \end{bmatrix}$$

where $\mathbf{h}^T$ is the ith row of the matrix H, $\mathbf{v} \sim N(0,R)$, and $R = \text{trace } \{\sigma_i^2\}$. This assumption implies that the measurements are independent. Furthermore, we will assume that there are k measurements of this form:

$$\mathbf{z}_k = \mathbf{H}_k \mathbf{x} + \mathbf{v}_k, \tag{3.39b}$$

or in component form

$$\mathbf{z}_k = \begin{bmatrix} z_1 \\ z_2 \\ \vdots \\ z_k \end{bmatrix}, \qquad H_k = \begin{bmatrix} h_1^T \\ h_2^T \\ \vdots \\ h_k^T \end{bmatrix}, \qquad \mathbf{v}_k = \begin{bmatrix} v_1 \\ v_2 \\ \vdots \\ v_k \end{bmatrix}$$

with $\mathbf{v}_k \sim N(0,R_k)$ and $R_k = \text{diag } \{\sigma_1^2, \sigma_2^2, \dots, \sigma_k^2\}$. The covariance matrix P_k for this system is

$$P_k = (H_k^T R_k^{-1} H_k)^{-1}, \tag{3.40a}$$

so that

$$P^{-1} = H_k^T R_k^{-1} H_k, \tag{3.40b}$$

while the estimate of $\mathbf{x}$ is given by

$$\hat{\mathbf{x}}_k = P_k H_K^T R^{-1} \mathbf{z}_k. \tag{3.41}$$

Now let us assume that we obtain an additional measurement of $\mathbf{x}$. Equation (3.39) can then be written in the form

$$\mathbf{z}_{k+1} = H_{k+1} \mathbf{x}_{k+1} + \mathbf{v}_{k+1}, \tag{3.42}$$

implying that we update $\mathbf{x}_k$ to $\mathbf{x}_{k+1}$. Therefore, we can adjoin this new observation to the previous observations, resulting in

$$\mathbf{z}_{k+1} = \begin{bmatrix} H_k \\ \mathbf{h}_{k+1}^T \end{bmatrix} \mathbf{x}_k + \mathbf{v}_{k+1}, \qquad \mathbf{v}_{k+1} \sim N\left(0, \begin{bmatrix} R_k & 0 \\ 0 & \sigma_{k+1}^2 \end{bmatrix}\right). \tag{3.43}$$

Similarly, the error covariance matrix P_{k+1}, based on the first $k+1$ measurements, is given by

$$P_{k+1} = (P_{k+1}^{-1} + H_{k+1}^T R_{k+1} H_{k+1})^{-1} \tag{3.44}$$

or, in terms of the $\mathbf{h}^T$'s,

$$P_{k+1} = \left(P_k^{-1} + \frac{h_{k+1} h_{k+1}^T}{\sigma_{k+1}^2}\right)^{-1}.$$

Equation (3.44) is the *error covariance update* equation. Using the matrix inversion lemma, Eq. (3.44) can be written in the form

$$P_{k+1} = P_k - P_k H_{k+1}^T (R_{k+1} + H_{k+1}^T P_k H_{k+1})^{-1} H_{k+1} P_k, \tag{3.45}$$

or again in term of the $\mathbf{h}^T$'s

$$P_{k+1} = P_k - P_k \mathbf{h}_{k+1}^T (\mathbf{h}_{k+1}^T P_k \mathbf{h}_{k+1} + \sigma_{k+1}^2)^{-1} \mathbf{h}_{k+1}^T P_k.$$

The *estimate update* equation is given by [57]

$$\hat{\mathbf{x}}_{k+1} = \hat{\mathbf{x}}_k + K_{k+1}(\mathbf{Z}_{k+1} - H_{K+1}\hat{\mathbf{x}}_k), \tag{3.46}$$

where

$$K_{k+1} = P_{k+1} H_{k+1}^T R_{k+1},$$

or

$$K_k = P_k^- H_k^T (H_k P_k^- H_k^T + R_k)^{-1}$$

where P_k^- is the *a priori* error covariance matrix.

In words, Eq.(3.46) states that the new estimate is obtained by adding to the old estimate $\hat{\mathbf{x}}$ a correction term which depends on the error between the new observation $\hat{\mathbf{z}}_{k+1}$ and the expected observation $\mathbf{H}_{k+1}\hat{\mathbf{x}}_k$; this difference is, as we have seen earlier, the residual error. The recursive equations (3.45) and (3.46) can be started by setting $P_0^{-1} = 0$ and $\hat{\mathbf{x}}_0 = \mathbf{0}$, which models the unknown prior statistics of $\mathbf{x}$.

3.5 BAYES ESTIMATION

In parameter estimation, we are concerned with the problem of determining or estimating values for system parameters. Specifically, if we denote the parameter by θ and its estimate by $\hat{\theta}$, where θ is a continuous random variable, we can assign a *cost* to all pairs $C[\theta, \hat{\theta}]$ over the range of interest. In many cases of engineering interest, it is realistic to assume that the cost depends only on the error of the estimate. We define this estimation error as [see also the definition in Section 3.3 and Eq.(3.15)]

$$\tilde{\theta} = \theta - \hat{\theta}. \tag{3.47}$$

This equation says that the error equals the true value minus the estimate. The expected value of the cost is given by the Bayes *risk* in the form

$$\mathscr{B}(\hat{\theta}) = \mathscr{E}\left\{C[\theta, \hat{\theta}(z)]\right\} = \int_{-\infty}^{\infty} \int_{-\infty}^{\infty} C[\theta, \hat{\theta}(z)] p(\theta, z) \, d\theta \, dz. \tag{3.48}$$

Equation (3.48) says that once we have specified the cost function and the *a priori* probability, we can minimize the risk. Writing $p(\theta, z)$ as $p(z|\theta)p(\theta)$, then Eq.(3.48) can be written as

$$\mathscr{B}(\hat{\theta}) = \int_{-\infty}^{\infty} \left[\int_{-\infty}^{\infty} C[\theta, \hat{\theta}(z)] p(z|\theta) \, dz \right] p(\theta) \, d\theta. \tag{3.49}$$

For a specified $\hat{\theta}$,

$$\mathscr{B} \triangleq \mathscr{E}\left\{C(\hat{\theta})\right\} = \mathscr{E}\left\{C(\theta - \hat{\theta})\right\}. \tag{3.50}$$

That is, the Bayes risk is the expected cost of an error in estimation. The risk can be minimized by proper choice of $\tilde{\theta}$. Therefore, the Bayes estimate is the estimate that minimizes the risk. From Eq.(3.49) we note that the inner integral is the conditional cost, given θ, and can be rewritten in the form

$$\mathscr{B}(\hat{\theta}|\theta) = \mathscr{E}\left\{C(\theta, \hat{\theta})|\theta\right\} = \int_{-\infty}^{\infty} C[\theta, \hat{\theta}(z)] \, p(z)|\theta) \, dz \tag{3.51}$$

representing the expected value of the cost for a given value of θ. Note that in Eq. (3.51) the expectation is over the random variable θ and the observed variables z.

Next, we wish to determine the risk which corresponds to the *mean square error* (MSE). Before we develop the MSE criterion, we rewrite Eq. (3.47) in the form

$$\mathscr{B}(\hat{\theta}) = \int_{-\infty}^{\infty} \left[\int_{-\infty}^{\infty} C[\theta, \hat{\theta}(z)] p(\theta|z) \, d\theta \right] p(z) dz. \tag{3.52}$$

where we have replaced the density $p(\theta,|z)$ with $p(\theta|z)p(z)$ for the observation z. If we elect to minimize $\hat{\theta}$, then

$$\mathscr{B}(\hat{\theta}|z) = \int_{-\infty}^{\infty} C[\theta, \hat{\theta}(z)] p(\theta|z) \, d\theta. \tag{3.53}$$

Substituting Eq. (3.47) into Eq. (3.52) we have for the mean square cost

$$C_{MS}(\theta, \hat{\theta}) = \|\theta - \hat{\theta}\|^2 = \sum_{i=1}^{l} (\theta_l - \hat{\theta}_i)^2 \tag{3.54}$$

the expression

$$\mathscr{B}_{MS}(\hat{\theta}|z) = \int_{-\infty}^{\infty} \int_{-\infty}^{\infty} \|\theta - \hat{\theta}\|^2 p(\theta|z) \, d\theta. \tag{3.55}$$

From Eq. (3.53) we have

$$\mathscr{B}_{MSE}(\hat{\theta}|z) = \int_{-\infty}^{\infty} \|\theta - \hat{\theta}\|^2 p(\theta|z) \, d\theta. \tag{3.56}$$

Therefore, the optimum estimator for the mean-square-cost criterion is given by [58]

$$\hat{\theta}_{MSE} = \int_{-\infty}^{\infty} \theta p(\theta|z) \, d\theta, \tag{3.57}$$

which is the conditional mean of θ given the observation z. This is a unique minimum, because the second derivative of the inner integral of Eq. (3.55) turns out to be equal to two.

3.6 CONCLUDING REMARKS

As we have noted in this chapter, there are three important classes of estimators that appear in the literature. These are:

1. *The Maximum Likelihood Estimator (MLE)* The maximum likelihood estimate is the value of the parameter that maximizes the probability density of the observed measurements, conditioned on the parameter. In other words, the ML estimate is the value of the parameter that makes the observed measurement the most probable measurement. Specifically, the ML estimate assumes the parameter to be nonrandom, so that one does not require any information on the statistics of the parameter. Furthermore, the ML estimate chooses the value of the parameter (or variable) which yields the largest probability of the measurement actually observed; it is the parameter value for which the observed measurement was most likely.

2. *The Maximum a Posteriori (MAP) Estimator* The maximum *a posteriori* estimate is the value of the parameter that has the largest probability (density), conditioned on the observed measurement. In other words, for the MAP estimate we choose the most probable value of the parameter, given that a particular measurement was observed. As noted earlier in this chapter, the MAP estimate differs from the ML estimate in that the MAP estimate requires prior statistics on the parameter. In general, the MAP and ML estimates need not be the same. The equation

$$p(\theta|z) = p(z|\theta)\, p(\theta)/p(z)$$

depicts the relationship between the two conditional densities that appear in the definitions of the ML and MAP estimates. From the above equation we see that if, in addition to the information required for the ML estimate, we have the probability densities for the parameter θ as well as for the measurement z, we can produce the MAP estimate. Typically, the density for z is not really necessary to determine the MAP estimate, as it merely normalizes the above equation, and does not change its dependence on θ. Finally, we note that if the probability density of the parameter θ is uniform (i.e., all values of θ are equal), then the two estimates ML and MAP will be the same.

3. *The Bayesian Minimum-Risk Estimator* In the Bayesian minimum-risk or minimum-cost estimator, one chooses a risk or cost function that one desires to keep as small as possible. One then uses the estimate which minimizes the expected value of the cost function. As we have seen, a common cost function is the squared error in the estimate, though other cost functions, such as the absolute value of the error, are also used. When squared error is the cost function, this type of estimator is called the Bayes *mean square error* (MSE) estimate.

A measure of quality to be considered in estimating the parameter $\hat{\theta}$ is the expectation of the estimate

$$\mathscr{E}\{\hat{\theta}(z)\} \triangleq \int_{-\infty}^{+\infty} \hat{\theta}(z)p(z|\theta)\, dz.$$

From this equation, we can draw the following conclusions:

1. If $\mathscr{E}\{\hat{\boldsymbol{\theta}}(\mathbf{z})\} = \boldsymbol{\theta}$ for all values of $\boldsymbol{\theta}$, then we say that the estimate is *unbiased*. This means that the average value of the estimates equals the quantity we are trying to estimate.
2. If $\mathscr{E}\{\hat{\boldsymbol{\theta}}(\mathbf{z})\} = \boldsymbol{\theta} + \mathbf{b}$, where $\mathbf{b}$ is not a function of $\boldsymbol{\theta}$, we say that the estimate has a known bias.
3. If $\mathscr{E}\{\hat{\boldsymbol{\theta}}(\mathbf{z})\} = \boldsymbol{\theta} + \mathbf{b}(\boldsymbol{\theta})$, we say that the estimate has an unknown bias. Here we see that since the bias depends on the unknown parameter $\boldsymbol{\theta}$, the bias cannot be subtracted out.

As we shall see in the next chapter, we deal mostly with unbiased estimators.

PROBLEMS

3.1 In many important distributions encountered in practice, the regression curves are straight lines with the equation $y = \alpha + \beta x$. Frequently, however, we wish to limit the functional form of $f(x)$, given $y = f(x)$. We may wish to have $f(x)$ represent only all polynomials of degree n, or only all linear functions. Our problem is then to find the best estimate of x or y in the sense of least squares. To obtain the best *linear* estimate of y the coefficients α and β must be so determined that

$$\mathscr{E}\{[y - f(x)]^2\} = \mathscr{E}\{(y - \alpha - \beta x)^2\} = \text{minimum}. \tag{1}$$

Letting $\mu_{20} = \sigma_x^2, \mu_{02} = \sigma_y^2$, and $\mu_{11} = \rho \sigma_x \sigma_y$, where the quantity ρ is the *correlation coefficient*, show that the two regression lines for x and y are

$$y = y_0 + \rho \sigma_y / \sigma_x (x - x_0), \tag{2}$$

$$x = x_0 + \rho \sigma_x / \sigma_y (y - y_0), \tag{3}$$

where each line passes through the point x_0 and y_0. [*Hint*: Take the partial derivatives of Eq. (1) first with respect to α and then with respect to β, and set the results equal to zero].

3.2 Consider the linear observation equation

$$\mathbf{z} = \mathbf{H}\mathbf{x} + \mathbf{v} \tag{1}$$

corrupted by additive noise $\mathbf{v}$. Furthermore, let the estimate of $\mathbf{x}$ be designated as $\hat{\mathbf{x}}$ and the error in the estimate by $\tilde{\mathbf{x}} = \hat{\mathbf{x}} - \mathbf{x}$. A useful quadratic cost function (to be discussed in more detail in Chapter 5, Section 5.2) is

$$J(\mathbf{z}) = 1/2(\mathbf{z} - \mathbf{H}\hat{\mathbf{x}})^T (\mathbf{z} - \mathbf{H}\hat{\mathbf{x}}). \tag{2}$$

The following are desired:

(a) Show that the *least-squares estimate* of $\hat{\mathbf{x}}$ is given by

$$\hat{\mathbf{x}} = (H^T H)^{-1} H^T \mathbf{z}. \tag{3}$$

[*Hint*: First, expand Eq. (2); then minimize $J(\mathbf{z})$ (i.e., take the partial derivative of $J(\mathbf{z})$ with respect to $\hat{\mathbf{x}}$ and set the result equal to zero)].
(b) Show that Eq. (3) averages the measurements $\mathbf{z}$ by assuming that $\mathbf{x}$ is a scalar and $\mathbf{z}$ is a k_1-vector of measurements. For this part, use the scalar form of Eq. (1) as

$$z_k = x + v_k, \qquad k = 1, \ldots, k_1,$$

with **H** is column vector of ones,

$$\mathbf{H}^T = [1 \quad 1 \cdots 1].$$

In essence, here we wish to estimate a scalar constant.

3.3 As an extension of Problem 3.2, show that the covariance P of estimation error is given by

$$P = \mathscr{E}\{\hat{\mathbf{x}}\hat{\mathbf{x}}^T\} = (\mathbf{H}^T\mathbf{H})^{-1}.$$

Note that all components of $\mathbf{v}$ in the observation equation

$$\mathbf{z} = \mathbf{H}\mathbf{x} + \mathbf{v}$$

are pairwise uncorrelated and have unit variance, that is,

$$\mathscr{E}\{v_i v_j\} = \delta_{ij} = \begin{cases} 1 & \text{if} \quad i = j, \\ 0 & \text{if} \quad i \neq j \end{cases}$$

and

$$\mathscr{E}\{\mathbf{v}\mathbf{v}^T\} = I.$$

[*Hint*: As in Problem 3.2, minimize $J(\mathbf{z}) = 1/2\,(\mathbf{z} - \mathbf{H}\hat{\mathbf{x}})^T(\mathbf{z} - \mathbf{H}\hat{\mathbf{x}}).$]

3.4 Consider a target, such as an aircraft, being tracked by two radars as shown in the diagram below. Let the range measurements $\mathbf{z}$ from the radars be r_a and r_b. The state of the target is given by $\mathbf{x}^T = [x \ y]$, and the state of the radars by $\mathbf{x}_a^T = [x_a \ y_a]$ and $\mathbf{x}_b^T = [x_b \ y_b]$. Neglecting measurement noise, the ranges are as follows:

$$r_a = \sqrt{(x - x_a)^2 + (y - y_a)^2},$$

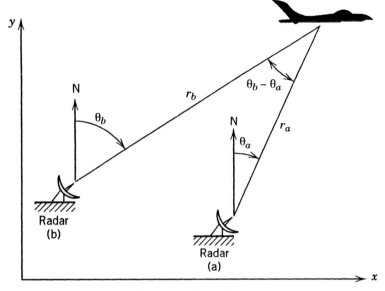

Figure Problem 3.4

$$r_b = \sqrt{(x - x_b)^2 + (y - y_b)^2}.$$

These form components of the nonlinear measurement equation

$$\mathbf{z} = \mathbf{h}(\mathbf{x}).$$

In order to solve this problem, linear sensitivity equations are required. Linearization can be made about some nominal point $\mathbf{x}_n$. Thus,

$$\Delta \mathbf{z} = \begin{bmatrix} r_a \\ r_b \end{bmatrix} = \begin{bmatrix} \dfrac{\partial z_a}{\partial x} \Delta x + \dfrac{\partial z_a}{\partial y} \Delta y \\ \dfrac{\partial z_b}{\partial x} \Delta x + \dfrac{\partial z_b}{\partial y} \Delta y \end{bmatrix} = \begin{bmatrix} \dfrac{x_n - x_a}{z_a} & \dfrac{y_n - y_a}{z_a} \\ \dfrac{x_n - x_b}{z_b} & \dfrac{y_n - y_b}{z_b} \end{bmatrix} \begin{bmatrix} \Delta x \\ \Delta y \end{bmatrix},$$

or

$$\Delta \mathbf{z} = H\, \Delta \mathbf{x}.$$

Show that

$$\mathrm{tr}\,(H^T H)^{-1} = 2/\sin^2(\theta_a - \theta_b),$$

where the trace tr of a square matrix is the sum of all the elements on its principal diagonal. *Hint:* The components of H can be written from the geometry of the problem as follows:

$$H = \begin{bmatrix} \sin \theta_a & \cos \theta_a \\ \sin \theta_b & \cos \theta_b \end{bmatrix}.$$

3.5 Consider the following observations of a scalar parameter θ (θ could represent a signal):

$$z_i = \theta + n_i, \qquad i = 1, 2, \ldots, I,$$

where the n_i's are independent and indentically distributed Gaussian random variables with mean zero and variance σ^2. The probability density function $p(z|\theta)$ is given by

$$p(z|\theta) = \prod_{i=1}^{I} p(z_i|\theta)$$

$$= \left(\frac{1}{\sigma\sqrt{2\pi}} \right) \prod_{i=1}^{I} e^{-(z_i - \theta)^2 / 2\sigma^2}$$

$$= (\sigma\sqrt{2\pi})^{-1} \exp \left[-\frac{1}{2\sigma^2} \sum_{i=1}^{I} (z_i - \theta)^2 \right].$$

Show that the maximum-likelihood estimate of the parameter θ turns out to be the *sample mean* and is given by

$$\hat{\theta}_{ML} = \frac{1}{I} \sum_{i=1}^{I} z_i.$$

[*Hint*: Take the partial derivative of $p(z|\theta)$ with respect to θ, and set the result equal to zero].

3.6 As an extension of Problem 3.4, let the observation z_i be given in the form

$$z_i = \theta + n_i, \qquad i = 1, 2, \ldots, N.$$

Assume that θ is Gaussian, $N(0, \sigma_\theta)$, and that the n_i are each independent Gaussian variables $N(0, \sigma_n)$. Then,

$$p(z/\theta) = \prod_{i=1}^{N} \frac{1}{\sqrt{2\pi}\sigma_n} \exp \left(-\frac{(z_i - \theta)^2}{2\sigma_n^2} \right),$$

$$p(\theta) = \frac{1}{\sqrt{2\pi}\sigma_\alpha} \exp \left(-\frac{\theta^2}{2\sigma_\alpha^2} \right)$$

Determine $p(\theta|\mathbf{z})$. (Note that $p(\mathbf{z})$ contributes to the constant needed to make

$$\int_{-\infty}^{+\infty} p(\theta|\mathbf{z})d\theta = 1.$$

In other words, $p(\mathbf{z})$ is simply a normalization constant).

3.7 Again, consider the observation model given in Problem 3.6:

$$z_i = \theta + n_i, \qquad i = 1, 2, \ldots, N. \tag{1}$$

Taking the logarithm of $p(z/\theta)$ and differentiating the expression

$$p(z/\theta) = \prod_{i=1}^{k} \frac{1}{\sqrt{2\pi}\sigma_n} \exp\left(-\frac{(z_i - \theta)^2}{2\sigma_n^2}\right). \tag{2}$$

$$p(\theta) = \frac{1}{\sqrt{2\pi}\sigma_\alpha} \exp\left(-\frac{\theta^2}{2\sigma_\alpha^2}\right) \tag{3}$$

show that: (a) $\hat{\theta}_{ML}(\mathbf{z}) = \frac{1}{N}\sum_{i=1}^{K} z_i$, and (b) $\hat{\theta}_{ML}(\mathbf{z})$ is unbiased; to do this, take the expectation of both sides of the result of part (1) of this problem.

3.8 The number of events in an experiment obey the Poisson law with mean value a and is given by

$$\Pr\left(n\,\text{events}\,|a = A\right) = (A^n/n!)e^{-A} \qquad n = 0, 1, \ldots \tag{1}$$

We want to observe the number of events and estimate the parameter a of the Poisson law. For this reason, we shall assume that a is a random variable with an exponential density

$$P_a(A) = \begin{cases} \lambda e^{-\lambda A} & \text{for } A > 0, \\ 0 & \text{elsewhere.} \end{cases} \tag{2}$$

The *a posteriori* density of a is given by

$$p_{a|n}(A|N) = \frac{\Pr(n = N|a = A)p_a(A)}{\Pr(n = N)} \tag{3}$$

Determine $P_{a|n}(A|N)$.

3.9 Using the Cramér–Rao inequality, Eq. (3.37), and Eq. (3.38), show that

$$\mathscr{E}\left\{\frac{\partial^2 \ln p(z|\theta)}{\partial \theta^2}\right\} = -\mathscr{E}\left\{\left[\frac{\partial \ln p(z|\theta)}{\partial \theta}\right]^2\right\}.$$

Note that

$$\mathscr{E}\left\{\left[\frac{\partial \ln p(z|e)}{\partial e}\right]^2\right\} = -\mathscr{E}\left\{\frac{\partial \ln p(z|\theta)^2}{\partial \theta}\right\}$$

$$= \int_{-\infty}^{\infty}\left[\frac{\partial \ln p(z|\theta)^2}{\partial \theta}\right] p(z|\theta)\, dz$$

$$= \int_{-\infty}^{\infty}\left[\frac{\partial \ln (z|\theta)^2}{\partial \theta}\right] p^{-1}(z|\theta)\, dz.$$

3.10 Given the observation equation

$$z_k = H_k x + v_k.$$

Assume that the noise v_k is Gaussian and the samples are independent and such that the density is given by

$$p(v_k) = K_v \exp\{v_k^T R_k^{-1} v_k\}.$$

Furthermore, the conditional density $p(z_N|x)$ is

$$p(z_N|x) = k \exp\left\{-\frac{1}{2}\sum_{i=0}^{N}(z_i - H_i \underline{x})^T R_i^{-1}(z_i - H_i \underline{x})\right\},$$

where k is a normalization constant. Determine the Fisher information matrix and the variance of the error in any unbiased estimate of x. Assumes that the expression in parenthesis in Eq.(3.38d) is integrable.

3.11 Consider the linear regression equation

$$Y = \Phi\theta + e,$$

where Y is an N-vector, Φ is an $N \times N$ matrix, θ is an N-vector, and e is Gaussian distributed with zero mean and covariance matrix $\lambda^2 I$ [i.e., $e \sim N(0, \lambda^2 I)$]. Assume that θ and λ^2 are known. The following are desired:

(a) Using the Cramér–Rao lower bound for unbiased estimates, calculate the Fisher information matrix. *Hint*: use the likelihood function

$$L(Y, \theta, \lambda^2) = \frac{1}{(2\pi)^{N/2}(\det \lambda^2 I_N)^{1/2}}\exp$$

$$\left[-\frac{1}{2}(\mathbf{Y} - \Phi\boldsymbol{\theta})^T (\lambda^2 I_N)^{-1} (\mathbf{Y} - \Phi\boldsymbol{\theta})\right]$$

where I_N is the identity matrix of order N, and its logarithm

$$\log L(\mathbf{Y}, \boldsymbol{\theta}, \lambda^2) = -\frac{1}{2\lambda^2}(\mathbf{Y} - \Phi\boldsymbol{\theta})^T (\mathbf{Y} - \Phi\boldsymbol{\theta}) - \frac{N}{2}\log 2\pi - \frac{N}{2}\log \lambda^2.$$

(b) From the results of part (a), write inequalities for $\mathrm{cov}(\boldsymbol{\theta})$ and $\mathrm{var}(\lambda^2)$.

3.12 Consider again the linear regression equation of Problem 3.11,

$$\mathbf{Y} = \Phi\boldsymbol{\theta} + \mathbf{e}, \qquad \mathbf{e} \sim N(0, R),$$

with R known and correlated errors. Find the Cramér–Rao lower bound on the covariance matrix of any unbiased estimator $\boldsymbol{\theta}$. *Hint:* Use

$$L(\mathbf{Y}, \boldsymbol{\theta}) = \frac{1}{(2\pi)^{N/2} (\det R)^{1/2}} \exp$$

$$\left[-\frac{1}{2}(\mathbf{Y} - \Phi\boldsymbol{\theta})^T R^{-1} (\mathbf{Y} - \Phi\boldsymbol{\theta})\right]$$

and the Cramér–Rao lower bound the right-hand side of the equation

$$\mathrm{cov}(\hat{\boldsymbol{\theta}}) \geqslant \left[\mathcal{E}\left(\frac{\partial \log L}{\partial \boldsymbol{\theta}}\right)^T \frac{\partial \log L}{\partial \boldsymbol{\theta}}\right]^{-1} = -\left[\mathcal{E}\frac{\partial^2 \log L}{\partial \boldsymbol{\theta}^2}\right]^{-1}.$$

3.13 As an example of **MAP** estimation, consider a Gaussian problem where $p(\mathbf{z}|\boldsymbol{\theta})$ is given by the expression

$$p(\mathbf{z}|\boldsymbol{\theta}) = K_2 \exp\left[-(\mathbf{z} - H\boldsymbol{\theta})^T (1/\sigma^2)(\mathbf{z} - H\boldsymbol{\theta})\right].$$

The product $p(\mathbf{z}|\boldsymbol{\theta})p(\boldsymbol{\theta})$ becomes

$$p(\mathbf{z}|\boldsymbol{\theta})p(\boldsymbol{\theta}) = K_3 \exp\left[-\tfrac{1}{2}(\mathbf{z} - H\boldsymbol{\theta})^T (1/\sigma^2)(\mathbf{z} - H\boldsymbol{\theta}) - (\tfrac{1}{2} V_0)(\boldsymbol{\theta} - \mu_0)^2\right].$$

Find $\hat{\boldsymbol{\theta}}_{\mathrm{MAP}}$ using Eq. (3.30). Note that $\boldsymbol{\theta}$ is Gaussian with mean μ_0 and variance V_0 so that

$$p(\boldsymbol{\theta}) = \frac{1}{\sqrt{2\pi V_0}} \exp\left[-\frac{(\boldsymbol{\theta} - \mu_0)^2}{2V_0}\right].$$

CHAPTER 4

THE KALMAN FILTER

The Kalman filtering theory is concerned with overcoming the difficulties of the Wiener filter enumerated in Section 2.7. In fact, the computation of the optimum filter becomes highly simplified, with generalized equations covering all cases, stationary or time-varying, one-dimensional or multidimensional. Although statistical data are required for the Kalman filter, they are presented in a much more simplified form than required for the Wiener problem. The Kalman filtering problem is treated entirely within the time domain. The theory accommodates both continuous-time and discrete-time linear systems, and the same equations are valid for filtering as for prediction problems.

As stated in Chapter 1, the Kalman filter is one of the most widely used estimation algorithms. This algorithm generates estimates of the variables of the system being controlled by processing available sensor measurements. This is done in real time by exercising a model of the system and using the difference between the model predictions and the measurements. Then, in conjunction with a closed-loop algorithm, the control signals are fine-tuned in order to bring the estimates into agreement with nominal performance, which is stored in the computer's memory. In essence, the Kalman filter consists of a linearized model of the system dynamics, employing statistical estimates of the system error sources in order to compute the time-varying gains for the processing of external measurement information. Consequently, the measurement information is used to generate corrections, and to improve the system compensation for critical error sources. Thus, if the system error dynamics and their associated statistics are exactly modeled in the filter, the optimum corrections for the avaiblable measurement information are generated.

4.1 THE CONTINUOUS-TIME KALMAN FILTER

We begin this chapter with certain important definitions and a discussion of the continuous-time Kalman filter for linear, time-varying systems. The continuous-time Kalman filter is used when the measurements are continuous functions of time. Linear, time-varying state models are commonly expressed through *state-space* methods. Intrinsic to any state model are three types of variables: (1) *input variables*, (2) *state variables*, and (3) *output variables*, all generally expressed as vectors. The state model identifies the dynamic and interaction of these variables. The aforementioned variables will now be defined more formally.

Input–Output Variables: The input and output variables characterize the interface between the physical system and the external world. The input reflects the excitations delivered to the physical system, whereas the output reflects the signal returned to the external world.

State Variables: The state variables represent meaningful physical variables or linear combinations of such variables. For example, the state vector is a set of n variables, whose values describe the system behavior completely.

The diagram in Figure 4.1 illustrates the general composition of a linear, time-varying state model.

From the figure, the following first-order, degree-n vector differential equation can be written:

$$\dot{\mathbf{x}}(t) = F(t)\,\mathbf{x}(t) + G(t)\,\mathbf{u}(t),$$

$$\mathbf{y}(t) = H(t)\,\mathbf{x}(t) + D(t)\,\mathbf{u}(t),$$

with initial state

$$\mathbf{x}(t_0) = \mathbf{x}_0,$$

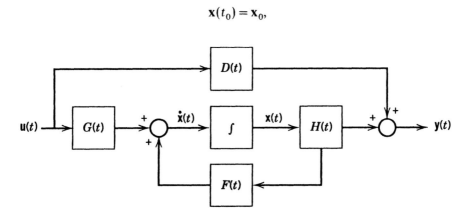

Figure 4.1 Representation for a general linear time-varying state model.

where $\mathbf{x}(t) \in R^n$ is the state vector (i.e., vector of the state variables), $\mathbf{u}(t) \in R^m$ is the system input vector, and $\mathbf{y}(t) \in R^r$ is system output vector. The matrices $F(t)$, $G(t)$, $H(t)$, and $D(t)$ have entries which are piecewise continuous, real-valued functions of time. The definitions of these matrices are as follows:

$F(t)$ is a feedback matrix of dimension $n \times n$, defining the internal interaction among the states.

$G(t)$ is a matrix of dimension $n \times m$, which is an input filter reflecting the interaction between the excitations of the external world and the state.

$H(t)$ is a matrix of dimension $r \times n$, acting as an output filter between the measurable output signals $\mathbf{y}(t)$ and the internal state variables $\mathbf{x}(t)$.

$D(t)$ is a matrix of dimension $r \times m$, and represents the feedforward system gains.

It is noted here that time-invariant state models are special cases of time-varying state models, with the time dependence deleted from the above system matrices.

Consider now an n-dimensional signal $\mathbf{x}(t)$, which is referred to as the state of the system or simply the *state vector*. (The state vector is not a measurable quantity, analogous to the one-dimensional input signal of the Wiener problem.) Then, the following time-varying linear system driven by white noise will be assumed to represent the process [2, 26, 48]:

$$\frac{d\mathbf{x}(t)}{dt} = F(t)\,\mathbf{x}(t) + G(t)\,\mathbf{w}\,(t), \qquad t \in [t_0, t_f],$$

$$\mathbf{x}(t_0) = \mathbf{x}_0, \qquad \mathscr{E}\{\mathbf{x}_0\} = \mu_x \tag{4.1}$$

where $\mathbf{x}(t) =$ a state vector of dimension $n \times 1$, which represents the error model states,

$F(t) =$ an $n \times n$ matrix, which describes the system and error model dynamics.

$G(t) =$ an $n \times r$ matrix, often called the noise gain matrix [this matrix scales the white-noise inputs and sums them with the desired combinations of the states $\mathbf{x}(t)$], which represents the effect of the input dynamics,

$\mathbf{w}(t) =$ a vector of stochastic inputs of dimensions $r \times 1$ (or zero-mean white-noise process).

The solution of the first-order time-varying vector differential equation (4.1) is given by

$$\mathbf{x}(t) = \Phi(t, t_0)\,\mathbf{x}(t_0) + \int_{t_0}^{t} \Phi(t, \tau)\,G(\tau)\,\mathbf{w}(\tau)\,d\tau \tag{4.2}$$

where the state transition matrix $\Phi(t, t_0)$ is a solution of the homogeneous matrix linear differential equation

$$\frac{d}{dt}\Phi(t, t_0) = F(t_0)\Phi(t, t_0) \qquad (4.3)$$

[handwritten annotations in margin: $\dot{x} = Ax$, $x = e^{At}$, $\dot{x} = Ae^{At}$, $Ae^{At} = Ae^{At}$]

with the initial (or boundary) condition

$$\Phi(t, t_0) = I,$$

where I is the identity matrix (see Section 4.4 for a detailed description of the state transition matrix).

Referring to Eq. (4.2), consider the second term on the right-hand side. This term represents the effect on the state vector of the input $\mathbf{u}(t)$ over the interval (t, t_0). In certain situations, the input is constant. For example, in inertial navigation systems, typical error sources such as bias errors, scale-factor errors, sensitive axis misalignments, g- and g^2-dependent drift rates, and so on, are modeled as constants and influence the propagation of navigation errors as elements of a constant $\mathbf{u}$, that is, $\mathbf{u}(t) = \mathbf{u} = $ constant. Therefore $\mathbf{u}$ may be moved to the right side of the integral, thus modifying Eq. (4.2) as follows:

$$\mathbf{x}(t) = \Phi(t, t_0)\,\mathbf{x}_0 + \left[\int_{t_0}^{t} \Phi(t, \tau)\,G(\tau)\,d\tau \right]\mathbf{u}.$$

As with the state transition matrix, the bracketed integral in the above equation may be viewed as a *sensitivity* matrix. This matrix determines the influence on $\mathbf{x}(t)$ of the constant drive term acting over the interval (t, t_0). For simplicity, we will let the symbol $S(t, t_0)$ denote this matrix, that is,

$$S(t, t_0) = \int_{t_0}^{t} \Phi(t, \tau)\,G(\tau)\,d\tau.$$

Suppose now there are available m measurements that are linearly related to the state and are corrupted by additive white noise:

$$\mathbf{z}(t) = H(t)\,\mathbf{x}(t) + \mathbf{v}(t), \qquad (4.4)$$

where $\mathbf{z}(t) = $ a vector of dimension $m \times 1$, the measurement or output vector,
   $H(t) = $ an $m \times n$ matrix, called the observation (or measurement) matrix,
   $\mathbf{v}(t) = $ a vector of stochastic observation errors of dimension $m \times 1$.

Equation (4.4) states that the output, which is a measurable quantity, is an m-dimensional vector $\mathbf{z}(t)$. This observation vector is composed of a known linear combination of the states vector with an m-dimensional noise vector $\mathbf{v}(t)$. The $m \times n$ observation matrix $H(t)$ represents the linear relationship that exists between the state and the observation vector. Assuming the prior statistics of the noise processes to be white, Gaussian, zero-mean processes, then the following apply:

$$\mathcal{E}\{\mathbf{w}(t)\} = \mathcal{E}\{\mathbf{v}(t)\} = \mathbf{0}, \qquad \mathcal{E}\{\mathbf{x}(0)\} = \mu_x(0),$$

process $\quad \mathcal{E}\{\mathbf{w}(t)\mathbf{w}^T(\tau)\} = Q(t)\delta(t - \tau),$

measurement $\quad \mathcal{E}\{\mathbf{v}(t)\mathbf{v}^T(\tau)\} = R(t)\delta(t - \tau),$ (4.5)

cross correlation $\quad \mathcal{E}\{\mathbf{v}(t)\mathbf{w}^T(\tau)\} = 0,$

where $\mathcal{E}\{\ \}$ = expectation operator,
  $\mu_x(0)$ = mean of $\mathbf{x}(0)$,
  $\delta(t)$ = Dirac delta function,
  $Q(t)$ = an $r \times r$ matrix, known as the covariance matrix of the state model uncertainties (system noise strength),
  $R(t)$ = an $m \times m$ matrix, known as the covariance matrix of the observation noise (also called measurement noise strength),

(The superscript T denotes matrix transposition.)

The Dirac delta function is defined in terms of an integral expression of the form

$$\int_{-\infty}^{\infty} \delta_D(t - \tau)\, dt = 1, \qquad \lim_{r \to 0} \int_{r - |\varepsilon|}^{t + |\varepsilon|} \delta_D(t - \tau)\, dt = 1,$$ (4.6)

so that for an arbitrary small positive ε,

$$\delta_D(t - \tau) = \begin{cases} \dfrac{1}{\varepsilon}, & \tau - \dfrac{\varepsilon}{2} < t < \tau + \dfrac{\varepsilon}{2}, \\ 0 & \text{otherwise.} \end{cases}$$ (4.7)

It should be noted from Eq. (4.5) that the white-noise sequences $\mathbf{w}(t)$ and $\mathbf{v}(t)$ are uncorrelated.

Given the above model, we update the best estimate of the state vector $\mathbf{x}(t)$ according to a linear combination of the measurements $\mathbf{z}(t)$ and the present state estimates $\hat{\mathbf{x}}(t)$ so as to minimize the performance index

$$\mathcal{E}\{[\mathbf{x}(t) - \hat{\mathbf{x}}(t)]^T [\mathbf{x}(t) - \hat{\mathbf{x}}(t)]\} = \text{minimum.}$$ (4.8)

The solution to this problem is the well-known Kalman–Bucy filter [41, 42]. The equation for the optimal estimator* (or optimal filter) is

$$A\hat{x} + L(y - C\hat{x})$$

$$\frac{d\hat{\mathbf{x}}(t)}{dt} = F(t)\hat{\mathbf{x}}(t) + K(t)[\mathbf{z}(t) - H(t)\hat{\mathbf{x}}(t)],$$ (4.9)

---

* The term "observer" appears often in the literature. An observer is an estimator with structure sufficiently well defined to guarantee consistent estimation. See Section 4.5.1 for more details on observers.

where the feedback control is formed from the best estimate of the state vector in accordance with the *seperation theorem* (see Section 5.1 for more details). Stated another way, the optimal Kalman filter estimates of the state $\hat{x}(t)$ are obtained from a weighted combination of predictions based upon the system model and corrections based upon the measurements. This optimal estimator, Eq. (4.9), is based on the correct information of initial conditions, noise covariances, and system (or coefficient matrices). Figure 4.2 depicts the matrix block diagram of the continuous-time Kalman filter.

The *Kalman gain* matrix $K(t)$ is an $n \times m$ matrix of the coefficients, which is determined by solving a nonlinear differential equation of the *Riccati* type. Now, define an error covariance matrix $P(t)$ by [1, 37]

$$P(t) = \mathscr{E}\{[\mathbf{x}(t) - \hat{\mathbf{x}}(t)][\mathbf{x}(t) - \hat{\mathbf{x}}(t)]^T\}. \tag{4.10}$$

The quantity $\mathbf{x}(t) - \hat{\mathbf{x}}(t)$ signifies the error in the estimate. Note that in Eq. (4.8), by minimum we mean minimizing the trace (i.e., $P(t) = \sigma_{\tilde{x}_1}^2 + \sigma_{\tilde{x}_2}^2 + \cdots + \sigma_{\tilde{x}_n}^2$, where $\tilde{\mathbf{x}}(t) = \mathbf{x}(t) - \hat{\mathbf{x}}(t)$ is the error in the estimate. With the initial condition $P(t)$ matrix,* $P(t_0)$, assumed known, the nonlinear matrix Riccati differential equation, also called the covariance equation, is [48, 58]

$$\frac{dP(t)}{dt} = F(t)\, P(t) + P(t)F^T(t)$$
$$- P(t)H^T(t)\, R^{-1}(t)\, H(t)\, P(t) \tag{4.11}$$
$$+ G(t)\, Q(t)\, G^T(t).$$

subject to the initial condition

$$P(t_0) = \mathrm{cov}\{\mathbf{x}(t_0), \mathbf{x}(t_0)\}.$$

$P(t)$ is a symmetric, positive definite matrix, which satisfies the matrix Riccati differential equation. Furthermore, we note that the propagation of the error covariance matrix $P(t)$ is independent of measurements; it depends only on the system dynamics $F(t)$ and system noise $Q(t)$. Consequently, if $H(t) = [0]$, no measurements are available, and Eq. (4.11) reduces to the linear covariance equation

$$AP + PA^T + GQG^T$$

$$\dot{P}(t) = F(t)\, P(t) + P(t)\, F^T(t) + G(t)\, Q(t)\, G^T(t). \tag{4.12}$$

Furthermore, we note that if $v(t)$ in Eq. (4.4) were identically zero (i.e., the measurements were *perfect*), then the covariance matrix equation (4.11) would be singular, since $R(t)$ would be a null matrix.

---

* The filter must be initialized by the user providing the initial estimate $\hat{x}(t_0)$ and its associated estimate error covariance matrix $P(t_0)$. A poor initialization will require more observations for the algorithm estimate to converge near the value of the state vector.

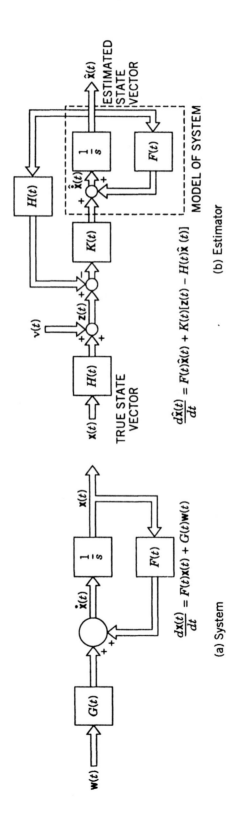

$$\frac{d\mathbf{x}(t)}{dt} = F(t)\mathbf{x}(t) + G(t)\mathbf{w}(t)$$

(a) System

$$\frac{d\hat{\mathbf{x}}(t)}{dt} = F(t)\hat{\mathbf{x}}(t) + K(t)[\mathbf{z}(t) - H(t)\hat{\mathbf{x}}(t)]$$

(b) Estimator

Figure 4.2 Matrix block diagram of the continuous-time Kalman filter system and estimator.

As stated earlier, the covariance propagation equation is only valid when certain limited restrictions are placed on the noise characteristics and the system. The flow diagram in Figure 4.3 is intended as a guide, illustrating how Eq. (4.12) can be used in the preparation of a computer program (see also Section 6.1 and Figure 6.2).

Some authors refer to Eq. (4.11) as the variance equation. This name can be explained as follows: Given two random variables X and Y with dimensions $n \times 1$ and $m \times 1$, respectively. Then the *covariance* of X and Y is defined by the

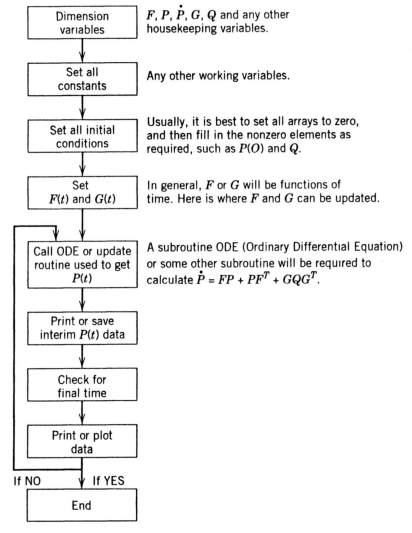

Figure 4.3 Flow diagram for computing the covariance matrix dP/dt.

$n \times m$ matrix [21]

$$\text{cov}(X, Y) = \mathscr{E}\{(X - \mathscr{E}\{X\})(Y - \mathscr{E}\{Y\})^T\}.$$

When $Y = X$, we have the *variance* matrix

$$\text{var}(X) = \text{cov}(X, X).$$

As stated in Section 2.4, the matrix $P(t)$ contains the second-order statistics of the system errors. That is, the diagonal elements are the variances of the components of the estimation error, and the off-diagonal elements are the covariances of the estimation errors. Therefore, the filter generates an $\hat{x}$ (estimate of the state) which minimizes the variances on the diagonal of $P(t)$. The optimal Kalman gain matrix is determined from the auxiliary relation

$$L = P C^T R^{-1}$$

$$K(t) = P(t) H^T(t) R^{-1}(t). \tag{4.13}$$

In the special case in which "all" the states are measured, that is, $H = I$ and $R = \text{diag}(r_1, r_2, \ldots, r_n)$, the Kalman filter gain takes the form

$$K(t) = P(t) H^T(t) R^{-1}(t) = \begin{bmatrix} \sigma^2_{\hat{x}_1}/r_1 & \sigma^2_{\hat{x}_1 \hat{x}_2}/r_2 & \cdots & \sigma^2_{\hat{x}_1 \hat{x}_n}/r_n \\ \sigma^2_{\hat{x}_2 \hat{x}_1}/r_1 & \sigma^2_{\hat{x}_2}/r_2 & \cdots & \sigma^2_{\hat{x}_2 \hat{x}_n}/r_n \\ \vdots & \vdots & \ddots & \vdots \\ \sigma^2_{\hat{x}_n \hat{x}_1}/r_1 & \sigma^2_{\hat{x}_n \hat{x}_2}/r_2 & \cdots & \sigma^2_{\hat{x}_n}/r_n \end{bmatrix}.$$

Finally, we note that the analysis leading to Eqs. (4.9), (4.11), and (4.13) is based on the assumption that $v(t)$ is white, that is, successive values for small intervals are essentially uncorrelated.

The significance or the effect of the various terms in the right-hand side of the general error covariance matrix equation, Eq. (4.11), are as follows:

$FP + PF^T$ These terms result from the behavior of the homogeneous (unforced) system without measurements. Furthermore, the sum of these terms represents the effect of the system dynamics on the propagation of the covariance matrix P.

GQG^T This term represents the increase of uncertainty due to process or driving noise input (this term is positive semidefinite). That is, this term tends to increase P.

$-PH^T R^{-1} HP$ This term represents the decrease of uncertainty as a result of measurements. It is the effect of the information provided by the measurements. Larger measurement noise will cause the error covariance matrix either to diminish less rapidly or to increase, depending on the system dynamics, the disturbances, and the initial value of P. A smaller noise will cause the filter estimates to converge on the true values more rapidly.

(Note that for convenience we have dropped the argument t.)

The term *positive definite* matrix was mentioned in the above discussion in connection with the covariance matrix $P(t)$. In the sequel, we shall have the occasion to use the terms *positive definite, positive semidefinite, negative definite,* and *negative semidefinite*. For the convenience of the reader, these terms will now be defined. Consider the following two corollaries [2]:

Corollary 1: If Q is a real, symmetric $n \times n$ matrix, then the eigenvalues $\lambda_1, \lambda_2, ..., \lambda_n$ of Q are all real.

Corollary 2: If $Q(\mathbf{v}, \mathbf{v}) = \langle \mathbf{v}, Q\mathbf{v} \rangle$ (the inner product $\langle \rangle$ is defined in Section 4.7) is the quadratic form induced by the matrix Q, then there is an orthogonal basis $\{\mathbf{f}_1, \mathbf{f}_2, ..., \mathbf{f}_n\}$ of R^n such that

$$Q(\mathbf{v}, \mathbf{v}) = \langle \mathbf{v}, Q\mathbf{v} \rangle = \sum_{i=1}^{n} \lambda_i \beta_i^2$$

for all $\mathbf{v}$ in R^n, where λ_i are the eigenvalues of Q and the β_i are the coordinates of $\mathbf{v}$ with respect to the $\mathbf{f}_i$, that is, $\mathbf{v} = \sum_{i=1}^{n} \beta_i \mathbf{f}_i$.

From these corollaries, the matrix Q will have the above properties as follows:

Positive definite if and only if all the eigenvalues $\lambda_1, \lambda_2, ..., \lambda_n$ of Q are positive, that is,

$$\lambda_i > 0 \qquad \text{for all } i.$$

Positive semidefinite if and only if all the eigenvalues $\lambda_1, \lambda_2, ..., \lambda_n$ of Q are nonnegative and at least on eigenvalue of Q is zero, that is,

$$\lambda_i \geq 0 \qquad \text{for all } i$$

and

$$\lambda_{i_1} > 0 \quad \text{for some} \quad i_1 \in \{1, 2, ..., n\}.$$

Negative definite if and only if all the eigenvalues $\lambda_1, \lambda_2, ..., \lambda_n$ of Q are negative, that is,

$$\lambda_i < 0 \qquad \text{for all } i$$

Negative semidefinite if and only if all the eigenvalues $\lambda_1, \lambda_2, ..., \lambda_n$ of Q are nonpositive and at least one eigenvalue of Q is zero, that is,

$$\lambda_i \leqslant 0 \qquad \text{for all } i$$

and

$$\lambda_i > 0 \qquad \text{for some } i_1 \in \{1, 2, ..., n\}.$$

Therefore, from the above definitions, we note that if Q is either positive or negative definite, then Q must be nonsingular, since Q is similar to a diagonal matrix which has zero entries on the main diagonal. Finally, if Q is either positive or negative semidefinite, then Q must be a singular matrix, since Q has a zero eigenvalue.

Several examples will now be given to illustrate the above discussion.

Example 1 Consider the double-integral plant (or system)

$$\dot{x}_1 = x_2,$$
$$\dot{x}_2 = w.$$

(see Figure 4.4.) The statistics of this system are as follows:

$$\mathscr{E}\{w(t)w^T(\tau)\} = 0,$$

$$\mathscr{E}\{v_1(t)\} = \mathscr{E}\{v_2(t)\} = \mathscr{E}\{v(t)\} = 0,$$

$$\mathscr{E}\{v(t)v^T(\tau)\} = \begin{bmatrix} 1 & 0 \\ 0 & 1 \end{bmatrix} \delta(t - \tau),$$

$$Q = 0, \qquad R = \begin{bmatrix} 1 & 0 \\ 0 & 1 \end{bmatrix}, \qquad P(0) = \begin{bmatrix} 4 & 0 \\ 0 & 4 \end{bmatrix}.$$

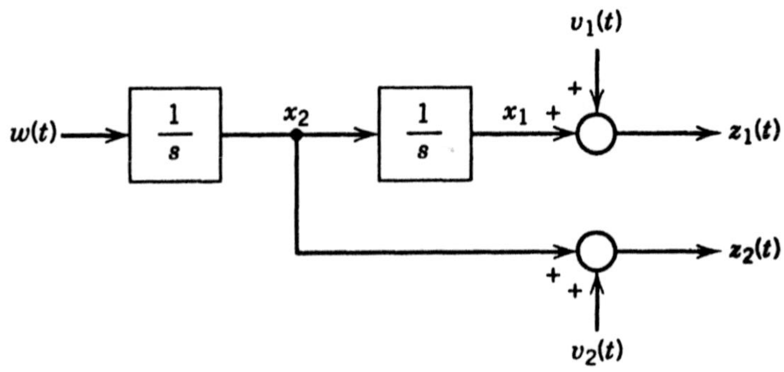

Figure 4.4 The double-integral plant.

Find the optimal estimator equations for the system (F, G, H).
    The state and observation equations are

$$\dot{\mathbf{x}}(t) = F\mathbf{x}(t) + G\mathbf{w}(t), \qquad \mathbf{z}(t) = H\mathbf{x}(t) + \mathbf{v}(t),$$

$$\dot{\mathbf{x}} = \begin{bmatrix} 1 & 0 \\ 0 & 1 \end{bmatrix}\mathbf{x} + \begin{bmatrix} 0 \\ 1 \end{bmatrix}\mathbf{w}, \qquad \mathbf{z} = \begin{bmatrix} 1 & 0 \\ 0 & 1 \end{bmatrix}\mathbf{x} + \mathbf{v}.$$

The estimation equations are, in matrix form,

$$K(t) = PH^T R^{-1} = \begin{bmatrix} P_{11} & P_{12} \\ P_{21} & P_{22} \end{bmatrix} \begin{bmatrix} 1 & 0 \\ 0 & 1 \end{bmatrix} \begin{bmatrix} 1 & 0 \\ 0 & 1 \end{bmatrix}$$

$$= \begin{bmatrix} P_{11} & P_{12} \\ P_{21} & P_{22} \end{bmatrix},$$

$$\dot{\hat{\mathbf{x}}} = \begin{bmatrix} 0 & 1 \\ 0 & 0 \end{bmatrix}\hat{\mathbf{x}} + \begin{bmatrix} P_{11} & P_{12} \\ P_{21} & P_{22} \end{bmatrix} \left(\begin{bmatrix} z_1 \\ z_2 \end{bmatrix} - \begin{bmatrix} \hat{x}_1 \\ \hat{x}_2 \end{bmatrix} \right),$$

[handwritten margin note: I think this is a mistake it is just F(t)]

where

$$\dot{p}_{11} = 2p_{12} - p_{11}^2 - p_{12}^2, \qquad p_{11}(0) = 4,$$
$$\dot{p}_{12} = p_{22} - p_{11}p_{12} - p_{12}p_{22}, \qquad p_{12}(0) = 0,$$
$$\dot{p}_{22} = -p_{12}^2 - p_{22}^2, \qquad p_{22}(0) = 4.$$

Solving the Riccati equation for p_{11}, p_{12}, and p_{22}, knowing that $K = PH^T$ $R^{-1} = P$ for this case, the estimator and variance matrix equations are easily obtained. It should be noted that even in simple examples, the nonlinear matrix Riccati equations are difficult to solve analytically.

Example 2 As an extension of the previous example, consider now the system

$$\mathscr{E}\{w_1(t)w_1(\tau)\} = \delta(t - \tau),$$

$$\mathscr{E}\{v_1(t)v_1(\tau)\} = \tfrac{1}{4}\delta(t - \tau),$$

$$P(0) = \begin{bmatrix} 1 & 0 \\ 0 & 1 \end{bmatrix}.$$

Here, as before, we wish to find the optimal estimator and determine the error covariance matrix as a function of time. The block diagram for this system is shown in Figure 4.5. The system is the same, except that no observation is taken on the state x_2. Furthermore, the noise covariance matrices for $w(t)$

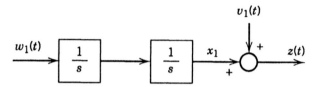

Figure 4.5   Variation of the double-integral plant.

and $v(t)$ are

$$\mathscr{E}\{w(t)\,w(\tau)\} = Q\,\delta(t-\tau) = \delta(t-\tau),$$

$$\mathscr{E}\{v(t)\,v(\tau)\} = \tfrac{1}{4}\delta(t-\tau),$$

$$P(0) = \begin{bmatrix} 1 & 0 \\ 0 & 1 \end{bmatrix}.$$

From the above, the system can be set into the form

$$\frac{dx}{dt} = \begin{bmatrix} 0 & 1 \\ 0 & 0 \end{bmatrix} x + \begin{bmatrix} 0 \\ 1 \end{bmatrix} w(t)$$

$$z(t) = [1 \ 0]\,x(t) + v(t).$$

The estimation equation is therefore

$$\frac{d\hat{x}}{dt} = \begin{bmatrix} 0 & 1 \\ 0 & 0 \end{bmatrix} \hat{x} + K(t)\{z(t) - [1 \ \ 0]\,\hat{x}(t)\},$$

$$\hat{x}(0) = 0,$$

where

$$K(t) = PH^T R^{-1} = 4 \begin{bmatrix} P_{11} \\ P_{12} \end{bmatrix},$$

and the error variance matrix is determined from the solution of

$$\dot{P}_{11} = 2P_{12} - 4(p_{11}^2 + p_{12}^2), \qquad\qquad P_{11}(0) = 1,$$

$$\dot{P}_{12} = P_{22} - 4(p_{11}\,p_{12} + p_{11}\,p_{12}), \qquad P_{12}(0) = 0,$$

$$\dot{P}_{22} = 1 - 4(p_{12}^2 + p_{22}^2), \qquad\qquad P_{22}(0) = 1.$$

Example 3   In unaided inertial navigation systems, gravity modeling errors can be one of the more significant sources of error. Let us now consider the model [61]

$$\delta\ddot{x} + \omega^2 x = \delta g + g\,\delta\varphi, \tag{1}$$

$$\delta\dot{g} + \beta\,\delta g = \sigma\sqrt{2\beta}\,w(t), \tag{2}$$

where δx = position error,
   $\omega^2 = g/R$,
   g = acceleration of gravity,
   R = earth's radius,
   δg = deflection of gravity,
   $\delta\varphi$ = inertial-platform tilt,
   $1/\beta$ = correlation time of δg (correlation time = distance/velocity)
   $w(t)$ = zero-mean, unity-PSD white-noise process.

For short periods of time (e.g., several hours), Eq. (1) represents approximately the propagation of the gravity deflections for a single-axis dynamic model driven by a correlated process [Eq. (2)], while Eq. (2) represents the deflection of gravity modeled as an exponential (or first-order Markov) noise. It should be noted, however, that in practice, higher-order models for δg are often used (e.g., second-order Markov). Furthermore, we will assume that the inertial platform is initially aligned to the true gravity vector. Therefore, any gravity deflection at $t = 0$ produces a platform tilt

$$\delta\varphi = -\delta g(0)/g, \tag{3}$$

so that δg will be a stationary process if we constrain

$$\mathscr{E}\{\delta g(0)^2\} = \sigma_{\delta g}^2. \tag{4}$$

Since we are considering only short-term propagations (gyroscope drifts are not being considered here), the dynamics of $\delta\varphi$ are simply

$$\delta\dot{\varphi} = 0. \tag{5}$$

Equations (1)–(5) can be placed in the form

$$\frac{d\mathbf{x}}{dt} = F\mathbf{x} + G\mathbf{w} \tag{6}$$

as follows:

$$\begin{bmatrix} \delta\dot{x} \\ \delta\ddot{x} \\ \delta\dot{g} \\ \delta\dot{\varphi} \end{bmatrix} = \begin{bmatrix} 0 & 1 & 0 & 0 \\ -g/R & 0 & 1 & g \\ 0 & 0 & -\beta & 0 \\ 0 & 0 & 0 & 0 \end{bmatrix} \begin{bmatrix} \delta x \\ \delta\dot{x} \\ \delta g \\ \delta\varphi \end{bmatrix} + \begin{bmatrix} 0 \\ 0 \\ \sigma_{\delta g}\sqrt{2\beta} \\ 0 \end{bmatrix} w(t) \tag{7}$$

$$\dot{\mathbf{x}} \quad = \qquad\qquad F \qquad\qquad \mathbf{x} \; + \quad G \quad \mathbf{w}$$

The covariance matrix for **x** is governed by the matrix Riccati equation

$$\frac{dP}{dt} = FP + PF^T + GQG^T, \qquad P = \mathscr{E}\{\mathbf{x}(t)\mathbf{x}^T(t)\}, \tag{8}$$

with the initial condition

$$P(0) = \begin{bmatrix} 0 & 0 & 0 & 0 \\ 0 & 0 & 0 & 0 \\ 0 & 0 & \sigma_{\delta g}^2 & \sigma_{\delta g}^2/g \\ 0 & 0 & -\sigma_{\delta g}^2/g & \sigma_{\delta g}^2/g^2 \end{bmatrix}. \tag{9}$$

Example 4 A final example will now be given. Figure 4.6 illustrates a simplified error propagation model for a single-axis, Schuler-loop inertial navigation system. Accelerometer measurements are integrated to obtain velocity and position, thereby defining the system dynamics. Specifically, the error inputs to the system are the velocity measurement error, position measurement error, platform tilt error, accelerometer bias error, and random gyroscope drift rate error. The mathematical model of this simplified inertial navigation system will

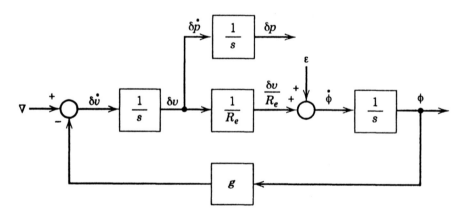

Figure 4.6 Single-axis inertial navigation system.

be formulated using state variables. The basic differential equations describing this system are [61]

$$\delta\dot{p} = \delta v,$$

$$\delta\dot{v} = -g\phi + \nabla,$$

$$\dot{\phi} = \frac{\delta v}{R_e} + \varepsilon,$$

where δp = position error,
$\qquad \delta v$ = velocity error,
$\qquad g$ = acceleration of gravity,
$\qquad \phi$ = platform tilt error,
$\qquad R_e$ = radius of the earth,
$\qquad \nabla$ = accelerometer bias error,
$\qquad \varepsilon$ = random gyroscope drift error,
$\quad 1/s$ = integrator.

The mathematical model of the errors in the inertial navigation system can be formulated using state variables. For Eqs. (4.1) and (4.4), the linear time-varying system is described by the vector differential equation

$$\dot{\mathbf{x}}(t) = F(t)\mathbf{x}(t) + G(t)\mathbf{w}(t),$$

and the discrete measurements are represented by the equation

$$\mathbf{z}(t_i) = H(t_i)\mathbf{x}(t_i) + \mathbf{v}(t_i),$$

where, as before, $\mathbf{w}(t)$ is a vector of white-noise inputs, and $\mathbf{v}(t)$ is a vector of white noise in the measurements.

Commonly, the random gyroscope drift rate is modeled as the sum of an exponentially time-correlated variable, ε_1 and a random constant $\varepsilon_2(\varepsilon = \varepsilon_1 + \varepsilon_2)$:

$$\dot{\varepsilon}_1 = -\frac{\varepsilon_1}{\tau_{\varepsilon_1}} + w_1,$$

$$\dot{\varepsilon}_2 = 0.$$

Furthermore, the accelerometer bias can be modeled as the sum of two distinct exponentially time-correlated variables, b_1 and $b_2(\nabla = b_1 + b_2)$:

$$\dot{b}_1 = -\frac{b_1}{\tau_{b_1}} + w_2,$$

$$\dot{b}_2 = -\frac{b_2}{\tau_{b_2}} + w_3.$$

Position and velocity measurements are provided to the filter. However, the system measurements are corrupted with exponentially time-correlated noises Δp and Δv in addition to the white-noise vector $v(t)$. Thus, Δp and Δv are modeled as follows [33]:

$$\Delta \dot{p} = -\frac{\Delta p}{\tau_{\Delta_p}} + w_4,$$

$$\Delta \dot{v} = -\frac{\Delta v}{\tau_{\Delta_v}} + w_5.$$

The various matrices for this simple example are listed below.

$$\mathbf{x} = [\delta p \quad \delta v \quad \phi \quad \varepsilon_1 \quad b_1 \quad \varepsilon_2 \quad b_2 \quad \Delta p \quad \Delta v]^T,$$

$$\mathbf{x} = [\delta \mathbf{p}_m \quad \delta \mathbf{v}_m]^T,$$

$$F = \begin{bmatrix} 0 & 1 & 0 & 0 & 0 & 0 & 0 & 0 & 0 \\ 0 & 0 & -g & 0 & 1 & 0 & 1 & 0 & 0 \\ 0 & 1/R_e & 0 & 1 & 0 & 1 & 0 & 0 & 0 \\ 0 & 0 & 0 & -1/\tau_{\varepsilon_1} & 0 & 0 & 0 & 0 & 0 \\ 0 & 0 & 0 & 0 & -1/\tau_{b_1} & 0 & 0 & 0 & 0 \\ 0 & 0 & 0 & 0 & 0 & 0 & 0 & 0 & 0 \\ 0 & 0 & 0 & 0 & 0 & 0 & -1/\tau_{b_2} & 0 & 0 \\ 0 & 0 & 0 & 0 & 0 & 0 & 0 & -1/\tau_{\Delta p} & 0 \\ 0 & 0 & 0 & 0 & 0 & 0 & 0 & 0 & -1/\tau_{\Delta v} \end{bmatrix},$$

$$G = \begin{bmatrix} 0 & 0 & 0 & 1 & 0 & 0 & 0 & 0 & 0 \\ 0 & 0 & 0 & 0 & 1 & 0 & 0 & 0 & 0 \\ 0 & 0 & 0 & 0 & 0 & 0 & 1 & 0 & 0 \\ 0 & 0 & 0 & 0 & 0 & 0 & 0 & 1 & 0 \\ 0 & 0 & 0 & 0 & 0 & 0 & 0 & 0 & 1 \end{bmatrix},$$

$$H = \begin{bmatrix} 1 & 0 & 0 & 0 & 0 & 0 & 0 & 1 & 0 \\ 0 & 1 & 0 & 0 & 0 & 0 & 0 & 0 & 1 \end{bmatrix},$$

$$Q = \begin{bmatrix} 0 & 0 & 0 & 0 & 0 & 0 & 0 & 0 \\ 0 & 0 & 0 & 0 & 0 & 0 & 0 & 0 \\ 0 & 0 & 0 & 0 & 0 & 0 & 0 & 0 \\ 0 & 0 & 0 & 2\sigma_{\varepsilon_1}^2/\tau_{\varepsilon_1} & 0 & 0 & 0 & 0 \\ 0 & 0 & 0 & 0 & 2\sigma_{b_1}^2/\tau_{b_1} & 0 & 0 & 0 \\ 0 & 0 & 0 & 0 & 0 & 0 & 0 & 0 \\ 0 & 0 & 0 & 0 & 0 & 2\sigma_{b_2}^2/\tau_{b_2} & 0 & 0 \\ 0 & 0 & 0 & 0 & 0 & 0 & 2\sigma_{\Delta p}^2/\tau_{\Delta p} & 0 \\ 0 & 0 & 0 & 0 & 0 & 0 & 0 & 2\sigma_{\Delta v}^2/\tau_{\Delta v} \end{bmatrix},$$

$$R = \begin{bmatrix} \sigma_{\delta_p}^2 & 0 \\ 0 & \sigma_{\delta_v}^2 \end{bmatrix}.$$

The state transition matrix of the above model must be determined before one proceeds with evaluating the model. For the simple case where the driving errors and noises are zero, that is, $\nabla = \varepsilon = 0$, the model is simplified to

$$\delta \dot{p} = \delta v,$$

$$\delta \dot{v} = -g\phi$$

$$\dot{\phi} = \frac{\delta v}{R_e},$$

so that the state transition matrix is given by

$$\Phi(\Delta T) = \begin{bmatrix} 1 & \dfrac{1}{\omega_s}\sin(\omega_s \Delta T) & R_e[\cos(\omega_s \Delta T) - 1)] \\ 0 & \cos(\omega_s \Delta T) & -\dfrac{g}{\omega_s}\sin(\omega_s \Delta T) \\ 0 & \dfrac{1}{R_e\omega_s}\sin(\omega_s \Delta T) & \cos(\omega_s \Delta T) \end{bmatrix},$$

where ΔT is the sampling period between the kth and $(k-1)$th samples, and ω_s is the Schuler frequency.

4.2   INTERPRETATION OF THE KALMAN FILTER

The equation for the Kalman filter is given by Eq. (4.9). From the terms on the right-hand side of Eq. (4.9) it can be seen that in the absence of any measurements, the optimal estimates of the states evolve in time according to the same dynamical relationships as in the actual system. Thus, it can be seen that if the measurements are interrupted or shut off, the state estimates can be found by solving the known equations of motion of the real system. This is intuitively reasonable, and is the heart of Kalman filtering theory. In particular, the second term on the right-hand side of Eq. (4.9) shows the effect of measurements on the state estimates. It is the difference between the actual measurement $z(t)$ and the expected measurement $H(t)\hat{x}(t)$ that drives the estimator. (The quantity in the brackets is also known as *innovations*.) Thus, when the estimator is doing well, the driving term to the Kalman filter is small. The Kalman gains $K(t)$ represent the weight given to the incoming measurements for tuning up the estimates of the state. From Eq. (4.13), $K(t)$ is proportional to the covariance of the errors in the estimate, $P(t)$, and inversely proportional to the covariance of the measurement errors $R(t)$. These are both reasonable results, since for a given $R(t)$, decreased confidence in the estimates—indicated by a larger $P(t)$—would call for a heavier weighting of the incoming measurements. Similarly, for given $P(t)$, an increase in $R(t)$ means the measurements are noisier, implying that they should be weighted less heavily.

A word now about white noise is appropriate (see Section 2.8 for a detailed mathematical description of white noise). First of all, the filter performs best for linear systems. Second, the system noises which drive the filter are assumed to be white and Gaussian. Under the restrictions of linearity, whiteness, and Gaussianness, the Kalman filter is the best filter of any conceivable form.

Whiteness implies that the noise value is not correlated in time (i.e., the correlation time is zero). It also implies that the noise has equal power at all frequencies. This is in analogy with white light, which contains all frequencies. Since this results in a noise with infinite power, a true white noise cannot exist. However, white noise can be approximated with a wideband noise having low power at frequencies about the system bandpass, and essentially constant power at all frequencies within the system bandpass. Within the bandpass of the system of interest, such noise looks identical to the fictitious white noise, and the mathematics involved in the Kalman filter is considerably simplified.

The properties of the filter that make it useful as an estimation model can be summarized as follows:

1. At a given time t, the filter generates an unbiased estimate $\hat{x}$ of the state vector x; that is, the expected value of the estimate is the value of the state vector at time t.
2. The estimate is a minimum-variance estimate.
3. The filter is recursive, meaning it does not store past data.
4. The filter is linear or it must be linearized. Linearization simplifies calculations, making them suitable for machine computation.

In applying the Kalman filtering theory, we make the following model assumptions:

1. The state vector $\mathbf{x}(t)$ exists at the time t in a random environment (i.e., system dynamics) that is Gaussian with zero mean and covariance matrix $Q(t)$.
2. The state vector, which is unknown, can be estimated using observations or data samples that are functions of the state vector.
3. An observation made at a point in time t is corrupted by uncorrelated, Gaussian noise, having a zero mean and covariance matrix $R(t)$.

In engineering applications, the engineer must make certain that the stability of Kalman filter is satisfied. By stability we mean that the asymptotic values of the estimated state and its error covariance matrix are independent of their respective *initial* values (note that filtering stability is one of the requirements for an unbiased estimate). Furthermore, if the chosen or designed filter has been tuned properly (see also Section 4.6), an *error budget* can be established. In error budget analysis, the sytem designer can determine which are the predominant error sources, so that he can utilize the best hardware that contribute smaller errors in the overall system accuracy. In this context, error budget analysis consists of repeated covariance analyses whereby the error sources in the truth model are tuned on individually in order to determine the separate effects of these error sources. The concept of the error budget can be summarized by considering the following points (see also Section 6.2.1):

1. It catalogs the contribution of the individual error sources to the total system error.
2. It provides clues to better design of the optimal (or suboptimal) filter.
3. It relies on the linearity of the covariance analysis formulation.
4. It can be used to assess the theoretical or technological performance limits of the system under design.
5. It can be considered as a special form of sensitivity analysis.

4.3 THE DISCRETE-TIME KALMAN FILTER

A Kalman filter can also be derived for discrete systems. For sampled data systems, the discrete-time form of the Kalman filter is of interest. In the discrete form, the measurements to improve the estimate of the state of the system are made at discrete intervals of time, even if the source of the measurements operates continuously. However, we note that some information is lost if a continuous source of measurement data is sampled at discrete points in time, but this situation can be avoided by prefiltering. On the other hand, measurements in certain avionics applications are available only at distinct points in time. For example, a radar is often used in a pulsed, and thus sampled, mode. The recursive nature of the discrete-time case implies that there is no need to store past measurements and older estimates for the purpose of computing present estimates. The recursive characteristic is important because the burden on the onboard computer memory capacity is considerably reduced. Since in practice the Kalman filter algorithm is implemented in a digital computer, and since the digital computer cannot perform the differentiation and integration operations required for solution of the continuous-time equations, a difference-equation formulation of the filter is appropriate. Therefore, the usefulness of the discrete-time case lies in the fact that digital computers accept only *discrete data*, or *data sequences*. The dynamical system of Eq. (4.1) may be written in discretized form, making use of the state-transition matrix defined by Eq. (4.3).

The linear, discrete-time equivalent of the continuous-time Kalman filter state model given in Section 4.1 can be generated from various schemes available for digital simulation. Some of these schemes are: (1) Euler's technique, (2) Simpson's method, and (3) trapezoidal method. For the present discussion, we will select the *forward Euler* approximation technique, which can be expressed as follows:

$$\dot{x}(t_k) \approx \frac{x(t_{k+1}) - x(t_k)}{t_{k+1} - t_k},$$

where $x(t_k)$ is thought of as the sampled data value of the continuous-time signal $x(t)$. Substituting the above equation into the continuous-time state model (4.1) results in

$$\frac{x(t_{k+1}) - x(t_k)}{t_{k+1} - t_k} \approx F(t_k)\, x(t_k) + G(t_k)\, u(t_k).$$

[Note that here we have replaced $w(t)$ with $u(t)$ in order to conform with the diagram and system equations given in the beginning of Section 4.1.]

Now, let $h = t_{k+1} - t_k = \Delta t$ be a fixed step size for all k. Consequently, the discrete-time model can be written in the form

$$x(t_{k+1}) \approx [1 + h F(t_k)]\, x(t_k) + h G(t_k)\, u(t_k).$$

Before we arrive at the final form for the discrete-time state model, certain preliminaries are in order. As it is common in the literature of estimation theory, let $x(t_k)$ be denoted $x(k)$. Then the matrix F can be written as

$$F(k) = I + h F(t_k)$$

and the matrix G as

$$G(k) = h G(t_k).$$

Therefore, we can write the linear time-variant (i.e., nonstationary) discrete-time state model as

$$x(k + 1) = F(k)\, x(k) + G(k)\, u(k),$$

$$y(k) = H(k)\, x(k) + D(k)\, u(k).$$

For time-invariant or slowly varying systems, the recursive discrete form of the Kalman filter may be written as follows [21, 48, 58]*:

$$\mathbf{x}(k + 1) = \Phi(k + 1, k)\, \mathbf{x}(k) + \Gamma(k)\, \mathbf{w}(k), \quad k = 0, 1, 2, \ldots, \tag{4.14}$$

$$\mathbf{w}(k) \sim N(0, Q(k)) \qquad \text{(white Gaussian sequence with } k > 0\text{)};$$

$$\mathbf{z}(k) = H(k)\mathbf{x}(k) + \mathbf{v}(k), \tag{4.15}$$

$$\mathbf{v}(k) \sim N(0, R(k)) \qquad \text{(white Gaussian sequence)};$$

$$\hat{\mathbf{x}}(k + 1 | k) = \Phi(k + 1, k)\, \hat{\mathbf{x}}(k | k), \tag{4.16}$$

$$P(k + 1 | k) = \Phi(k + 1, k)\, P(k | k)\, \Phi^T(k + 1, k) + \Gamma(k)Q(k + 1)\Gamma^T(k), \tag{4.17}$$

$$\hat{\mathbf{x}}(k | k) = \hat{\mathbf{x}}(k | k - 1) + K(k)[\mathbf{z}(k) - H(k)\hat{\mathbf{x}}(k | k - 1)]; \tag{4.18}$$

$$P(k | k) = [I - K(k)H(k)]\, P(k | k - 1) \tag{4.19a}$$

or

$$P(k | k) = P(k | k - 1) - K(k)\, H(k)\, P(k | k - 1); \tag{4.19b}$$

$$K(k) = P(k | k - 1)\, H^T(k)\, [H(k)\, P(k | k - 1)\, H^T(k) + R(k)]^{-1}, \tag{4.20}$$

where the superscript T denotes matrix transposition. Equation (4.14) is called the state equation. It is a first-order difference equation in k, relating one value of $\mathbf{x}$, $\mathbf{x}(k)$, to the next value $\mathbf{x}(k + 1)$. The vector $\mathbf{x}(k)$ represents the parameter, or state vector, whose components we will try to estimate. When the components of the state vector are time-dependent, Eq. (4.14) structures their dynamical behavior. Most of the generality of the recursive model is due to the use of this equation. Therefore, Eq. (4.14) is a model for the true state of affairs, the values of the state vector at time k. Equation (4.15), called the observation equation, is a model of the measurement process. $\mathbf{z}(k)$ represents the measurement (or observation) vector whose components are the individual scalar measurements made at time k. Furthermore, Eq. (4.15) relates these measurements to the state vector via the observation matrix $H(k)$ for $k \geqslant 1$. The random measurement noise is represented by $\mathbf{v}(k)$ for $k \geqslant 1$.

It should be pointed out that the observation equation contributes no dynamics to the model. Equation (4.17) represents the covariance propagation. In Section 4.2 we discussed the importance of the Kalman gain $K(k)$ in connection with the continuous-time case. Consider now Eq. (4.18). This equation says that our best estimates at $t = k$, using all observations up to and including

---

* The notation $\mathbf{w}(k) \sim N(0, Q(k))$ means that the noise has zero mean and covariance $Q(k)$. In general, if a random variable x is Gaussian with mean m and variance σ^2, then we can write $\mathbf{x} \sim N(m, \sigma^2)$. (see also Section 2.4)

k, is equal to our predicted estimate plus the error (difference) between what was observed and what we claim should have been observed multiplied by some weighting factor (or gain) $K(k)$. The key to this equation is indeed the weighting factor. In essence, it determines how much we will alter or change our estimate of the state based on the new observation. Therefore, (1) if the elements of $K(k)$ are small, we have considerable confidence in our model, and (2) if they are large, we have considerable confidence in our observation measurements.

$K(k)$ changes with time, and we wish to assure that it is always optimum. It is highly dependent upon our characterization of the measurement noise $R(k)$. Thus, if $R(k) = 1$, the elements of $K(k)$ strictly decrease in absolute value and represent the classical least-squares method. From another point of view, we note from Eq. (4.20) that, if a measurement is processed for which the measuresurement noise is small relative to the prior errors [that is, if R is small relative to $H(k) P(k|k-1) H^T(k)$], then the gain is larger, and if the measurement is noisy relative to the prior errors [that is, if R is large relative to $H(k) P(k|k-1) H^T(k)$], then the gain is small and the filter pays little attention to the measurement.

In applying the linear filter to a specific system, the dynamical model $[\Phi, \Gamma, H]$, noise statistics (Q, R), and a priori data $[\hat{x}(0), P(0)]$ must be specified. As in the continuous-time case, the prior statistics of the noise processes $w(t)$ and $v(t)$ are assumed to be zero-mean and white. Thus,

$$\mathscr{E}\{w(k)\} = \mathscr{E}\{v(k)\} = 0, \qquad \mathscr{E}\{x(0)\} = \mu_x(0),$$

$$\mathscr{E}\{w(k)w^T(j)\} = Q(k)\delta_{kj},$$

$$\mathscr{E}\{v(k)v^T(j)\} = R(k)\delta_{kj}, \tag{4.21}$$

$$\mathscr{E}\{w(k)v^T(j)\} = 0, \qquad \forall j, k,$$

where $\mu_x(0)$ is the mean and δ_{kj} is the Kronecker delta defined as

$$\delta_{kj} = \begin{cases} 0 & \text{for } k \neq j, \\ 1 & \text{for } k = j. \end{cases} \tag{4.22}$$

The system noise and measurement noise covariance matrices can also be expressed as symmetric matrices in the form

$$Q(k) = \begin{bmatrix} q_{11} & q_{12} & \cdots & q_{1s} \\ q_{12} & q_{22} & \cdots & q_{2s} \\ \vdots & \vdots & & \vdots \\ q_{1s} & q_{2s} & \cdots & q_{js} \end{bmatrix} = \begin{bmatrix} \sigma_1^2 & \rho_{12}\sigma_1\sigma_2 & \cdots & \rho_{1s}\sigma_1\sigma_s \\ \rho_{12}\sigma_1\sigma_2 & \sigma_2^2 & \cdots & \rho_{2s}\sigma_2\sigma_s \\ \vdots & \vdots & & \vdots \\ \rho_{1s}\sigma_1\sigma_s & \rho_{2s}\sigma_2\sigma_s & \cdots & \sigma_s^2 \end{bmatrix},$$

where ρ_{ij} is the cross-correlation coefficient for the ith and jth variables [see also Chapter 6, Eq. (6.3)]. A similar expression can be written for $R(k)$, with the exception that the off-diagonal terms will be zero.

The noise strength matrix $Q(k)$ is given by

$$\mathscr{E}\{w(t_{k+1})w^T(t_{k+1})\} = Q(t_k) = \int_{t_k}^{t_{k+1}} \Phi(t_{k+1}, \tau) Q(\tau) \Phi^T(t_{k+1}, \tau)\, d\tau.$$

Note that here we substituted t_{k+1} for k. For time-invariant or slowly varying matrices $F(t)$, $G(t)$, and $G(t)Q(t)G^T(t)$, a first-order approximation of $Q(k)$ is given by

$$Q(k) \approx G(t_k) Q(t_k) G^T(t_k) [t_{k+1} - t_k].$$

Given a set of sequential observations $\mathbf{Z}(k) = \{z(1), z(2), ..., z(k)\}$, we wish to determine an estimate of $\mathbf{x}(j)$, which we shall represent by $\hat{\mathbf{x}}(j|k)$, which is the estimate of the state at the jth sample, based on observations to the kth sample. Then the estimation error will be denoted by [33, 43]

$$\tilde{\mathbf{x}}(j|k) = \mathbf{x}(j) - \hat{\mathbf{x}}(j|k).$$

In view of the above discussion, refer to the estimation process as *smoothing* or *interpolation* if $j < k$, *filtering* if $j = k$, and *prediction* or *extrapolation* if $j > k$.

It can be shown that in the limit as the sample time $T_s \to 0$, that is, for infinitesimally small sampling intervals, the above difference equations approach the continuous differential covariance Eq. (4.11). Furthermore, the continuous Kalman gains, as defined by Eq. (4.13), are related to the discrete-time gains of Eq. (4.20) by

$$K(t) = \lim_{T_s \to 0} \frac{K(k)}{T_s}. \tag{4.23a}$$

Similarly, the continuous-time and discrete-time measurement noise covariances are related by

$$R(t) = \lim_{T_s \to 0} \frac{R(k)}{T_s}. \tag{4.23b}$$

From the above discussion, the covariance matrix $\dot{P}(t)$ can also be shown to be related to the discrete-time covariance matrix by the expression

$$\lim_{\Delta t \to 0} \frac{P(k) - P(k-1)}{\Delta t} = \dot{P}(t). \tag{4.24}$$

The system noise strength matrix Q can be interpreted as follows: if one were to increase Q, this would indicate one of two things: (1) that stronger noises are driving the dynamics, or (2) increased uncertainty in the ability of the model to describe the *true* dynamics accurately. Consequently, the rate of growth of the elements of the covariance matrix $P(t)$ will also increase, which also means that the filter gains will increase, thus weighting the measurements more heavily. Therefore, it must be remembered that by increasing Q, we in effect are putting less confidence in the output of the filter's own dynamics model. Similary, increasing R indicates that the measurements are subjected to a stronger corruptive noise, and therefore should be weighted less by the filter.

The variables appearing in Eqs. (4.14)–(4.20) are defined as follows:

$\mathbf{x}(k)$ — The *true* state vector; it consists of all parameters or errors that are estimated by the filter (e.g., position, velocity, attitude).

$\Phi(k+1, k)$ — The state transition matrix; it transforms a given state at a time t_k to another state at time t_{k+1}. This matrix is used to propagate the state covariance matrix. It is also used in the calculation of the state-vector estimate at the present point in time, t_k, from the state-vector estimate at a past point in time, t_{k-1}.

$\Gamma(k)$ — The system noise coefficient matrix.

$\mathbf{w}(k)$ — The system noise.

$\mathbf{v}(k)$ — The measurement noise.

$\mathbf{z}(k)$ — The measurement or observation vector at time t_k. The measurement $\mathbf{z}(k)$ is processed by the filter to yield the best estimate of the error state $\mathbf{x}(k)$. Also, the $\mathbf{z}(k)$ components are linear combinations of the components of the true state vector $\mathbf{x}(k)$ which have been corrupted by uncorrelated Gaussian noise $\mathbf{v}(k)$.

$H(k)$ — The measurement or observation matrix. It relates the components of the true state vector to the measurements. Moreover, the measurement matrix is computed at each measurement time and contains the partial derivatives of the measurement equations. This matrix depends, to a great extent, on the type of measured data needed for the filter implementation.

$\hat{\mathbf{x}}(k|k)$ — The state estimate at t_k, given $\mathbf{z}_k = \{z_1, z_2, ..., z_k\}$. This is an unbiased* (zero-mean) minimum variance estimate of the "true" state vector $\mathbf{x}(k)$ at time t_k.

$\hat{\mathbf{x}}(k+1|k)$ — The state estimate at t_{k+1}, given $\mathbf{z}_k$. The quantity $\mathbf{z}(k) - H(k)\hat{\mathbf{x}}(k|k-1)$ in Eq. (4.18) is the difference between the actual measurement $\mathbf{z}(k)$ and the predicted measurement. As stated earlier, this quantity is referred to as "innovations".

$Q(k+1)$ — The system noise covariance matrix; it relates errors in the state propagation to the uncertainty of the current estimate expressed in the covariance matrix. This matrix must be input in advance by the user; its value is important, since it affects the filter's performance.

---

* An *unbiased* estimate is one whose expected value is the same as that of the quantity being estimated. A *minimum-variance*, unbiased estimate has the property that its error variance is less than or equal to that of any other unbiased estimate. Furthermore, the minimum-variance estimate is always the conditional mean of the state vector, regardless of its probability density function (see also Section 2.2.1).

$R(k+1)$ The measurement noise covariance matrix. This matrix is determined by the accuracy of the device (e.g., navigational) used to generate the input. This accuracy, or lack of it, affects the performance of
the filter. The diagonal elements of $R(k+1)$ are the variances of the
sensor measurements. This matrix must be provided by the user.

$P(k|k)$ The error covariance matrix. That is, this matrix represents the
covariance of the difference between the true state vector $\mathbf{x}(k)$ and
the estimated state vector $\hat{\mathbf{x}}(k)$:

$$P(t_k) = \mathscr{E}\{[\mathbf{x}(t_k) - \hat{\mathbf{x}}(t_k)][\mathbf{x}(t_k) - \hat{\mathbf{x}}(t_k)]^T\}.$$

In minimizing the diagonal elements of $P(t)$, each of the filter
estimation rms errors is minimized. The error sources may be
initial uncertainties in $\hat{\mathbf{x}}(0)$, the process noise $\mathbf{w}(t)$, or the measurements noise $\mathbf{v}(t)$.

$K(k+1)$ The weighting or Kalman gain matrix at t_{k+1}. More specifically,
the Kalman gains $K(k+1)$ represent the weight given to the incoming measurements for "tuning up" the estimates of the state.
$K(k+1)$ is chosen so that each element in the diagonal of $P(k|k)$
is a minimum [that is, it minimizes the trace of $P(k|k)$]. Also, this
matrix determines the relative magnitude of corrections which will
be made to the state parameters to allow for the measurement
residuals. The size of the matrix $[H(k) P(k|k-1) H^T(k) + R(k)]^{-1}$
is determined by the number of components in $\mathbf{z}(k)$, i.e., the number of observed (or measured) quantities. Algorithms for avoiding a
direct computation of the inverse of this matrix are discussed in
reference [21].

Figure 4.7 shows the discrete-time Kalman filter estimator.

The Kalman filter operates in a *predict–correct* manner. Consider the equation

$$\hat{\mathbf{x}}(k) = \Phi\hat{\mathbf{x}}(k-1) + K(k)[\mathbf{z}(k) - H\Phi\hat{\mathbf{x}}(k-1)],$$

where $\hat{\mathbf{x}}(k)$ is the best estimate for signal $\mathbf{x}(k)$ at the kth instant of time. The first
term, $\Phi\hat{\mathbf{x}}(k-1)$, predicts the signal estimate at time k by projecting the signal
estimate at time $k-1$ into the future through the use of the system's state
transition matrix. Next, the result of this prediction is corrected by comparing
the projected estimate with the observation as weighted by the Kalman gain
matrix $K(k)$. Our aim here is to derive the *one-stage* predictor algorithm.

Towards this end, consider the prediction stage

$$\hat{\mathbf{x}}(k|k-1) = \Phi\hat{\mathbf{x}}(k-1|k-1)$$

and

$$P(k|k-1) = \Phi P(k-1|k-1)\Phi^T + Q(k-1).$$

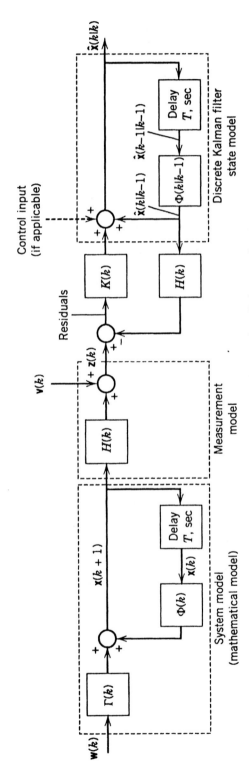

Figure 4.7 Block diagram of the discrete-time Kalman estimator.

118

The results of prediction are then used to *update*, or *correct*, the estimation procedure using Eqs. (4.18), (4.19a), and (4.20). The prediction problem starts with

$$\hat{\mathbf{x}}(k+1|k) = \Phi\hat{\mathbf{x}}(k)$$

where $\hat{\mathbf{x}}(k) = \hat{\mathbf{x}}(k|k)$ is the current signal estimate, and $\hat{\mathbf{x}}(k+1|k)$ is the projection of this estimate one time period into the future. Specifically, this equation implies that given the filtered signal, the best estimate of the future value of the signal is the projection using the system's state transition matrix Φ. Now, ignoring noise in this prediction scheme, and using Eq. (4.18) in the above equation for $\hat{\mathbf{x}}(k+1|k)$, we have

$$\hat{\mathbf{x}}(k+1|k) = \Phi\hat{\mathbf{x}}(k|k-1) + G(k)[\mathbf{z}(k) - H(k)\hat{\mathbf{x}}(k|k-1)],$$

where the prediction gain matrix is given by

$$G(k) = \Phi K(k).$$

For the prediction problem, we require the error covariance matrix $P(k+1|k)$. The error covariance matrix is derived by writing the equation for $P(k|k-1)$ given above, indexed into the future by one time period, giving

$$P(k+1|k) = \Phi P(k|k)\Phi^T + Q(k).$$

Finally, using Eq. (4.19a) to obtain $P(k|k)$ in terms of $P(k|k-1)$ and substituting Eq. (4.19a) into the above equation results in

$$P(k+1|k) = [\Phi - G(k)H(k)]\,P(k|k-1)\,\Phi^T + Q(k),$$

which is the one-stage prediction error covariance matrix.

The conventional discrete-time Kalman filter algorithm indicated by Eq. (4.14)–(4.20) can also be written as follows (arguments are dropped in system matrices for simplicity):

State prediction:

$$\hat{\mathbf{x}}(k|k-1) = \Phi(k|k-1)\,\hat{\mathbf{x}}(k-1|k-1),$$

$$\hat{\mathbf{x}}(0|0) = \hat{\mathbf{x}}(0).$$

Observation prediction:

$$\hat{\mathbf{z}}(k|k-1) = H\hat{\mathbf{x}}(k|k-1).$$

Innovations:

$$v(k) = \mathbf{z}(k) - \hat{\mathbf{z}}(k|k-1).$$

Covariance prediction:

$$P(k|k-1) = \Phi P(k-1|k-1)\Phi^T + GQG^T,$$

$$P(0|0) = P(0).$$

Innovations covariance:

$$S(k) = HP(k|k-1)H^T + R.$$

Kalman gain:

$$K(k) = P(k|k-1)H^T S(k)^{-1}.$$

State update:

$$\hat{\mathbf{x}}(k|k) = \hat{\mathbf{x}}(k|k-1) + K(k)v(k).$$

Covariance update:

$$P(k|k) = P(k|k-1) - K(k)S(k)K^T(k).$$

[Note that $K(k)$ is also given as $K(k) = P(k|k)H^T R^{-1}$.]

The above equations can be coded using the MATLAB* program. In particular, this program is useful for rapid prototyping and debugging of algorithms, such as the Kalman filter algorithm. It should be pointed out, however, that the MATLAB functions are slow. Therefore, debugged and successful algorithms should be coded in a high-order computer language, if speed and flexibility are important to the system analyst. Higher-order languages recommended are the following: FORTRAN, Ada, C, or C^{++}.

At this point, we will summarize the time-update and measurement-update algorithms. Consider the discrete global model of the form

$$\mathbf{x}(k) = \Phi(k, k-1)\mathbf{x}(k-1) + \Gamma(k)\mathbf{w}(k) \tag{4.25a}$$

$$\mathbf{z}(k) = H(k)\mathbf{x}(k) + \mathbf{v}(k) \tag{4.25b}$$

$$k = 1, 2, \ldots,$$

where the following usual assumptions are invoked: $\mathbf{x}_0 \sim N(\bar{\mathbf{x}}_0, P_0), \mathbf{w}(k) \sim N(0, Q), \mathbf{v}(k) \sim N(0, R)$; and $\mathbf{x}(k) \in R^n, \mathbf{w}(k) \in R^p, \mathbf{z}(k) \in R^m, \Phi \in R^{n \times n}, \Gamma \in R^{n \times p}$, and $H \in R^{m \times n}$. Also, $\mathbf{w}(k)$ and $v(k)$ are white-noise processes independent of each other and of the initial state estimate $\mathbf{x}_0$. We assume that P_0, Q, R are symmetric and positive definite matrices.

The measurement, $\mathbf{z}(k)$, is processed by the Kalman filter to yield the best linear estimate of the state vector $\mathbf{x}(k)$. The estimation scheme has two distinct stages, namely, the time-update estimate and error covariance matix, and the corresponding measurement-update quantities. They are as follows:

1. Time Update (Propagation):[†]

$$\hat{\mathbf{x}}(t_k^-) = \Phi(t_k, t_{k-1})\hat{\mathbf{x}}(t_{k-1}^+) \tag{4.26}$$

$$P(t_k^-) = \Phi(t_k, t_{k-1})P(t_{k-1}^+)\Phi^T(t_k, t_{k-1}) + Q(t_{k-1}). \tag{4.27}$$

---

*MATLAB is a commercially available software package for use on a personal computer. For more information, the reader is referred to Appendix B.

2. Measurement Update:[†]

$$\hat{\mathbf{x}}(t_k^+) = \hat{\mathbf{x}}(t_k^-) + K(t_k)[z(t_k) - H(t_k)\hat{\mathbf{x}}(t_k^-)], \qquad (4.28)$$

$$P(t_k^+) = P(t_k^-) - K(t_k) H(t_k) P(t_\mathbf{K}^-), \qquad (4.29)$$

where

$$K(t_k) = P(t_k^-) H^T(t_k)[H(t_k)P(t_k^-) H^T(t_k) + R(t_k)]^{-1} \qquad (4.30)$$

with initial conditions (i.e., recursion initialization quantities)

$$\hat{\mathbf{x}}(t_0) = \mathscr{E}\{\mathbf{x}(t_0)\} = \hat{\mathbf{x}}_0,$$

$$P(t_0) = \mathscr{E}\{[\mathbf{x}(t_0) - \hat{\mathbf{x}}_0][\mathbf{x}(t_0) - \hat{\mathbf{x}}_0]^T\} = P_0.$$

Each of the above stages contains difference equations for propagation of a state estimate and its error covariance matrix. The efficiency and simplicity of the Kalman filter algorithm make it attractive for use in real-time estimation problems involving small, airborne digital computers. Figure 4.8 illustrates the propagation and time update timing sequence for the discrete-time case [Eqs. (4.27) and (4.28)]. From this figure, we note that the standard deviation increases before a measurement is taken and decreases after an update to the system is made.

In certain cases, when no *a priori* information about the system is available, *batch* processing can be carried out. More specifically, in batch processing the observation vector **z** is the vector of all measurements that are available, and

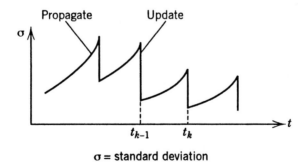

σ = standard deviation

Figure 4.8 Propagation and update signature in the σ–t plane.

[†]It is easily verified that in the time-update stage the required computational effort of the filter to process a single measurement is proportional to $O(n^3)$, while in the measurement-update stage the computational complexity is proportional to $O(mn^2)$. The notation has been changed here in order to conform with that of Figure 4.9 below. The notation given above is preferred by most engineers, however.

thus all measurements are simultaneously incorporated into the estimate. Therefore, in recursive processing, the observation vector $\mathbf{z}$ is partitioned into components as follows [48]:

$$
\mathbf{z} = \begin{bmatrix} \mathbf{z}_1 \\ \hline \mathbf{z}_2 \\ \hline \vdots \\ \mathbf{z}_k \end{bmatrix}.
$$

For example, two simultaneously measurements of position can be incorporated simultaneously in batch.

As noted earlier, the Kalman filter provides an estimate of the state of the system at the current time based on all measurements obtained up to and including the present time. Figure 4.9 shows the timing diagram of the discrete-time Kalman filter.

Equations (4.14)–(4.20) constitute the discrete-time Kalman filter algorithm. It should be noted that the measurements $\mathbf{z}(t)$ are assumed to be discrete samples in time. Thus, as stated above, the filter uses all available measurements, regardless of their precision, to improve the accuracy of the overall data system. A flowchart illustrating the computational order of Eqs. (4.14)–(4.20) is shown in Figure 4.10, and Figure 4.11 shows how the Kalman filter enters into a typical navigation loop [61].

The conventional sequential discrete-time Kalman filter algorithm can be implemented in real time with reasonable ease by most computer processors. Its algebraic simplicity and computational efficiency have contributed to its widespread application. However, it is well recognized that the Kalman filter can be numerically unstable due to the asymmetric nature of its covariance update equation

$$
P(k/k) = [I - K(k)H(k)]P(k|k-1). \tag{4.19a}
$$

Under certain conditions, the subtraction operation in this equation can cause the covariance matrix $P(k|k)$ to be asymmetric (and also lose positive definiteness) because of finite-precision computation, which leads to numerical instability and eventual filter divergence. Moreover, Eq. (4.19a) involves the product of a nonsymmetric matrix and a symmetric matrix, which makes it less attractive to use. However, we note that this expression is better behaved numerically when many measurements of accuracy comparable to that of the prior information are processed.

An alternative form of the covariance update equation which is commonly used is

$$
P(k|k) = P(k|k-1) - K(k)H(k)P(k|k-1). \tag{4.19b}
$$

This form is efficient, but in certain applications it may involve the small difference of large numbers when the measurements are rather accurate, which may lead into loss of positive definiteness, especially on finite-work computers.

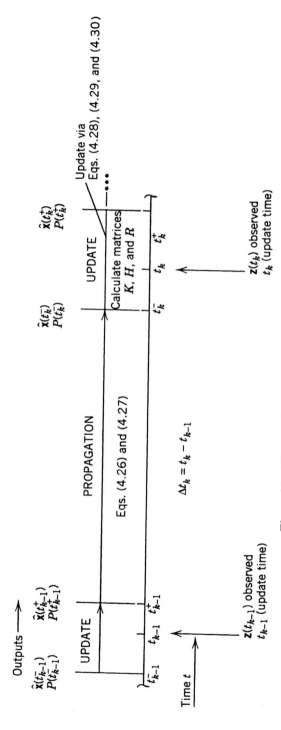

Figure 4.9 Discrete-time Kalman filter algorithm sequence.

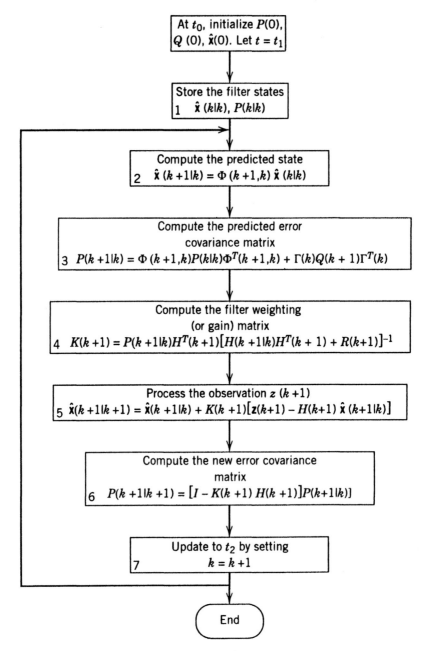

Figure 4.10

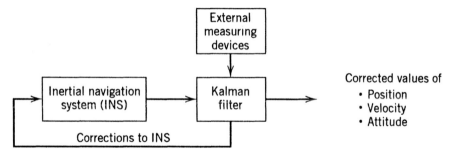

Figure 4.11 Block diagram of the Kalman Filter in a typical navigation loop.

Another method for avoiding divergence, which is preferred by some users, is to force the symmetry of $P(k|k)$ by averaging it with its transpose:

$$P(k|k) = \frac{P(k|k) + P^T(k|k)}{2}.$$

In order to alleviate the instability and/or divergence problem, several alternatives have been proposed in the literature. Some of these methods are summarized below:

1. *Stabilized Kalman Algorithm:* In to ensure that $P(k|k)$ is positive definite, one common approach is to rewrite Eq. (4.19a) in a quadratic formulation as

$$P(k|k) = [I - K(k)H(k)]P(k|k-1)$$

$$[I - K(k)Hk)]^T + K(k)R(k)K^T(k), \qquad (4.31)$$

known as *Joseph's** form, which effectively avoids roundoff errors in machine computations. Joseph's form is better behaved numerically when a measurement which is very accurate relative to the prior information is processed. One drawback of this equation is that is requires a greater number of computations. Namely, it requires almost three times as many arithmetic operations as that given by Eq. (4.19a).

Assuming that $\Gamma(k) \equiv I$ in Eq. (4.17), we can write this equation in the form

$$P(k+1|k) = \Phi(k+1, k)P(k|k)\Phi^T(k+1, k) + Q(k).$$

---

*Joseph's form has the advantage of less sensitivity to computer implementation errors: to first order, it is unaffected by errors in the gain matrix K. However, it is less desirable in that it requires increased computer processing time.

Using Eq.(4.19b) in the above equation, we obtain

$$P(k+1|k) = [\Phi(k+1,k) - G(k)H(k)]P(k|k-1)\Phi^T(k+1,k) + Q(k) \quad (4.32)$$

where $G(k) = \Phi(k+1,k)K(k)$. This is the so-called *one-stage* prediction algorithm derived earlier

2. *Carlson Square-Root Formulation*: In this formulation, the error covariance matrices are represented in terms of their upper triangular square roots. Namely,

$$P(t_k^-) \triangleq S(t_k^-)S^T(t_k^-),$$

$$P(t_k^+) \triangleq S(t_k^+)S^T(t_k^+),$$

where $S(t)$ is the square-root matrix of $P(t)$ and assures that the $P(t)$ generated will always be symmetrical and positive semidefinite [13, 51, 56, and 57]. This square-root filter formulation provides twice the effective precision of the conventional filter, but it has the disadvantage that n square roots are necessary for each sequential measurement update. The superscripts $-$ and $+$ refer to the times just before and just after the incorporation of a measurement. In Carlson's upper triangular square-root algorithm, the scalar measurements are incorporated sequentially according to the following equations [13]:

$$P_f(-) = S_f(-)S_f^T(-), \tag{1a}$$

$$\mathbf{f} = S_f^T(-)h_i \qquad \text{for } i = 1, ..., M \tag{1b}$$

$$\alpha_0 = \sigma_i^2, \tag{1c}$$

$$b_0 = 0 \text{ for } j = 1, ..., N_f \tag{1d}$$

$$\alpha_j = \alpha_{j-1} + f_j^2, \tag{1e}$$

$$\mathbf{b}_j = \mathbf{b}_{j-1} + S_{fj}(-)f_j, \tag{1f}$$

$$S_{fj}(+) = [S_{fj}(-) - \mathbf{b}_{j-1}f_j/\alpha_{j-1}]\sqrt{\alpha_{j-1}/\alpha_j}, \quad \text{for } j = 1, ..., N_f \tag{1g}$$

$$K = \mathbf{b}_N/\alpha_N, \tag{1h}$$

$$\hat{X}_f(+) = \hat{X}_f(-) + K \Delta Z_i, \qquad \text{for } i = 1, ... M \tag{1i}$$

$$P_f(+) = S_f(+)S_f^T(+), \tag{1j}$$

where $P_f(\cdot) =$ filter covariance matrix before $(-)$ or after $(+)$ measurement incorporation,

$S_f(\cdot) =$ the upper triangular square root matrix of the filter covariance (using the Cholesky decomposition; for more details see Section 4.7),

S_{jj} = the jth column of S_f,
f_j = the jth element of $\mathbf{f}$,
Δz_i = the ith measurement residual,
h_i = the measurement gradient vector the ith measurement.
σ_i^2 = variance of the noise in the ith measurement,
N_f = dimension of the filter state,
M = number of measurements,
$\mathbf{K}$ = Kalman gain vector.

3. *Bierman U–D Factorization Algorithm*: The U–D factorization method developed by Bierman [9] and by Thornton and Bierman [69] is an extension of the square-root filter method. In this formulation, the covariance matrix is factored into an upper triangular matrix U with ones along the diagonal, a diagonal matrix D, and the matrix U^T, so that

$$P(t_i^-) = U(t_i^-)D(t_i^-)U^T(t_i^-),$$

$$P(t_i^+) = U(t_i^+)D(t_i^+)U^T(t_i^+).$$

where the $-$ and $+$ signs refer to the incorporation of measurements before and after, respectively. Thus, the U–D algorithm reformulates the Kalman measurement update equations in terms of the U and D matrices. Experience with this method has shown that it improves both the accuracy and the computational efficiency of the Kalman filter. A requirement for the U–D algorithm is that the measurements be normalized so that the covariance matrix of the measurement noise, R, is diagonal and the measurements may be applied singly as scalars. The Bierman UDU^T square-root formulation is computationally more expedient: in addition to providing the requisite stability in recursive computations so important in real-time applications over long mission times, it avoids unnecessary inversions of covariance matrices. Finally, this method has all the advantages of the square-root formulation without the need of computing square roots. For more details on the U–D method, see Section 4.7.

4. *Square-Root Filtering*: Perhaps the most recognized method for overcoming the numerical instability mentioned above in the Kalman filter is the so-called *square-root* filtering, in which the covariance matrix P is written as the product of a lower triangular matrix S (called the square root of P and sometimes written as $P^{1/2}$) and its transpose S^T, namely $SS^T = P$, where the superscript T denotes the transpose, and the Kalman covariance updates are determined in terms of the matrix S. This guarantees the positive semidefiniteness of the matrix P, as SS^T always possesses this property. Furthermore, this algorithm is recommended for its greatly improved dynamic range (compared to the covariance Kalman filter algebra). The square-root algorithm can be divided into three parts as follows: (a) noise-free integration of S between measurements, (b) a triangulation algorithm to incorporate integrated noise in S, and (c) the update algorithm.

(a) *Integration:* We have

$$dS/dt = FS,$$

where the matrix F incorporates the deterministic equations of the model. Next, we form the equation

$$S = S + \frac{dS}{dt} \Delta t,$$

where Δt is the integration time step. Note that here and in the development that follows, the equal sign represents a *replacement operation.*

(b) *Triangularization Algorithm:* The purpose of the triangularization algorithm is to *add* the integrated square-root noise matrix q to the integrated square-root covariance matrix. Now, if q is adjoined to S to form an $N \times 2N$ matrix

$$A = [S \quad q], \tag{A-1}$$

then it is seen that if $q^T q = Q$, one has

$$AA^T = P + Q, \tag{A-2}$$

which is the desired form. The triangularization algorithm in effect post multiplies A by an orthogonal matrix in such a way that the right half of A is reduced to zero. Note that Eq.(A-2) is preserved if A is post multiplied by any orthogonal matrix. Thus, the noise has in effect been incorporated back into the left half of A (namely, S), which is the desired result. The algorithm does not explicitly find the orthogonal matrix, but proceeds by elementary row operations as follows:

$$K = 1 \qquad \text{(row counter)}, \tag{A-3}$$

$$\sigma = \text{sign}(A_{KK}) \left(\sum_{j=k}^{2N} A_{KJ}^2 \right)^{1/2}. \tag{A-4}$$

$$\text{If } K = N, \text{ go to step (A-10). Otherwise} \tag{A-5}$$

$$\beta = \frac{1}{\sigma}(\sigma + A_{KK}), \tag{A-6}$$

$$A_{KK} = A_{KK} + \sigma. \tag{A-7}$$

Do the next two steps for $L = K + 1, 1, N$:

$$Y = \beta \sum_{J=K}^{2N} A_{KJ} A_{LJ}; \tag{A-8}$$

for $J = K, 1, 2N,$

$$A_{LJ} = A_{LJ} - YA_{KJ}. \tag{A-9}$$

$$A_{KK} = -\sigma \text{ or } A_{KJ} \text{ if } J \text{ exceeds } K. \tag{A-10}$$

The execution of the algorithm is complete if $K = N$; otherwise step K and go to (A-4).

(c) *Measurement Update*: For each measurement z, there is a corresponding measurement matrix H and a measurement noise variance R. The computation is therefore as follows:

$$M = S^T H^T, \tag{U-1}$$

$$a = \frac{1}{M^T M + R}, \tag{U-2}$$

$$K = S M a, \tag{U-3}$$

$$\Delta X = Kz, \tag{U-4}$$

$$\gamma = \frac{1}{1 + (Ra)^{1/2}}, \tag{U-5}$$

$$S = S - \gamma K M^T. \tag{U-6}$$

In the foregoing discussion, the vector M is a useful intermediate which saves computation. The other parameters are as follows:

$a =$ reciprocal of the expected variance of the measurement from the state prediction
$K =$ Kalman gain matrix,
$\Delta X =$ correction to the state vector,
$\gamma =$ factor which makes Eq. (U-6) work

Early development and applications of square-root filtering can be found in the works of Battin [5], Schmidt et al. [59], McGee [50], and Smith et al. [63]. Later work by Bierman [9] has extended these results to develop and implement the so-called *square-root information filter* (SRIF), in which square-root techniques are used on the information filter as opposed to the covariance filter. (See also Chapter 8.) Reference [43] contains a survey of various square-root filtering methods. The conventional Kalman filtering algorithm is referred to as the covariance filter, since the original error covariances are propagated and updated. The name *information filter* (or *information matrix*) is taken from the idea that the inverse of the error covariance matrix, defined as

$$\mathscr{I} = P^{-1},$$

indicates the accuracy or the amount of information present in the measurements. For instance, in the covariance matrix method, the measurement update equation for the covariances is given by

$$P(+) = [I - KH] P(-)$$

$$= P(-) - P(-)H^T[HP(-)H^T + R]^{-1}HP(-),$$

while the corresponding information-filter equation has the following simpler form:

$$P^{-1}(+) = P^{-1}(-) + H^TR^{-1}H,$$

where $(+)$ indicates "after the update" and $(-)$ "before the update." This equation requires the inversion of two $n \times n$ matrices, but the inverse of the covariance could be updated without further inversion. Therefore, from the above definition, that is $\mathscr{I} = P^{-1}$, the last equation can be written as

$$\mathscr{I}(+) = \mathscr{I}(-) + H^TR^{-1}H.$$

Although the covariance and information filters are algebraically equivalent, they possess different numerical properties. For example, in situations where there is either poor or no *a priori* information available (which is a cause for numerical difficulties in the covariance filter), the information filter has superior performance. This could be the case in navigation applications, for example, where lack of *a priori* information on the noise processes leads to difficulties in initializing the covariance filter. Finally, the SRIF is a square-root version of the information filter; it mitigates numerical problems by employing square-root algorithms and also allows for lack of *a priori* information by using the information filter. In SRIF mechanizations, the square-root of the information matrix (i.e., the square root of the inverse of the covariance matrix) is updated. The properties of the square-root information filter (SRIF) are summarized below.

(a) The SRIF works in the information rather than the covariance domain (information is inversely proportional to covariance).
(b) It accomodates large *a priori* covariances in a precise, numerically stable manner.
(c) The numerical stability is due to the square-root structure and Householder (orthogonal) transformations.
(d) Implementation is penalized when the number of measurements is small.
(e) The filter is economical when explicit state estimates and covariances are not needed frequently.
(f) The algorithm is more complex to implement; however, the storage requirements are comparable to those of the Kalman and U–D methods.

Example   The discrete Kalman filter will now be applied to tracking an aircraft by means of a radar. Specifically, we will consider the radar tracking problem, with discrete measurements. Typically, the tracking radar measures the slant range r and azimuth angle θ as illustrated in Figure 4.12.

The range r is given as

$$r = [(\mathbf{x} - \mathbf{x}_0)^T(\mathbf{x} - \mathbf{x}_0) + (\mathbf{y} - \mathbf{y}_0)^T(\mathbf{y} - \mathbf{y}_0)]^{1/2}$$

or simply (assuming $\mathbf{x}_0 = 0$ and $\mathbf{y}_0 = 0$)

$$r^2 = x^2 + y^2, \tag{1}$$

while the azimuth angle θ is given by

$$\tan \theta = \frac{y}{x}. \tag{2}$$

Differentiating Eq. (1), we obtain

$$\dot{r} = \frac{x\dot{x} + y\dot{y}}{r} \tag{3}$$

From the diagram we note that $x = r \cos \theta$, $y = r \sin \theta$, $\dot{x} = V, \dot{y} = 0$, so that

$$\dot{r} = V \cos \theta. \tag{4}$$

Again, differentiating Eq. (4) we have

$$\ddot{r} = \dot{V} \cos \theta - V\dot{\theta} \sin \theta. \tag{5}$$

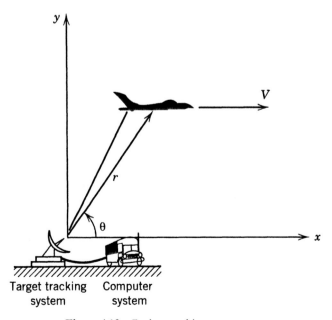

Figure 4.12   Radar tracking geometry.

Next, we differentiate Eq. (2). The result is

$$\dot\theta \sec^2 \theta = \frac{x\dot y - \dot x y}{x^2},$$

$$\dot\theta = \frac{x\dot y - \dot x y}{x^2 \sec^2 \theta} = \frac{x\dot y - \dot x y}{r^2},$$

$$\dot\theta = -\frac{V}{r}\sin\theta. \tag{6}$$

A second differentiation of Eq. (6) yields

$$\ddot\theta = -\left(\frac{\dot V r - V\dot r}{r^2}\right)\sin\theta - \frac{V}{r}\dot\theta\cos\theta. \tag{7}$$

Equations (5) and (7) are nonlinear. Commonly, linearization is done about the present best estimate. In the present case, linearization is possible by assuming that $\dot V = 0$, and that r and θ do not change much during each scan of the radar. The state vector for this problem is given by

$$\mathbf{x}(t) = [r \quad \dot r \quad \theta \quad \dot\theta]^T.$$

Before the discrete Kalman filter is used, certain preliminaries are in order. The system model can be written in the form

$$\dot{\mathbf{x}} = F\mathbf{x}(t) + \mathbf{w}(t), \tag{8}$$

where $\mathbf{w}(t)$ is the process noise. Therefore,

$$\frac{d}{dt}\begin{bmatrix} r \\ \dot r \\ \theta \\ \dot\theta \end{bmatrix} = \begin{bmatrix} 0 & 1 & 0 & 0 \\ 0 & 0 & 0 & 0 \\ 0 & 0 & 0 & 1 \\ 0 & 0 & 0 & 0 \end{bmatrix}\begin{bmatrix} r \\ \dot r \\ \theta \\ \dot\theta \end{bmatrix} + \begin{bmatrix} 0 \\ w_1(t) \\ 0 \\ w_2(t) \end{bmatrix} \tag{9}$$

In order to compensate for the unmodeled dynamics, and to allow for small changes in $\dot r$ and $\dot\theta$ during the scan time T, we have added the random disturbances $w_1(t)$ and $w_2(t)$. Thus,

$$\frac{d}{dt}\dot r = w_1(t), \tag{10a}$$

$$\frac{d}{dt}\dot\theta = w_2(t). \tag{10b}$$

From Eq. (9) the state transition matrix can be obtained as follows:

$$\Phi = e^{FT} = I + FT + \frac{(FT)^2}{2!} + \cdots + \frac{(FT)^n}{n!} + \cdots.$$

Neglecting second-order and higher terms, the state transition matrix assumes the simple form

$$\Phi = \begin{bmatrix} 1 & T & 0 & 0 \\ 0 & 1 & 0 & 0 \\ 0 & 0 & 1 & T \\ 0 & 0 & 0 & 1 \end{bmatrix}.$$

The measurements $\mathbf{z}(t)$ are available at each radar scan time T. Consequently,

$$\mathbf{z}(t) = H(k)\mathbf{x}(k) + \mathbf{v}(k)$$

where

$$H(k) = \begin{bmatrix} 1 & 0 & 0 & 0 \\ 0 & 0 & 1 & 0 \end{bmatrix}$$

and

$$\mathbf{v}(k) \sim N(0, R) \sim N\left(0, \begin{bmatrix} \sigma_r^2/T & 0 \\ 0 & \sigma_\theta^2/T \end{bmatrix}\right).$$

The state equation can thus be written in the form

$$\mathbf{x}(k+1) = \Phi(k+1, k)\mathbf{x}(k) + \mathbf{w}(k)$$

$$= \begin{bmatrix} 1 & T & 0 & 0 \\ 0 & 1 & 0 & 0 \\ 0 & 0 & 1 & T \\ 0 & 0 & 0 & 1 \end{bmatrix} \mathbf{x}(k) + \mathbf{w}(k).$$

Assume now that we have at time $t = k$ a predicted value $\hat{\mathbf{x}}(k|k-1)$ and its covariance matrix $P(k|k-1)$ for the state of the system $\mathbf{x}(k)$. The state estimate of the system $\hat{\mathbf{x}}(k|k)$ and the covariance matrix $P(k|k)$ are given by the expressions

$$\hat{\mathbf{x}}(k|k) = \hat{\mathbf{x}}(k|k-1) + K(k)[\mathbf{z}(k) - H(k)\hat{\mathbf{x}}(k|k-1)],$$

$$P(k|k) = [I - K(k)H(k)]P(k|k-1),$$

where

$$K(k) = P(k|k-1) H^T(k) [H(k) P(k|k-1) H^T(k) + R(k)]^{-1}.$$

From the Kalman filter theory, $\hat{\mathbf{x}}(k|k)$ represents the optimum (best) estimate of the state. Therefore, the equation for $\hat{\mathbf{x}}(k|k)$ states that our best estimate at $t = k$, using all observations up to and including k, is equal to our predicted estimate plus the error difference between what was observed and what we claim should have been observed, multiplied by the Kalman gain (or weighting) matrix $K(k)$; it determines how much we will change our estimate of the state in view of the new observation. If the elements of $K(k)$ are small, it means that we have considerable confidence in our model; on the other hand, if the elements of $K(k)$ are large, then we have considerable confidence in our observation measurements.

$K(k)$ is highly dependent upon our characterization of the measurement noise $R(k)$; in particular, if $R(k) = I$, the elements of $K(k)$ decrease in absolute value and represent the classical least-squares method (discussed earlier in this book). Finally, given the initial state vector $\mathbf{x}(0)$ and covariance matrix $P(0)$,

$$\mathbf{x}(0) = \begin{bmatrix} r_0 \\ 0 \\ \theta_0 \\ 0 \end{bmatrix}, \qquad P(0) = \frac{1}{T} \begin{bmatrix} \sigma_r^2 & 0 & 0 & 0 \\ 0 & \sigma_{\dot{r}}^2 & 0 & 0 \\ 0 & 0 & \sigma_\theta^2 & 0 \\ 0 & 0 & 0 & \sigma_{\dot{\theta}}^2 \end{bmatrix},$$

the best estimates of r, $\dot{r}$, θ, and $\dot{\theta}$ can be obtained (the initial conditions of r_0 and θ_0 can be obtained from the physics of the problem).

Example   The above example can be extended to the case where the tracking sensor measures target bearing (or azimuth) α, and elevation ε, and range ΔR. Consider now a moving target whose position vector $\mathbf{R}_T$ and velocity $\mathbf{V}_T$ are expressed in an inertial coordinate system as illustrated in Figure 4.13. From the diagram, the target's state vector in terms of position and velocity can be expressed as

$$\mathbf{x} = [R_{Tx} \quad R_{Ty} \quad R_{Tz} \quad V_{Tx} \quad V_{Ty} \quad V_{Tz}]^T.$$

The line-of-sight (LOS) measurement vector L, measured by the sensor in the inertial space, and which points along the measured LOS to the target, can now be defined as follows:

$$L = [\Delta R \sin\alpha \cos\varepsilon \quad \Delta R \cos\alpha \cos\varepsilon \quad \Delta R \sin\varepsilon]^T.$$

where $\Delta R = |R_T - R_S|$,
        R_s = sensor position.

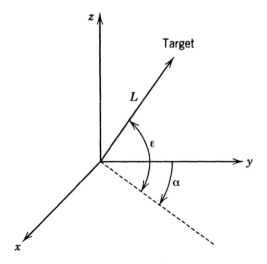

Figure 4.13 Target coordinate system.

The state vector and associated covariance matrix P are updated to the current time t_k in accordance with the expressions

$$\hat{\mathbf{x}}(k|k-1) = \Phi(k-1)\,\hat{\mathbf{x}}(k-1|k-1),$$

$$P(k|k-1) = \Phi(k-1)\,P(k-1|k-1)\,\Phi^T(k-1) + Q(k-1),$$

where the state transition matrix is given by

$$\Phi(k-1) = \begin{bmatrix} 1 & 0 & 0 & \Delta t & 0 & 0 \\ 0 & 1 & 0 & 0 & \Delta t & 0 \\ 0 & 0 & 1 & 0 & 0 & \Delta t \\ 0 & 0 & 0 & 1 & 0 & 0 \\ 0 & 0 & 0 & 0 & 1 & 0 \\ 0 & 0 & 0 & 0 & 0 & 1 \end{bmatrix}, \qquad \Delta t = t_k - t_{k-1},$$

and the measurement covariance matrix $R(k)$ by

$$R(k) = \begin{pmatrix} \sigma_r^2 & 0 & 0 \\ 0 & \sigma_\alpha^2 & 0 \\ 0 & 0 & \sigma_\varepsilon^2 \end{pmatrix}$$

where σ_r^2 is the variance associated with the range measurement, σ_α^2 is the variance associated with the azimuth, and σ_ε^2 is the variance associated with the elevation.

We noted earlier in this example that the initial state $x(0)$ and initial error covariance matrix $P(0)$ were given. Therefore, in order to start the computational or estimation procedure, an initial estimate of the state, $\hat{x}(0)$, with its initial error covariance matrix $P(0)$, must be specified. Furthermore, the initial values of the covariance matrices $R(k)$ and $Q(k)$ must be specified. The initial value of the measurement noise covariance matrix $R(k)$ is specified by the characteristics of the particular tracking radar system used. That is, it can be obtained from a model of actual radar noise characteristics. The process noise covariance matrix $Q(k)$ associated with the random forcing function (or process noise vector) $w(k)$ is a function of estimated state parameters and compensates for the model inaccuracies. That is, a good initial choice can be determined experimentally or from the physics of the problem. For more details on the covariance matrices $R(k)$ and $Q(k)$, see the next section.

4.3.1 Real-World Model Errors

The results of Sections 4.2 and 4.3 can be summarized by considering a general example. Specifically, we will consider the real-world (or *truth*) model navigation errors of an inertial navigation system. The state of, say, a navigation system is commonly described in terms of many different parameters (states) representing error sources that affect system performance. System error dynamics are modeled as a first-order linear vector–matrix differential equation of the form

$$\frac{dx(t)}{dt} = Fx + w + u,$$

where $x =$ system *state* vector with n elements (states),
   $F =$ state *dynamics* matrix with $n \times n$ elements,
   $w = n \times 1$ vector uncorrelated (white) *process noise* that affects certain system states,
   $u = n \times 1$ *control* vector representing adjustments applied to the navigation system outputs, based on the kalman filter's estimates of the system errors.

Normally, one implements a discrete-time (sampled data) version of the above equation, which is defined at time $t = t_n$ ($n = 0, 1, 2, \ldots$) as

$$x_{n+1} = \Phi_n x_n + w_n + u_n.$$

Implementation of the Kalman filter estimation algorithm requires initial information on the real-world model. Specifically, the filter requires the following:

- a statistical description of the *initial* system error states x_0 at time $t = 0$,
- a statistical description of the *process* noise w_n,
- a statistical description of the *measurement* noise v_n.

These statistical properties will now be discussed:

Initial System Error Statistics   The system error state vector $\mathbf{x}_n$ can be described statistically in terms of the *system error covariance matrix* P_n defined as [see also Eq. (4.10)]

$$P_n = \mathscr{E}\{\mathbf{x}_n \mathbf{x}_n^T\}.$$

The *diagonal* (or *trace*) elements of P_n respresent the *variance* (or mean square values) of the system error states, and the *off-diagonal* elements represent the *covariances* (or cross-correlations) between pairs of error states. Therefore, the square roots of the diagonal elements represent the rms (root mean square) system errors. In particular, for error state x_i, the rms error at time t_n is given by the relation

$$\text{rms}[x_i]_n = \sqrt{[P_{ii}]_n},$$

where $[P_{ii}]_n$ is the ith element of P_n. In other words, the square roots of its diagonal terms yield the time histories of standard deviations (or 1σ values equal to rms values if the processes are zero mean) of errors in the estimates of the quantities of interest. Stated another way, the trace elements of the error covariance matrix, that is, the elements P_{ii} on the diagonal of P are the variances of the state parameters:

$$P_{ii} = \sigma_i^2,$$

where σ_i is the standard deviation of the ith parameter [see also Eq. (6.2)].

The error covariance matrix P provides a statistical representation of the uncertainty in current estimate of the state and the correlation between the individual elements of the state. At time $t = 0$, P_n can be diagonalized (that is, the error cross-correlations are assumed to be zero initially), and its diagonal elements are set equal to the *initial mean-square uncertainties* of the system error states.

One option is to allow certain diagonal and off-diagonal elements of P to be initialized to their steady-state values. As discussed previously, the real-world model of system error states is used as a baseline in order to evaluate the performance of the Kalman estimation filter. Thus, it is desirable to reduce the time required for the real-world model to compute the *steady-state* values of the system error covariance matrix P. One way to achieve steady-state conditions faster is to model the initial cross-correlations between key error states. Another way is to set the initial variances of the key error states to their approximate steady-state values.

Finally, the time history of P can be generated from a covariance analysis program, depicting the covariance of the true estimation errors caused by the filter model under design. Table 4.1 presents a suggestion on how the system

Table 4.1

System Error Category	State Number	Error Source	Diagonal Elements of Covariance Matrix P Initial Values[a]	Units[b]
Aircraft dynamics errors		Position errors		
	1, 2	X, Y Axes	$(150)^2$	m^2
	3	Z Axis	$(3)^2$	m^2
		Velocity errors		
	4, 5	X, Y Axes	$(5)^2$	$(m/sec)^2$
	6	Z Axis	$(0.6)^2$	$(m/sec)^2$
		Computed platform tilts		
	7, 8	X, Y Axes	$(60)^2$	$arcmin^2$
	9	Z Axis	$(120)^2$	$arcmin^2$
	10	Computer azimuth error	$(120)^2$	$arcmin^2$
Inertial instrument errors	11–13	Gyro bias drifts	$(0.001)^2$	$(deg/h)^2$
	14–16	Accelerometer bias errors	$(20)^2$	$(10^{-6}g)^2$
External instrument errors		Gravity-correlated noise		
		North vertical deflection	$(24.24)^2$	$(10^{-6}g)^2$
	17	East vertical deflection	$(24.24)^2$	$(10^{-6}g)^2$
	18	Vertical anomaly	$(35.0)^2$	$(10^{-6}g)^2$
	19			
	20–22	Gravity bias errors	0	$(10^{-6}g)^2$
		Doppler Azimuth Errors		
	23	Bias	$(0.001)^2$	rad^2
	24	Correlated Noise	$(0.003)^2$	rad^2
		Barometric Altimeter Errors		
	25	Bias	$(60)^2$	m^2
	26	Correlated Noise	0	m^2

[a] Each diagonal element of P is the square of the rms value.
[b] All units must be input as indicated.

Note: • Correlation times for the gyro, accelerometer, Doppler radar azimuth, and barometric altimeter error noises will vary between 10 and 60 min, depending on the application.

• Gravity noise values are proportional to the aircraft ground speed V_g. The gravity noise correlation distance d can be obtained form the relation

$$d = V_g \tau,$$

where τ is the correlation time. Typical correlation distances vary from 20 n.m. (37 km) to 60 n.m. (111 km).

designer can arrange the state vector **x** and initial covariacne matrix P for a real-world error model.

Process Noise Statistics   The process noise vector $\mathbf{w}(k)$ can be described statistically in terms of the *process noise covariance matrix* $Q(k)$, defined as follows [see also Eq. (4.21)]:

$$Q(k) = \mathcal{E}\{\mathbf{w}(k)\,\mathbf{w}^T(k)\}.$$

The diagonal elements of $Q(k)$ represent the variances (or mean square values) of the process noise terms. The process noise covariance matrix is related to the process noise power spectral density (PSD) matrix Q by the equation

$$Q(k) = Q\,\Delta T,$$

where ΔT is the interval between sample times t_k ($k = 0, 1, 2, \ldots, n$). Normally, a program simulation requires process noise Q-values as input; the program then will automatically convert the to $Q(k)$-values by multiplying them by ΔT. A similar table to the initial covariance matrix P can be made for the initial process noise PSD matrix Q for the real-world error model. Note that some of the diagonal elements of this matrix will be zero, depending on the parameter in question.

From the discussion in Section 4.2 and 4.3 the matrix $Q(k)$ allows us to determine the maximum amount of confidence we will place in our optimum estimates of $\hat{\mathbf{x}}(k|k)$, since $P(k|k)$ is the covariance matrix of errors associated with our optimum estimates. $Q(k)$ is user-specified and may be changed at the discretion of the user. The elements of $Q(k)$ are determined empirically by varying the noise strength values until 63% or more of the estimation error values are within $\pm 1\sigma$ of the filter model. Finally, $Q(0) = 0$ implies that one is willing to accept maximum confidence.

The process noise PSD matrix Q can be determined as shown in Table 4.2. In designing error models the initial rms values $\sigma_i(t_0)$ for the correlated noise states can be equal to the *steady-state* rms values $\sigma_i(t_{ss})$. Thus, one can determine Q_i in the above equations. (See also the Q-matrix of the single-axis inertial navigation example in Section 4.1.)

Measurement Noise Statistics   The measurement noise vector $\mathbf{v}_n$ in Eq. (4.15) can be described statistically in terms of the *measurement noise covariance matrix* R_n, defined as [see also Eq. (4.21)]

$$R_n = \mathcal{E}\{\mathbf{v}_n\,\mathbf{v}_n^T\}.$$

For a given error mode, R_n is a diagonal matrix and must be provided by the user. As discussed in Section 4.2, if $R(t)$ is large, $K(t)$ will be small. This means that since one would tend to put little confidence in very noisy measurements, then one should weight them lightly. Furthermore, if the dynamic system noise covariance matrix Q is large, then $P(k|k-1)$ [Eq. (4.19b)] will be large, and so

Table 4.2

Form of System Error State x_i	ith Diagonal Element of the Noise PSD Matrix Q
$x_i(t) = $ *Time Correlated Noise* With steady-state rms value $\sigma_i(t_{ss})$ and correlation time τ_i	$Q_i = \dfrac{2}{\tau_i}\sigma_i^2(t_{ss})$*
$x_i(t) = $ *Distance Correlated Noise* With steady-state rms value $\sigma(t_{ss})$, correlation distance d_i, and aircraft ground speed V_g	$Q_i = \dfrac{2V_g}{d_i}\sigma_i^2(t_{ss})$
$x_i(t) = $ *Random Walk Process* With initial rms value $\sigma_i(t_0)$ and rms growth-rate factor G_i, that is, $\sigma_i(t) = \sigma_i(t_0) + \sqrt{G_i}\, t$	$Q_i = G_i$

* Note that for states that are modeled as a first-order Markov process, the Q_i term is described in terms of the P_0 term and correlation time τ_i in such a manner as to yield stationary processes. Thus, $Q_i = 2P_0/\tau_i$, since $P_{0_{ii}} = \sigma_i^2$.

will $K(t)$. In this case, one is not certain of the output of the system model in the filter, and therefore would weight the measurements heavily. For a given $R(k)$, decreased confidence in the estimates, indicated by a larger $P(k)$, would call for a heavier weighting of the incoming measurements. Similarly, for a given $P(k)$, an increase in $R(k)$ means that the measurements are noisier, implying that they should be weighted less heavily.

Finally, the matrix R is a database-controlled matrix which is not recomputed for each measurement time. Note that a poor choice of R will adversely affect the performance of the filter.

From the discussion of the above example, it is clear that the Kalman filter requirements can be summarized as follows:

- *Exact* knowledge of the system dynamics (i.e., matrices F, G) and the measurement process matrix H.
- *Exact* knowledge of the initial covariances.
- *Exact* knowledge of the error source statistics (Q, R). (See also Section 4.6.)
- Computing capability—(i.e., storage and throughput (or speed))

In connection with the discussion of the example given, we note that the Kalman gain matrix $K(k)$, which is computed at each measurement time, is an $n \times m$ matrix which relates measurement residuals ($m \times 1$ matrices) to corrections in the state. As $K(k)$ goes to infinity, the error covariance matrix $P(k)$ tends to the system dynamics covariance matrix $Q(k)$. In other words, as more information is obtained about the state vector (through observation), uncertainty of the estimate approaches the uncertainty of the environment in which the state vector exists.

The diagram in Figure 4.14 illustrates some of the requirements enumerated above.

It should be stated that the covariance matrix sequences $Q(k)$, $R(k)$, $\mathbf{w}(k)$, $\mathbf{v}(k)$ comprise the main statistical information. Although $\mathbf{w}(k)$ and $\mathbf{v}(k)$ are statistically independent, the components of $\mathbf{w}(k)$ or $\mathbf{v}(k)$ need not be independent. That is, $Q(k)$ and $R(k)$ need not be diagonal.

4.4   THE STATE TRANSITION MATRIX

Consider the linear, continuous-time system whose first-order vector differential equation is given by

$$\dot{\mathbf{x}}(t) = F(t)\mathbf{x}(t) + G(t)\mathbf{w}(t),$$

$$\mathbf{x}(t_0) = \mathbf{x}_0. \tag{4.33}$$

The solution of Eq. (4.33) can be obtained by the usual techniques for differential equations [assuming that the matrix $F(t)$ is constant], and is [see Eq. (4.2)]

$$\mathbf{x}(t) = e^{F(t-t_0)}\mathbf{x}(t_0) + \int_{t_0}^{t} e^{F(t-t_0)}\, G(\tau)\mathbf{w}(\tau)\, d\tau, \tag{4.34}$$

where t_0 is the initial time and $\mathbf{x}(t_0)$ is the initial state. We note that this equation consists of two terms: the first term gives the dependence on the initial state $\mathbf{x}(t_0)$, while the second term gives the dependence of the input vector $\mathbf{w}(t)$. If we assume for the moment that $\mathbf{w}(t) \equiv \mathbf{0}$, then the system can be

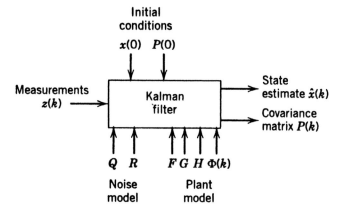

Figure 4.14   Block diagram of the Kalman filter.

expressed as

$$\mathbf{x}(t) = e^{F(t-t_0)} \mathbf{x}(t_0) \tag{4.35}$$

That is, the initial state $\mathbf{x}(t_0)$ is transformed to the state $\mathbf{x}(t)$ by means of the $n \times n$ matrix [52]*

$$\Phi(t, t_0) = \Phi(t - t_0) = e^{F(t-t_0)} \qquad \forall\, t \geqslant t_0. \tag{4.36}$$

For this reason, the matrix $\Phi(t, t_0)$ is called the *state transition matrix* (also the *fundamental matrix*). Equation (4.34) can now be written as [2]

$$\mathbf{x}(t) = \Phi(t, t_0)\,\mathbf{x}(t_0) + \int_{t_0}^{t} \Phi(t, \tau)\ G(\tau)\mathbf{w}(\tau)\, d\tau, \tag{4.37}$$

The state transition matrix is a solution of the homogeneous linear differential equation [52]

$$\dot{\Phi}(t, t_0) = \frac{d}{dt}\Phi(t, t_0) = F(t)\,\Phi(t, t_0) \tag{4.38}$$

with the initial condition $\Phi(t_0, t_0) = I$, where I is the identity matrix. That is,

$$\Phi(t_0, t_0) = \begin{bmatrix} 1 & 0 & \cdots & 0 \\ 0 & 1 & \cdots & 0 \\ \vdots & \vdots & & \vdots \\ 0 & 0 & \cdots & 1 \end{bmatrix} = I.$$

For instance, if $t_0 = 0$, and $F(t)$ is a 2×2 matrix, then

$$e^{Ft} = \Phi(t) = \begin{bmatrix} \Phi_{11}(t) & \Phi_{12}(t) \\ \Phi_{21}(t) & \Phi_{22}(t) \end{bmatrix}.$$

In the case of time-invariant systems, the state transition matrix is frequently expressed in the form $\Phi(t - t_0)$, since both variables t and t_0 appear as a difference in Eq. (4.36). In terms of its elements, Eq. (4.38) can be written in the form[†]

$$\dot{\Phi}_{ik} = \sum_{\nu-1}^{n} f_{i\nu}\,\Phi_{\nu k}, \qquad i, k = 1, 2, \ldots, n. \tag{4.39}$$

---

*We also have $\Phi(t, t_0) = \exp\left(\int_{t_0}^{t} F(\tau)\, d\tau \right)$.

[†] $F(t) = [f_{i,j}(t)]$.

This is a system of n^2 differential equations for the n^2 elements $\Phi_{ik}(t, t_0)$. The corresponding initial condition is

$$\Phi(t_0, t_0) = e^{F\,0} = I. \tag{4.40}$$

Next, we consider the unforced $[\mathbf{w}(t)]$ finite-dimensional system

$$\dot{\mathbf{x}}(t) = F(t)\,\mathbf{x}(t), \qquad t \in T. \tag{4.41}$$

Starting form the initial state $\mathbf{x}(t_0) = \mathbf{x}_0$, the solution of Eq. (4.41) has the unique state trajectory $\mathbf{x}(t)$ given by

$$\mathbf{x}(t) = \Phi(t, t_0)\,\mathbf{x}_0, \tag{4.42}$$

where, as before, the matrix $\Phi(t, t_0)$ is the unique matrix satisfying

$$\frac{\partial}{\partial t}\Phi(t, t_0) = F(t)\,\Phi(t, t_0) \qquad \text{for} \quad t \geqslant t_0 \tag{4.43}$$

and $\Phi(t_0, t_0) = I$. We note that the state transition matrix can be expressed as

$$\Phi(t, t_0) = \begin{bmatrix} x_1^1(t) & x_1^2(t) & \cdots & x_1^n(t) \\ x_2^1(t) & x_2^2(t) & \cdots & x_2^n(t) \\ \vdots & \vdots & & \vdots \\ x_n^1(t) & x_n^2(t) & \cdots & x_n^n(t) \end{bmatrix}, \tag{4.44}$$

where the columns are solutions of the unforced part of the system (4.33)

Equation (4.42) describes the free, unforced, "natural" motion of the state vector $\mathbf{x}(t)$ of a linear system, and represents the linear transformation which maps the initial state $\mathbf{x}_0$ at time t_0 into the state $\mathbf{x}(t)$ at time t. That is, $\Phi(t, t_0): X \to X$ is the matrix which takes the state of a system at time t_0 and yields the state to which the system will move under zero input by time t [44]. If the matrix $F(t)$ is constant, that is, $F(t) = F$, then

$$\Phi(t, t_0) = e^{F(t-t_0)} \qquad \forall\, t \geqslant t_0. \tag{4.36}$$

In general, if (t_0, t_1, t_2) are some points on the t-axis, then point t_2 on the trajectory can be reached as follows:

$$\mathbf{x}(t_1) = \Phi(t_1, t_0)\,\mathbf{x}(t_0),$$

$$\mathbf{x}(t_2) = \Phi(t_2, t_1)\,\mathbf{x}(t_1),$$

so that

$$\mathbf{x}(t_2) = [\Phi(t_2, t_1)\,\Phi(t_1, t_0)]\,\mathbf{x}(t_0),$$

or one can go directly from $\mathbf{x}(t_0)$ to $\mathbf{x}(t_2)$:

$$\mathbf{x}(t_2) = \Phi(t_2, t_0)\,\mathbf{x}(t_0). \tag{4.45}$$

Figure 4.15 illustrates this property [52]*.

The state transition matrix $\Phi(t_1, t_0)$ possesses an inverse form $\Phi(t_0, t_1)$. Designating an arbitrary point t_1 with t, then

$$\Phi^{-1}(t, t_0) = \Phi(t_0, t). \tag{4.46a}$$

From this property, Eq. (4.37) can also be written as

$$\mathbf{x}(t) = \Phi(t, t_0)\left\{\mathbf{x}_0 + \int_{t_0}^{t} \Phi^{-1}(\tau, t_0)\, G(\tau)\, \mathbf{u}(\tau)\, d\tau\right\}.$$

For time-invariant or slowly varying systems,*

$$\Phi^{-1}(t, t_0) = e^{F(t-t_0)} = e^{F(t-t_0)}. \tag{4.46b}$$

Again, if we assume that the matrix $F(t) = F$, constant, then from the Peano –Baker formula

$$\Phi(t, t_0) = \lim_{k \to \infty} \Phi_{k+1}(t, t_0) = I + \int_{t_0}^{t} F(\tau)\, d\tau + \int_{t_0}^{t} F(\tau) \int_{t_0}^{\tau} F(\tau_1)\, d\tau_1\, d\tau + \cdots \tag{4.47}$$

we have

$$\Phi(t, t_0) = I + (t - t_0)F + \frac{1}{2!}(t - t_0)^2 F^2 + \frac{1}{3!}(t - t_0)^3 F^3 + \cdots. \tag{4.48}$$

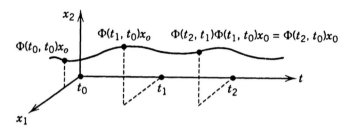

Figure 4.15   The transition property of the matrix $\Phi(t_2, t_0)$.

---

*The time interval $t - t_0$ is the propagation time interval; in actual applications, it is commonly between 5 and 10 sec. However, we note that for $\Delta t = 0.05$, the series decays rapidly, so that only the first three terms in Eq. (4.48) need to be computed.

This equation follows from the Taylor series expansion of the exponential function

$$e^x = 1 + x + \frac{1}{2!}x^2 + \frac{1}{3!}x^3 + \cdots.$$

Therefore, we can define the matrix exponential [2, 58]

$$e^A = 1 + A + \frac{1}{2!}A^2 + \frac{1}{3!}A^3 + \cdots. \tag{4.49}$$

for any $n \times n$ matrix. This series converges because

$$\left\| \sum_{i=m}^{m+k} \frac{A^i}{i!} \right\| \leqslant \sum_{i=m}^{m+k} \frac{\|A^i\|}{i!} \leqslant \sum_{i=m}^{m+k} \frac{\|A\|^i}{i!}$$

for all nonnegative integers m and k.

The foregoing discussion can be illustrated by a simple example. Consider the system

$$\dot{\mathbf{x}}(t) = \begin{bmatrix} -1 & 0 \\ 0 & -2 \end{bmatrix} \mathbf{x}(t) + \begin{bmatrix} 1 & 0 \\ 0 & 2 \end{bmatrix} \mathbf{w}(t).$$

Then, from Eqs. (4.48) and (4.49), we have

$$e^{Ft} = \begin{bmatrix} 1 & 0 \\ 0 & 1 \end{bmatrix} + \begin{bmatrix} -1 & 0 \\ 0 & -2 \end{bmatrix} \frac{t}{1!} + \begin{bmatrix} 1 & 0 \\ 0 & 4 \end{bmatrix} \frac{t^2}{2!} + \begin{bmatrix} -1 & 0 \\ 0 & -8 \end{bmatrix} \frac{t^3}{3!} + \cdots$$

$$= \begin{bmatrix} 1 - \dfrac{t}{1!} + \dfrac{t^2}{2!} - \dfrac{t^3}{3!} + \cdots & 0 \\ 0 & 1 - \dfrac{2t}{1!} + \dfrac{(2t)^2}{2!} - \dfrac{(2t)^3}{3!} + \cdots \end{bmatrix},$$

or in closed form

$$e^{Ft} = \begin{bmatrix} e^{-t} & 0 \\ 0 & e^{-2t} \end{bmatrix}.$$

The Peano–Baker formula is extremely difficult to compute, except for some special cases. One important case is that where, $F(t) = F$, is a constant matrix (i.e., independent of t). The Peano–Baker series can then be expressed as

$$\Phi(t, t_0) = e^{F(t-t_0)} = \sum_{k=0}^{\infty} F^k \frac{(t-t_0)^k}{k!}. \tag{4.50}$$

There are several ways in which one can solve for e^{Ft}:

1. Series.
2. Laplace transforms.
3. Diagonalization.
4. Decomposing F into commuting matrices.

Let us now briefly discuss the merits of each method.

The series method appeals to the series definition [52]

$$e^{Ft} = \Phi(t, 0) = \sum_{n=0}^{\infty} F^n \frac{t^n}{n!}.$$

It can be seen that this method is long and tedious unless F is a diagonal matrix, or some power of F vanishes (i.e., $F^k = 0 \Rightarrow F^n = 0$, $n \geqslant k$).

The use of the Laplace transform is shown in the following example.

Example 1 A form of the state transition matrix often encountered in physical problems is the following:

$$F = \begin{bmatrix} 0 & 1 & 0 \\ 0 & 0 & 1 \\ 0 & 0 & 0 \end{bmatrix},$$

$$e^{Ft} = I + \begin{bmatrix} 0 & 1 & 0 \\ 0 & 0 & 1 \\ 0 & 0 & 0 \end{bmatrix} t + \begin{bmatrix} 0 & 0 & 1 \\ 0 & 0 & 0 \\ 0 & 0 & 0 \end{bmatrix} \frac{t^2}{2} + \cdots = \begin{bmatrix} 1 & t & t^2/2 \\ 0 & 1 & t \\ 0 & 0 & 1 \end{bmatrix}.$$

The Laplace transform formulation of the homogeneous equation $\dot{\mathbf{x}}(t) = F(t) \mathbf{x}(t)$ with the initial condition $\mathbf{x}(0) = \mathbf{x}_0$, is as follows [2]:

$$s\mathbf{x}(s) - \mathbf{x}_0 = F\mathbf{x}(s),$$

$$\mathbf{x}(s) = (sI - F)^{-1} \mathbf{x}_0. \tag{4.51a}$$

Taking the inverse Laplace transform, we have $\mathbf{x}(s) = \mathscr{L}[\mathbf{x}(t)]$ and

$$\mathbf{x}(t) = \mathscr{L}^{-1}[\mathbf{x}(s)] = \mathscr{L}^{-1}[(sI - F)^{-1}]\mathbf{x}_0 \quad \Rightarrow \quad e^{Ft} = \mathscr{L}^{-1}(sI - F)^{-1},$$

or simply

$$\Phi(t) = \mathscr{L}^{-1}(sI - F)^{-1} \equiv e^{Ft}. \tag{4.51b}$$

Therefore, the solution is

$$\mathbf{x}(t) = e^{Ft}\mathbf{x}_0. \tag{4.52}$$

We assume here that the inverse of the matrix $sI - F$ exists, which is actually the case if $\det(sI - F) \neq 0$.

This method is the most straightforward and the easiest to apply.

Example 2 Let

$$F = \begin{bmatrix} 0 & w \\ -w & 0 \end{bmatrix}, \quad \text{so that} \quad sI - F = \begin{bmatrix} s & -w \\ w & s \end{bmatrix},$$

$$\Phi(t) = e^{Ft} = \mathcal{L}^{-1} \begin{bmatrix} \dfrac{s}{s^2 + w^2} & \dfrac{w}{s^2 + w^2} \\[2mm] \dfrac{-w}{s^2 + w^2} & \dfrac{s}{s^2 + w^2} \end{bmatrix} = \begin{bmatrix} \cos wt & \sin wt \\ -\sin wt & \cos wt \end{bmatrix}.$$

Example 3 A similar system to the second-order Markov process discussed in Section 2.9 is the simple one-dimensional harmonic oscillator problem governed by the equation

$$m\ddot{x} + kx = w(t), \tag{1}$$

where m is the mass, k is the spring constant, and $w(t)$ is the external applied force (positive in the $+x$ direction and negative in the $-x$ direction). For convenience, we will assume that the ratio k/m is unity, and designate $w(t)/m = w'(t)$. Then the system assumes the simpler form

$$\ddot{x} + x = w'(t). \tag{2}$$

Equation (2) can be written in state-space notation as follows: Let

$$x_1(t) = x(t), \tag{3}$$

$$x_2(t) = \dot{x}(t). \tag{4}$$

The state equations are

$$\dot{x}_1 = x_2, \tag{5}$$

$$\dot{x}_2 = -x_1 + w'(t). \tag{6}$$

Using the linear system equation

$$\frac{d\mathbf{x}}{dt} = F\mathbf{x} + \mathbf{w}, \tag{7}$$

we have

$$\begin{bmatrix} \dot{x}_1 \\ \dot{x}_2 \end{bmatrix} = \begin{bmatrix} 0 & 1 \\ -1 & 0 \end{bmatrix} \begin{bmatrix} x_1 \\ x_2 \end{bmatrix} + \begin{bmatrix} 0 \\ w' \end{bmatrix}. \tag{8}$$

The solution to the problem can be obtained taking the Laplace transform of both sides of Eq. (7) as follows:

$$s\mathbf{x}(s) = F\mathbf{x}(s) + \mathbf{w}(s),$$

or

$$\mathbf{x}(s) = (Is - F)^{-1} \mathbf{w}(s). \tag{9}$$

The inverse transform of Eq. (9) gives

$$\Phi(t, t_0) = \mathcal{L}^{-1}(Is - F)^{-1} = \begin{bmatrix} s & -1 \\ 1 & s \end{bmatrix}^{-1} = \frac{\begin{bmatrix} s & +1 \\ -1 & s \end{bmatrix}}{s^2 + 1}, \tag{10}$$

yielding (compare this result with Example 2)

$$\Phi(t, t_0) = \begin{bmatrix} \cos(t - t_0) & \sin(t - t_0) \\ -\sin(t - t_0) & \cos(t - t_0) \end{bmatrix}. \tag{11}$$

Therefore, the solution for $\mathbf{x}(t)$ in terms of Eq. (4.2) or (4.37) is given by

$$\mathbf{x}(t) = \Phi(t, t_0) \mathbf{x}(t_0) + \int_{t_0}^{t} \Phi(t, \tau) G(\tau) \mathbf{w}(\tau) \, d\tau$$

$$\begin{bmatrix} x_1(t) \\ x_2(t) \end{bmatrix} = \begin{bmatrix} \cos(t - t_0) & \sin(t - t_0) \\ -\sin(t - t_0) & \cos(t - t_0) \end{bmatrix}$$

$$+ \int_{t_0}^{t} \begin{bmatrix} \cos(t - \tau) & \sin(t - \tau) \\ -\sin(t - \tau) & \cos(t - \tau) \end{bmatrix} \begin{bmatrix} 0 \\ w'(\tau) \end{bmatrix} d\tau. \tag{12}$$

Consider now the system of linear differential equations given by

$$\dot{\mathbf{x}} = A\mathbf{x} + B\mathbf{u}.$$

we have

$$\mathcal{L}\{\dot{\mathbf{x}}\} = \mathcal{L} \begin{bmatrix} \dot{x}_1 \\ \vdots \\ \dot{x}_n \end{bmatrix} = \begin{bmatrix} \mathcal{L}\{\dot{x}_1\} \\ \vdots \\ \mathcal{L}\{\dot{x}_n\} \end{bmatrix}$$

$$= \begin{bmatrix} sX_1(s) - x_{10} \\ \vdots \\ sX_n(s) - x_{n0} \end{bmatrix} = sX(s) - x_0.$$

Assuming the matrices A and B to be constant, then we have

$$sX(s) - x_0 = AX(s) + BU(s),$$

$$(sI - A)X(s) = BU(s) + x_0.$$

Solving for $X(s)$, we have

$$X(s) = (sI - A)^{-1} BU(s) + (sI - A)^{-1} x_0.$$

Again, we assume here that the inverse of the matrix $sI - A$ exists, which is actually the case if $\det(sI - A) \neq 0$. Now, the determinant can be written in the form

$$\det(sI - A) = \begin{vmatrix} s - a_{11} & -a_{12} & \cdots & -a_{1n} \\ -a_{21} & s - a_{22} & \cdots & -a_{2n} \\ \vdots & & & \\ -a_{n1} & -a_{n2} & \cdots & s - a_{nn} \end{vmatrix}.$$

If we set $U(s) \equiv 0$, we have the simple equation

$$X(s) = (sI - A)^{-1} x_0.$$

As a result, the solution in the time domain is, for $u(t) \equiv 0$ and $t_0 = 0$,

$$x(t) = e^{At} x_0,$$

where $\mathcal{L}\{e^{At}\} = (sI - A)^{-1}$. In the general case, where $t \neq 0$, we have as before

$$x(t) = e^{A(t - t_0)} x(t_0),$$

where $\Phi(t, t_0) = e^{A(t - t_0)}$.

Diagonalization is perhaps the most tedious of the above methods to compute e^{Ft}. Suppose now that the eigenvalues of F are real and distinct. Let $\lambda_1, \lambda_2, \ldots, \lambda_n$ denote these eigenvalues, and let Λ be the matrix of the eigenvalues. Then

$$\Lambda = \begin{bmatrix} \lambda_1 & 0 & \cdots & 0 \\ 0 & \lambda_2 & \cdots & 0 \\ \vdots & \vdots & & \vdots \\ 0 & 0 & \cdots & \lambda_n \end{bmatrix}. \tag{4.53}$$

If P is an $n \times n$ nonsingular matrix of eigenvectors of F, then we can form

$$\Lambda = P^{-1}FP, \tag{4.54}$$

or, in terms of the *Jordan canonical form J*,

$$J = P^{-1}FP. \tag{4.55}$$

The solution of the homogeneous equation is then [2]

$$\dot{\mathbf{x}}(t) = F\mathbf{x}(t), \qquad \mathbf{x}(0) = \xi,$$

$$\mathbf{x}(t) = Pe^{\Lambda t}P^{-1}\xi,$$

$$= P\begin{bmatrix} e^{\lambda_1 t} & 0 & \cdots & 0 \\ 0 & e^{\lambda_2 t} & \cdots & 0 \\ \vdots & \vdots & & \vdots \\ 0 & 0 & \cdots & e^{\lambda_n t} \end{bmatrix}P^{-1}\xi,$$

where we have utilized the standard reverse relationship. Again, in terms of the Jordan canonical form, the solution is

$$\mathbf{x}(t) = Pe^{Jt}P^{-1}\xi. \tag{4.56}$$

Decomposing the matrix F into a sum of commuting matrices, when applicable, is probably one of the most useful methods for computing e^{Ft}. The decomposition is based on the fact that if $F = A + B$, where $AB = BA$, then (see Problem 4.11)

$$e^{Ft} = e^{(A+B)t} = e^{At}e^{Bt}.$$

As an illustration of this method, consider the following example:

Example 4   Let

$$F = \begin{bmatrix} A & 0 \\ 0 & B \end{bmatrix} = \begin{bmatrix} A & 0 \\ 0 & 0 \end{bmatrix} + \begin{bmatrix} 0 & 0 \\ 0 & B \end{bmatrix}.$$

Therefore,

$$e^{Ft} = \begin{bmatrix} e^{At} & 0 \\ 0 & e^{Bt} \end{bmatrix}.$$

The above four methods can be compared in this example. Let

$$F = \begin{bmatrix} \sigma & w \\ -w & \sigma \end{bmatrix}.$$

We wish to compute e^{Ft} by these methods.

1. *Infinite Series*:

$$e^{Ft} = \sum_{n=0}^{\infty} \frac{t^n}{n!} \begin{bmatrix} \sigma & w \\ -w & \sigma \end{bmatrix}^n = \sum_{n=0}^{\infty} \frac{t^n}{n!} \left\{ \sigma \begin{bmatrix} 1 & 0 \\ 0 & 1 \end{bmatrix} + w \begin{bmatrix} 0 & 1 \\ -1 & 0 \end{bmatrix} \right\}^n.$$

Since

$$\begin{bmatrix} 1 & 0 \\ 0 & 1 \end{bmatrix} \quad \text{and} \quad \begin{bmatrix} 0 & 1 \\ -1 & 0 \end{bmatrix}$$

commute, we can use the binomial expansion, so that

$$\left\{ \sigma \begin{bmatrix} 1 & 0 \\ 0 & 1 \end{bmatrix} + w \begin{bmatrix} 0 & 1 \\ -1 & 0 \end{bmatrix} \right\}^n = \sum_{i=0}^{\infty} \sigma^i w^{n-i} \binom{n}{i} \begin{bmatrix} 0 & 1 \\ -1 & 0 \end{bmatrix}^{n-i},$$

where

$$\binom{n}{i} = \frac{n!}{(n-i)!\,i!}.$$

Therefore,

$$e^{Ft} = \sum_{n=0}^{\infty} \frac{t^n}{n!} \sum_{i=0}^{\infty} \sigma^i w^{n-i} \binom{n}{i} \begin{bmatrix} 0 & 1 \\ -1 & 0 \end{bmatrix}^{n-i}$$

$$= \sum_{i=0}^{\infty} \sum_{n=0}^{\infty} \frac{t^n}{n!} \sigma^i w^{n-i} \binom{n}{i} \begin{bmatrix} 0 & 1 \\ -1 & 0 \end{bmatrix}^{n-i}.$$

Letting $k = n - 1$, we have

$$e^{Ft} = \sum_{i=0}^{\infty} \frac{(\sigma t)^i}{i!} \sum_{k=0}^{\infty} \frac{(wt)^k}{k!} \begin{bmatrix} 0 & 1 \\ -1 & 0 \end{bmatrix}^k = e^{\sigma t} \begin{bmatrix} \cos wt & \sin wt \\ -\sin wt & \cos wt \end{bmatrix}.$$

2. *Laplace Transformation*:

$$e^{Ft} = \mathcal{L}(sI - F)^{-1} = \begin{bmatrix} s - \sigma & -w \\ w & s - \sigma \end{bmatrix}^{-1} = \frac{1}{(s-\sigma)^2 + w^2} \begin{bmatrix} s - \sigma & w \\ -w & s - \sigma \end{bmatrix}$$

$$= \begin{bmatrix} e^{\sigma t} \cos wt & e^{\sigma t} \sin wt \\ -e^{\sigma t} \sin wt & e^{\sigma t} \cos wt \end{bmatrix}.$$

3. *Diagonalization*: To diagonalize a matrix, we must find a matrix P for which [2]

$$\Lambda = P^{-1}FP, \tag{4.57}$$

where

$$\Lambda = \begin{bmatrix} \lambda_1 & 0 \\ 0 & \lambda_2 \end{bmatrix}.$$

This implies that $P\Lambda = FP$. If we let the ith column of P be P_i, then

$$\lambda_i P_i = FP_i.$$

Therefore, P is the matrix which has columns that are the eigenvectors of F, and Λ is a diagonal matrix which has the characteristic values as its entries. Now,

$$e^{Ft} = Pe^{\Lambda t}P^{-1},$$

where

$$\Lambda = \begin{bmatrix} \lambda_1 & 0 \\ 0 & \lambda_2 \end{bmatrix} = \begin{bmatrix} \sigma + j\omega & 0 \\ 0 & \sigma - j\omega \end{bmatrix},$$

$$e^{\Lambda t} = \begin{bmatrix} e^{(\sigma + j\omega)t} & 0 \\ 0 & e^{(\sigma - j\omega)t} \end{bmatrix} = e^{\sigma t}\begin{bmatrix} e^{j\omega t} & 0 \\ 0 & e^{-j\omega t} \end{bmatrix},$$

$$P^{-1} = \begin{bmatrix} j & -1 \\ -1 & j \end{bmatrix}\left(-\frac{1}{2}\right).$$

Therefore,

$$e^{Ft} = \begin{bmatrix} j & 1 \\ 1 & j \end{bmatrix}\begin{bmatrix} e^{j\omega t} & 0 \\ 0 & e^{-j\omega t} \end{bmatrix}\begin{bmatrix} j & -1 \\ -1 & j \end{bmatrix}\frac{e^{\sigma t}}{(-2)} = e^{\sigma t}\begin{bmatrix} \cos\omega t & \sin\omega t \\ -\sin\omega t & \cos\omega t \end{bmatrix}.$$

4. *Matrix Decomposition.* Writing $F = B + C$, where B and C commute, and e^{Bt} and e^{Ct} are known, we have

$$F = \begin{bmatrix} \sigma & 0 \\ 0 & \sigma \end{bmatrix} + \begin{bmatrix} 0 & \omega \\ -\omega & 0 \end{bmatrix}.$$

Therefore,

$$e^{Ft} = \exp\left(\begin{bmatrix} \sigma & 0 \\ 0 & \sigma \end{bmatrix}_t\right)\exp\left(\begin{bmatrix} 0 & \omega \\ -\omega & 0 \end{bmatrix}_t\right) = \begin{bmatrix} e^{j\omega t} & 0 \\ 0 & e^{-j\omega t} \end{bmatrix}\begin{bmatrix} \cos\omega t & \sin\omega t \\ -\sin\omega t & \cos\omega t \end{bmatrix}$$

$$= e^{\sigma t}\begin{bmatrix} \cos\omega t & \sin\omega t \\ -\sin\omega t & \cos\omega t \end{bmatrix}.$$

For purposes of digital computation, the state transition matrix can be expanded as a power series [see Eq. (4.48)]

$$\Phi(t) = e^{Ft} = \sum_{n=0}^{\infty} F^n \frac{t^n}{n!} = I + Ft + F^2 \frac{t^2}{2!} + F^3 \frac{t^3}{3!} + \cdots, \qquad (4.58a)$$

where I is the identity matrix. At $t = 0$,

$$\Phi(0) = e^{F0} = I + F0 + F^2 \frac{0^2}{2!} + \cdots = I.$$

Using the abbreviation $X = Ft$, Eq. (4.58a) can be written as

$$\Phi(t) = I + \frac{X}{1!} + \frac{X^2}{2!} + \cdots + \frac{X^n}{n!} + \cdots, \qquad t \geqslant 0. \qquad (4.58b)$$

The series (4.58b) coverges for finite $t \geqslant 0$. Letting now

$$\beta_1 = I,$$

$$\beta_2 = I\frac{X}{1} = \beta_1 \frac{X}{1},$$

$$\beta_3 = \left(I\frac{X}{1}\right)\frac{X}{2} = \beta_2 \frac{X}{2}, \qquad (4.59)$$

$$\vdots$$

$$\beta_n = \left(I\frac{X^{n-s}}{(n-2)!}\right)\frac{X}{n-1} = \beta_{n-1}\frac{X}{n-1}.$$

Forming the partial sums, we have

$$\Phi_1 = I = \beta_1,$$

$$\Phi_2 = I + \frac{X}{1!} = \Phi_1 + \beta_2,$$

$$\Phi_3 = I + \frac{X}{1!} + \frac{X^2}{2!} = \Phi_2 + \beta_3, \qquad (4.60)$$

$$\vdots$$

$$\Phi_n = I + \frac{X}{1!} + \frac{X^2}{2!} + \cdots + \frac{X^{n-1}}{(n-1)!} = \Phi_{n-1} + \beta_n.$$

The computation will stop as soon as each element of β_n is smaller than a value, say, ε. In order to limit the computation time in the case of bad convergence, it is recommended that an upper bound K_{max} be placed for the corresponding terms in the series. According to reference [54], the series expansion of $\Phi(t)$ [Eq. (4.58b)] leads to large computation times when the series has bad convergence and/or the dimension of $\Phi(t)$ is large. Bad convergence occurs when the norm of the vector X,

$$\|X\| = \|Ft\| = \sum_i \sum_j |f_{i,j}| t,$$

takes on large values. Plant [54] suggests the following procedure in order to avoid bad convergence: let

$$N = \frac{Ft}{\|Ft\|} \qquad (\text{i.e.,} \quad \|N\| = 1),$$

where N is again an $n \times n$ matrix. From the definition $X = Ft$, we have

$$\Phi(t) = e^{N\|X\|}. \tag{4.61}$$

The norm of $\|X\|$ is a scalar number ($\|X\| > 0$), which can be decomposed as

$$\|X\| = k + y,$$

where $k = $ a positive number ($k = 0, 1, 2, ...$),
$\quad y = $ a positive number between 0 and 1.

With this decomposition, $\Phi(t)$ can be written in the form

$$\Phi(t) = e^{N(k+y)} = e^{Nk} e^{Ny}.$$

Since $\|N\| = 1$, and $0 \leqslant y < 1$, the factor e^{Ny} can be computed for the case of good series convergence. For a series with 12 terms, reference [54] gives a value of $\varepsilon = 10^{-8}$.

The treatment of discrete-time systems is closely related to the continuous-time systems discussed above. Consider the discrete-time system

$$x(k+1) = \Phi(k+1, k) x(k) + \Gamma(k) w(k) \tag{4.14}$$

with statistics

$$\text{cov}(w(k), w(j)) = Q(k, j),$$

$$\mathscr{E}\{w(k)\} = \mu_w(k).$$

Analogous to the continuous-time case, the solution of Eq. (4.14) is given by [58]

$$\mathbf{x}(j) = \Phi(j,0)\,\mathbf{x}(0) + \sum_{i=0}^{j-1} \Phi(j,i+1)\,\Gamma(i)\,\mathbf{w}(i). \qquad (4.62)$$

We will now summarize the properties of the state transition matrix. Given the state transition matrix $\Phi(t, t_0)$ of the system $\dot{\mathbf{x}}(t) = F(t)\,\mathbf{x}(t)$, $t \in T$, we have

$$|\Phi(t, t_0)| \neq 0 \qquad \forall\, t_0, t \in (t_1, t_2) \bigcap T,$$

so that $\Phi(t, t_0)$ is nonsingular for every t in (T_1, T_2);

$$\Phi(t_0, t_2) = \Phi(t_0, t_2)\,\Phi(t_1, t_2), \qquad \text{(semigroup property)},$$

$$\Phi^{-1}(t_1, t_2) = \Phi(t_2, t_1),$$

$$\Phi^{-1}(t) = \Phi(-t) \quad \text{and} \quad \Phi(t - \tau) = \Phi^{-1}(\tau - t) = \Phi(-\tau)\,\Phi(t);$$

$$(e^{Ft})^{-1} = \Phi^{-1}(t, 0) = \Phi(0, t) = e^{-Ft}$$

[since $\Phi(0, t_0) = e^{-Ft_0}$ for all t_0, then e^{Ft} is invertible];

$$\Phi(t_0, t_0) = e^{F0} = I \qquad \text{(identity transformation)},$$

$$\Phi(t_0, t_1)\,\Phi(t_1, t_0) = I;$$

$$\det \Phi(t, t_0) = \det \Phi(t_0, t_0)\, \exp\left(\int_{t_0}^{t} \text{trace}\, F(\tau)\, d\tau \right),$$

where

$$\det \Phi(t_0, t_0) = \det I = 1 \quad \text{and} \quad \text{trace}\, F = \sum_{i=1}^{n} [f]_{ii};$$

$$\frac{d}{dt}\Phi(t, t_0) = \dot{\Phi}(t, t_0) = F(t)\Phi(t, t_0), \qquad \Phi(t, t_0) = e^{F(t - t_0)},$$

$$\frac{d}{dt} e^{Ft} = F\, e^{Ft}.$$

In particular,

$$\frac{\partial}{\partial t}\Phi(t_0, t) = \frac{\partial}{\partial t}\Phi^{-1}(t, t_0) = -\Phi^{-1}(t, t_0)\left[\frac{\partial}{\partial t}\Phi(t, t_0) \right]\Phi^{-1}(t, t_0)$$

$$= -\Phi^{-1}(t, t_0)\, [F(t)\,\Phi(t, t_0)]\, \Phi^{-1}(t, t_0)$$

$$= -\Phi^{-1}(t, t_0)\, F(t)$$

$$= -\Phi(t_0, t)\, F(t).$$

Reference [58, pp.140–142] gives a detailed program listing for computing the state transition matrix, based on the Sylvester expansion theorem. However, a simpler computational FORTRAN program based on Eq. (4.49), which uses a Taylor series expansion, is presented below. If more accuracy is needed in calculating e^{Ft}, then the Padé approximation may appeal to the user.

```
C    **********************************************************
C    *
C    * SUBROUTINE: MEXP.FOR
C    *
C    * PURPOSE: COMPUTE THE MATRIX EXPONENTIAL EXP(Ft)
C    *          USING THE TAYLOR SERIES APPROACH FOR THE
C    *          STATE TRANSITION MATRIX
C    *
C    **********************************************************
```

$$e^{Ft} = \sum_{k=0}^{\infty} \frac{F^k}{k!} t^k$$

$$= I + \frac{F}{1!}t + \frac{F^2}{2!}t^2 + \frac{F^3}{3!}t^3 + \cdots.$$

```
      SUBROUTINE   MEXP   (D, F, A, T1, T2, T3, EXPTA, N, KMAX, RE, AE,
      KLM)

CCCC  NOTE THAT ALL MATRICES ARE SQUARE WITH DIMEN-
C     SION N x N. THE DIMENSION IS VARIABLE IN THE SUBROUTINE,
C     BUT MUST BE DEFINED IN THE MAIN PROGRAM.

      DOUBLE PRECISION K, T1, T2, T3, EXPTA, PK, T, TK, D1, D2, D3

      DIMENSION A(N, N),  F(N, N),  EXPTA(N, N),  T1(N, N),  T2(N, N),
      T3 (N, N)

CCCC  TO REDUCE RUN TIME, REDEFINE AE, RE, AND D AS
      DOUBLE PRECISION

      D1 = DBLE(AE)
      D2 = DBLE(RE)
      D3 = DBLE(D)

      T = 0.
      PK = 1.

      DO 10 J = 1, N
      DO 10 I = 1, N
         T1(I, J) = T*DBLE(A(I, J))

10          EXPTA(I, J) = T1(I, J)
      IF(KLM.EQ.1) THEN
            DO 15 I = 1, N
15             EXPTA (I, I) = EXPTA (I, I) + D3
      ENDIF
```

```
18          PK = PK + 1.
            TK = T/PK
            DO 30 I = 1, N
            DO 30 J = 1, N
                T2 (I, J) = 0.
                DO 30 L = 1, N
30                  T2 (I, J) = T2 (I, J) + DBLE (F(I, L))* T1 (L, J)
            DO 20 J = 1, N
            DO 20 I = 1, N
                IF (KLM.EQ.1) THEN
                    T1(I, J) = T2(I, J)*TK
                ELSE
                    T1(I, J) = (T2(I, J) + T2 (J, I))* TK
                ENDIF
20              EXPTA (I, J) = EXPTA (I, J) + T1(I, J)
            DO 70 J = 1, N
            DO 70 I = 1, N
                ER = DABS (T1(I.J)/(D1 + D2*DABS(EXPTA(I, J))))
                IF(ER.GT.1.0)THEN
                    IF (PK.LT.FLOAT(KMAX))THEN
                        GOTO 18
                    ELSE
                        STOP 'MAXIMUM NUMBER OF
                        ITERATIONS EXCEEDED!'
                    ENDIF
                ENDIF
70          CONTINUE
            K = IDINT(PK)
            WRITE(*.'("NUMBER OF ITERATIONS IS: ".I3)')K
            DO 120 I = 1, N
            DO 120 J = 1, N
120             T3(I, J) = EXPTA(I, J)
            CALL DRUK(T3, N, N)
            RETURN
            END
            SUBROUTINE DRUK (A, N, M)
            DIMENSION A(N, M)
101         FORMAT (* TRANSPOSE*)
102         FORMAT (1X, 12E11.4)
            IF (N.GT.M) GO TO 2
            DO 1 I = 1, N
1           PRINT 102, (A(I, J), J = 1, M)
            RETURN
2           PRINT 101
            DO 3 I = 1, M
3           PRINT 102, (A(J, I), J = 1, N)
            RETURN
```

Because the *norm* (mentioned earlier) plays an important role in optimal control theory, we conclude this section with a summary of its characteristics. We begin by noting that a mathematical form appearing often in optimal control problems is the *inner product*,* which has the form

$$\langle \mathbf{x}, \mathbf{y} \rangle = \mathbf{x}^T \mathbf{y} = \sum_{i=1}^{n} x_i\, y_i = x_1\, y_1 + x_2\, y_2 + \cdots + x_n\, y_n$$

provided that $\mathbf{x}$ and $\mathbf{y}$ are real. The inner product of a vector $\mathbf{x} = (x_1, x_2, \ldots, x_n)^T$ with itself is $\langle \mathbf{x}, \mathbf{x} \rangle$ and is called the *norm* of $\mathbf{x}$. It is denoted by

$$\| x \| = \langle \mathbf{x}, \mathbf{x} \rangle.$$

A frequently used norm is the *Euclidean norm* defined by

$$\| \mathbf{x} \| = \sqrt{\mathbf{x}^T \mathbf{x}} = \left(\sum_{i=1}^{n} x_i^2 \right)^{1/2}.$$

In general, a *vector norm* $\| \mathbf{x} \|$ is a real function in the subspace of the vector $\mathbf{x}$. If X is a linear space, then the function $\| \cdot \| : X \to R$ is said to be a norm if the following conditions are satisfied:

1. $\| \mathbf{x} \| \geqslant 0 \ \forall \, \mathbf{x} \neq 0$, and $\mathbf{x} \in R^n$;
2. $\| \mathbf{x} \| = 0$ if and only if $\mathbf{x} = \mathbf{0}$.
3. $\| \alpha \mathbf{x} \| = |\alpha| \, \| \mathbf{x} \|$ for all scalars $\alpha \in K$.
4. $\| \mathbf{x} + \mathbf{y} \| = \| \mathbf{x} \| + \| \mathbf{y} \| \ \ \forall \, \mathbf{x}, \mathbf{y} \in R^n$.
5. $|\langle \mathbf{x}, \mathbf{y} \rangle| \leqslant \| \mathbf{x} \| \cdot \| \mathbf{y} \| \ \ \forall \, \mathbf{x}, \mathbf{y} \in R^n$.

In addition to the above Euclidean norm, the following vector norms are of interest:

1. $\| \mathbf{x} \| = \max |x_i|$ with $\mathbf{x} = [x_1, x_2, \ldots, x_i, \ldots, x_n]^T \in R^n$;
2. $\| \mathbf{x} \| = \sum_{i=1}^{n} |x_i|$;
3. $\| \mathbf{x} \| = [(T\mathbf{x})^* \, T\mathbf{x}]^{1/2} = (\mathbf{x}^* T^* \, T\mathbf{x})^{1/2} = (\mathbf{x}^* Q\mathbf{x})^{1/2}$,

where the asterisk (*) denotes the *conjugate* (or *adjoint*) matrix, and T is a linear transformation. Specifically, the * mapping on matrices has the same

---

* The inner product is also called a "scalar" (or "dot") product, since it yields a scalar function.

properties as the transpose. It has

1. $(A^*)^* = A$ for all matrices A.
2. $(A + B)^* = A^* + B^*$.
3. $(AB)^* = B^* A^*$.

Finally, it can be shown that the norm function $\|\cdot\|$ is continuous and obeys the inequality

$$\|\mathbf{x}\| - \|\mathbf{y}\| \leqslant \|\mathbf{x} - \mathbf{y}\|.$$

4.5 CONTROLLABILITY AND OBSERVABILITY

Two important concepts associated with the Kalman filter are: (1) *controllability* and (2) *observability*. Controllability and observability arise in a fundamental way in attempting to stabilize unstable models of linear systems in filtering theory and in optimal control theory. In the discussion of these concepts, it is convenient to consider the system input noise $\mathbf{w}(t)$ as a control input, which may be selected arbitrarily, and the measurement noise $\mathbf{v}(t)$, both assumed to be zero. Consider the unforced, linear, time-varying system

$$\dot{\mathbf{x}}(t) = F(t)\mathbf{x}(t) + G(t)\mathbf{w}(t) \tag{4.63}$$

$$\mathbf{z}(t) = H(t)\mathbf{x}(t), \tag{4.64}$$

where $F(t)$, $G(t)$, $H(t)$ are $n \times n$, $n \times m$, and $q \times n$ matrix functions. respectively. This system is often referred to as the (F, G, H) system. Controllability and observability will now be defined.

Controllability. Controllability is the ability to transfer an arbitrary initial state to an arbitrary final state in finite time. Seen from a simple point of view, controllability implies the ability of the control input to affect each state variable. Using the message models Eq. (4.1) or Eq. (4.14), a system is said to be controllable at time t_0, if there exists a $t_f > t_0$ and a control $\mathbf{w}(t)$ defined on $[t_0, t_f]$—which depend, in general, on t_0 and $\mathbf{x}(t_0)$—such that $\mathbf{x}(t_f) = \mathbf{0}$ [1]. Furthermore, if a control can be found for every state $\mathbf{x}(t_0)$, then we say that the message model is completely controllable at t_0, and if this is true for every t_0, then the message model is *completely controllable*. Obviously, if a system is controllable on $[t_0, t_f]$, then it is controllable on any interval which contains $[t_0, t_f]$. For a linear, stationary, continuous-time system of dimension n in which the matrices F and G are constant, the system is completely controllable if and only if the criterion

$$\text{rank } [G \mid FG \mid F^2 G \mid \cdots \mid F^{n-1} G] = n \tag{4.65}$$

is satisfied [i.e., Eq. (4.65) is of full rank]. For the discrete-time, case,

$$\text{rank } [\Gamma | \Phi\Gamma | \Phi^2\Gamma | \cdots | \Phi^{n-1}\Gamma] = n. \qquad (4.66)$$

Observability. Observability is the ability to uniquely determine the state of a free linear dynamic system from observations of linear combinations of the output of the system in finite time. In discussing observability, it is necessary to consider the observation and the message models together. Furthermore, it will be assumed that both $\mathbf{w}(t)$ and $\mathbf{v}(t)$ are identically zero. Consequently, an unforced system is said to be *completely observable* on $[t_0, t_f]$ if, for given t_0 and $t_f > t_0$, every state $\mathbf{x}(t_0)$ can be determined from the knowledge of $\mathbf{z}(t) = H(t)\mathbf{x}(t)$ on $[t_0, t_f]$. If for every t_0 there exists a $t_f > t_0$ (which may depend on t_0) such that the system is completely observable on $[t_0, t_f]$, then we say that the system is completely observable. For a system with constant matrices, a continuous-time system is completely observable if and only if [21]

$$\text{rank } [H^T | F^T H^T | (F^T)^2 H^T | \cdots | (F^T)^{n-1} H^T] = n, \qquad (4.67)$$

and for the discrete-time case

$$\text{rank } [H^T | \Phi^T H^T | (\Phi^T)^2 H^T | \cdots | (\Phi^T)^{n-1} H^T] = n. \qquad (4.68)$$

The observability test is often used to determine how many state variables in a proposed model can be separately estimated from the output of the measuring devices. However, if any state variable cannot be estimated, a different model must be sought.

In practical applications of the Kalman filter, it is conceivable that one of the states in the output may not be observable. In such a case, two options are available: (1) reduce the state order of the filter, or (2) use another measurement sensor. For both controllability and observability, we can say that a completely controllable and observable system has the following properties:

- The optimum filter is asymptotically stable.
- The effect of the initial guess for the covariance matrix becomes of secondary importance as more data are processed.

Notice that these definitions require only that the input be able to drive the state from one value to another in a finite time. Furthermore, no constraint is made on the path followed in moving between the two states, nor on the type of input used.

Example In order to illustrate the concept of controllability, consider the following simple system:

$$\dot{x}(t) = \begin{bmatrix} -1 & 0 & 0 \\ 0 & 2 & 0 \\ 0 & 0 & 3 \end{bmatrix} x(t) + \begin{bmatrix} 1 & 0 \\ 0 & 1 \\ 1 & 1 \end{bmatrix} w(t),$$

$$z(t) = \begin{bmatrix} 1 & 1 & 1 \\ 2 & 2 & 2 \end{bmatrix} x(t),$$

$$\text{rank } [G \vdots FG \vdots F^2 G] = |\text{rank} \begin{bmatrix} 1 & 0 & -1 & 0 & 1 & 0 \\ 0 & 1 & 0 & 2 & 0 & 4 \\ 1 & 1 & 3 & 3 & 9 & 9 \end{bmatrix} = 3.$$

Therefore, the system is completely controllable. An example of a system that is not completely controllable is

$$\dot{x}(t) = \begin{bmatrix} a & 1 & 0 \\ 0 & a & 0 \\ 0 & 0 & b \end{bmatrix} x(t) + \begin{bmatrix} 1 & 0 \\ 0 & 0 \\ 0 & 1 \end{bmatrix} w(t),$$

$$z(t) = \begin{bmatrix} 1 & 0 & 1 \\ 0 & 1 & 1 \end{bmatrix} x(t),$$

$$\text{rank } [G \vdots FG \vdots F^2 G] = \text{rank} \begin{bmatrix} 1 & 0 & a & 0 & a^2 & 0 \\ 0 & 0 & 0 & 0 & 0 & 0 \\ 0 & 1 & 0 & b & 0 & b^2 \end{bmatrix} = 2 \neq 3.$$

A system which is neither controllable nor observable is of the form

$$\dot{x}(t) = \begin{bmatrix} -3 & 4 \\ -2 & 3 \end{bmatrix} x(t) + \begin{bmatrix} 1 \\ 1 \end{bmatrix} u(t),$$

$$z(t) = \begin{bmatrix} -1 & 2 \end{bmatrix} x(t).$$

Controllability and observability can be determined using Eqs. (4.65) and (4.67). Thus,

$$\text{rank } [G \vdots FG] = \begin{bmatrix} 1 & 1 \\ 1 & 1 \end{bmatrix}$$

has rank 1; therefore the system is not controllable.

$$\text{rank } [H^T \vdots F^T H^T] = \begin{bmatrix} -1 & -1 \\ 2 & 2 \end{bmatrix}$$

has rank 1; therefore the system is not observable.

This system is also unstable. In order to determine the instability, we must examine the system eigenvalues. The system eigenvalues can be determined as follows:

$$\det(\lambda I - F) = \det \begin{bmatrix} \lambda + 3 & -4 \\ 2 & \lambda - 3 \end{bmatrix} = \lambda^2 - 1, \qquad \lambda = +1, -1,$$

so that the system is unstable. However, this system can be stabilized if we use an output feedback of the form $\mathbf{u}(t) = k\mathbf{z}(t)$. This is equivalent to the system

$$\dot{\mathbf{x}}(t) = \begin{bmatrix} -3 & 4 \\ -2 & 3 \end{bmatrix} \mathbf{x}(t) + \begin{bmatrix} 1 \\ 1 \end{bmatrix} k[-1 \quad 2]\,\mathbf{x}(t)$$

$$= \begin{bmatrix} -3-k & 4+2k \\ -2-k & 3+2k \end{bmatrix} \mathbf{x}(t).$$

From the preceding equation, we note that if we let $k = -2$, then

$$\dot{\mathbf{x}}(t) = \begin{bmatrix} -1 & 0 \\ 0 & -1 \end{bmatrix} \mathbf{x}(t)$$

so that the system is stabilized. In conslusion, we can say that the unstable mode can be stabilized by means of output feedback.

In connection with the above discussion, it should be pointed out that a very important aspect of the Kalman filter theory is the so-called *Kalman duality theorem*.

Kalman Duality Theorem. The system (F, G, H) is observable (respectively, reachable) if and only if its *dual* (F*, G*, H*) is reachable (respectively, observable).

This is a very important result. Specifically, it says that whenever we have algorithms to test reachability (or controllability) of any system, it requires no extra effort to handle observability; that is, in order to test the observability of a system, we simply apply the reachability criterion to the dual system. For more details on the duality theorem, the reader should consult reference [52, Chapter 4].

The concepts of controllability and observability can be summarized by the general structure of the control system depicted in the Figure 4.16.

Consider the continuous linear dynamic system (or plant) defined as follows:

$$\frac{\mathbf{d}x(t)}{dt} = A(t)\mathbf{x}(t) + B(t)\mathbf{u}(t),$$

$$\mathbf{y}(t) = C(t)\mathbf{x}(t) + D(t)\mathbf{u}(t),$$

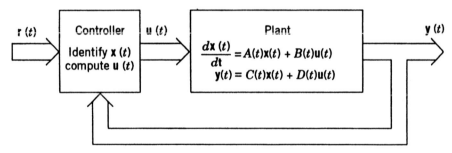

Figure 4.16   Typical control system.

where $\mathbf{x}(t)$ = system state vector = $[x_1(t), x_2(t), ..., x_n(t)]^T$,

$\mathbf{u}(t)$ = input vector = $[u_1(t), u_2(t), ..., u_m(t)]^T$,

$\mathbf{y}(t)$ = output vector = $[y_1(t), y_2(t), ..., y_p(t)]^T$,

$\mathbf{r}(t)$ = input vector to the controller,

$\mathbf{A}(t)$ = system matrix,

$\mathbf{B}(t)$ = input distribution matrix,

$\mathbf{C}(t)$ = measurement (or observation) matrix,

$\mathbf{D}(t)$ = noise coefficient matrix.

The function of the controller is twofold:

1. *Observation*: The controller must identify the state of the plant by observation of the output.
2. *Control*: The controller must guide the state of the plant along a desired trajectory by producing suitable control signals.

In certain control systems, the controller needs only partial information about the state of the plant. Furthermore, successful control of the output may often be achieved without complete control of the state. It should be pointed out, however, that much of the optimal control engineering is based on the following two assumptions:

1. *Complete Observability*: The initial state $\mathbf{x}(t_0)$ of the plant may be identified by observing the output $\mathbf{y}(t)$ for a finite time.
2. *Complete Controllability*: The plant may be transferred from an arbitrary initial state $\mathbf{x}(t_0)$ to any other state $\mathbf{x}(t_1)$ by applying a suitable control $\mathbf{u}(t)$ for a finite time.

4.5.1 Observers

From the discussion given in Section 4.1, it is sometimes desired to obtain a good estimate of the state $\mathbf{x}(t)$, given a knowledge of the output $\mathbf{y}(t)$, the input

$\mathbf{u}(t)$, and the system matrices, A, B, C, and D. This problem is referred to as the *state reconstruction problem*. That is, the state is reconstructed from in-put–output measurements. Another dynamic system, called the *observer*, is to be constructed. Its input will depend on $\mathbf{y}(t)$ and $\mathbf{u}(t)$, and its state (output) should be a good approximation to $\mathbf{x}(t)$. The continuous measurements of $\mathbf{u}(t)$ and $\mathbf{y}(t)$ over the interval $[t_0, t_1]$ drive a linear dynamical system, known as a *dynamic observer*, whose output $\mathbf{x}(t)$ approximates (i.e., asymptotically tracks) the original plant (or system) state vector. The structure of the dynamic ob-server mirrors the usual state model equations and depend implicitly on the known plant matrices A, B, C, and D. Specifically, the idea of an observer was formulated for reconstructing the state vector of an observable, deterministic linear system, from *exact* measurements of the state vector [47]. Observer concepts have found extensive use in the application of deterministic feedback control problems, where the control law may depend on knowledge of all the system states, but only limited combinations of the states are measurable.

The observers that will be presently discussed are called *full-state* or *identity observers*, since the total state vector is reconstructed. That is, a dynamic observer builds around a replica of the given plant to provide an on-line, continuous estimate of the system state. From intuition, in order for the replica to be a dynamic observer, the state $\hat{\mathbf{x}}(t)$ must asymptotically approach $\mathbf{x}(t)$ for large t. In order for this to happen, the error

$$\lim_{t \to +\infty} [\hat{\mathbf{x}}(t) - \mathbf{x}(t)] = 0 \tag{1}$$

for large t. Consider initially the deterministic, open-loop, linear time-invariant state model written in the usual multi-input, multi-output (MIMO) form

$$\frac{d\mathbf{x}(t)}{dt} = A\mathbf{x}(t) + B\mathbf{u}(t), \qquad \mathbf{x}(0) = x_0,$$

$$\mathbf{y}(t) = C\mathbf{x}(t) + D\mathbf{u}(t), \tag{2}$$

$$\mathbf{z}(t) = H\mathbf{x}(t),$$

where $\mathbf{x}(t) = n \times 1$ state vector $(x \in R^n)$,

$\mathbf{y}(t) = m \times 1$ output vector $(y \in R^m)$,

$\mathbf{z}(t) = r \times 1$ observation or measurement vector $(z \in R^r)$,

$\mathbf{u}(t) = l \times 1$ control input vector $(u \in R^l)$,

$A = n \times n$ constant matrix,

$B = n \times l$ constant matrix,

$C = m \times n$ constant matrix,

$D = m \times l$ constant matrix,

$H = r \times n$ observation matrix.

The open-loop system with observer is diagrammed in Figure 4.17.

Assuming for simplicity that the matrix $D = [0]$, the deterministic linear time-invariant observer model will have the form [47]

$$\frac{d\hat{\mathbf{x}}(t)}{dt} = M_1\hat{\mathbf{x}}(t) + M_2\mathbf{u}(t) + M_3\mathbf{z}(t), \tag{3}$$

where M_1, M_2, and M_3 are (as yet unspecified) constant matrices of dimension $n \times n$, $n \times l$, and $n \times r$, respectively. Define now the estimation error as

$$\mathbf{e}(t) \triangleq \hat{\mathbf{x}}(t) - \mathbf{x}(t). \tag{4}$$

Subtracting Eq. (2) from Eq. (3) and using the relation $\mathbf{z}(t) = H\mathbf{x}(t)$ yields

$$\frac{d\mathbf{e}(t)}{dt} = M_1\hat{\mathbf{x}}(t) + M_2\mathbf{u}(t) + M_3H\mathbf{x}(t) - A\mathbf{x}(t) - B\mathbf{u}(t)$$

$$= M_1\mathbf{e}(t) + (M_1 - A + M_3H)\,\mathbf{x}(t) + (M_2 - B)\,\mathbf{u}(t). \tag{5}$$

In practice, we would like to have the error dynamics be independent of the input control $\mathbf{u}(t)$ and present value of $\mathbf{x}(t)$. In order to eliminate the dependence on $\mathbf{u}(t)$, we set

$$M_2 = B. \tag{6}$$

Furthermore, we would like to remove the explicit dependence on $\mathbf{u}(t)$, so that we can set

$$M_1 = A - M_3H. \tag{7}$$

As a result, Eq. (5) reduces to the simpler form

$$\frac{d\mathbf{e}(t)}{dt} = \hat{\mathbf{x}}(t) - \dot{\mathbf{x}}(t) = (A - M_3H)\,\mathbf{e}(t). \tag{8}$$

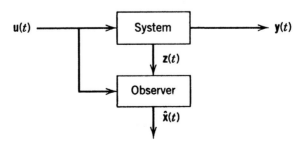

Figure 4.17   Open-loop system with observer.

The solution of Eq. (8) is simply

$$\mathbf{e}(t) = \hat{\mathbf{x}}(t) - \mathbf{x}(t) = e^{(A - M_3 H)} [\hat{\mathbf{x}}(0) - \mathbf{x}(0)]. \tag{9}$$

Now, if we choose M_3 such that the eigenvalues of $A - M_3 H$ all have negative real parts, then

$$\lim_{t \to +\infty} \mathbf{e}(t) = \mathbf{0}, \tag{10}$$

which means that the estimate produced by the observer is asymptotically stable.

Therefore, we can summarize the linear time-invariant observer for the system Eq. (2) as follows:

$$\frac{d\hat{\mathbf{x}}(t)}{dt} = (A - M_3 H) \hat{\mathbf{x}}(t) + M_3 \mathbf{z}(t) + B\mathbf{u}(t)$$

$$= A\hat{\mathbf{x}}(t) + M_3 [\mathbf{z}(t) - H\hat{\mathbf{x}}(t)] + B\mathbf{u}(t), \tag{11}$$

where the term $M_3[\mathbf{z}(t) - H\hat{\mathbf{x}}(t)]$ determines the degree to which $d\hat{\mathbf{x}}(t)/dt$ is proportional to the difference between the system output measurement $\mathbf{z}(t)$ and the estimated output $H\hat{\mathbf{x}}(t)$. The following requirements are imposed on the dynamics of the system:

1. $\hat{\mathbf{x}}(t) \to \mathbf{x}(t)$ for large t; that is, all the eigenvalues of $A - M_3 H$ lie in the open left-hand complex plane.
2. The rate at which $\hat{\mathbf{x}}(t)$ approaches $\mathbf{x}(t)$ is determined by the control design engineer by suitable choice of M_3.

It should be pointed out, however, that since information about some of the states is directly obtainable from $\mathbf{z}(t)$, the above method contains a certain amount of redundancy. As stated earlier, the observers described above are used to estimate the state $\mathbf{x}(t)$. If a constant feedback matrix K is used with $\hat{\mathbf{x}}(t)$ instead of $\mathbf{x}(t)$, the resulting system as shown in Figure 4.18. Obviously, the composite system is of order $2n$. By proper selection of K, n of the closed-loop eigenvalues can be specified. That is, a feedback system with desired poles can be designed, proceeding as if all states were measurable. In Figure 4.18, F is a feedforward matrix, $\mathbf{r}(t)$ is an input vector, and the control law takes the form

$$\mathbf{u}(t) = F\mathbf{r}(t) - K\hat{\mathbf{x}}(t). \tag{12}$$

Therefore, the closed-loop system is stable if the state feedback law (12) stabilizes the system (2).

The above results can be extended to stochastic observer systems. For more details the reader is referred to reference [70].

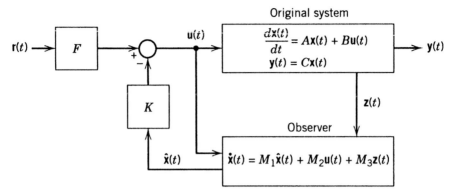

Figure 4.18 Implementation of the control **u** using an observer and measurement **z**.

As stated earlier, oftentimes only the output variables or some subset thereof are available for measurement. However, many feedback design schemes require complete state information. In order to remedy this situation, control design engineers insert dynamic observer structures in the feedback loop, and use the state estimate in the feedback control law of the type indicated by Eq. (12). Insertion of a dynamic observer in the feedback path, however, creates additional system dynamics, additional eigenvalues, and additional natural frequencies. Because of the *eigenvalue separation theorem*, the additional dynamics do not interfere with the desired system behavior. The eigenvalue separation theorem states that the characteristic polynomial of the feedback system with a dynamic observer equals the product of the characteristic polynomial of the observer and that of the state feedback control without the observer. Consequently, the dynamic behavior of the observer does not interfere with the desired eigenstructure of the controlled plant. The diagram in Figure 4.19 illustrates the need for the eigenvalue separation theorem.

Therefore, a feedback system with the desired poles can be designed, proceeding as if all the states were measurable. Then, a separate design of the observer can be used to provide the desired observer poles. The feedback

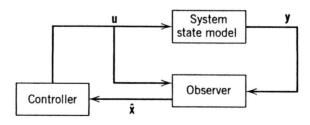

Figure 4.19 The eigenvalue separation theorem.

control law (12), which stabilizes the system (2), can be written in the form

$$\mathbf{u}(t) = F\mathbf{r}(t) - K\hat{\mathbf{x}}(t)$$

$$= [F\mathbf{r}(t) - K\mathbf{x}(t)] - K\mathbf{e}(t), \tag{13}$$

where $\mathbf{e}(t) = \hat{\mathbf{x}}(t) - \mathbf{x}(t)$. Substituting Eq. (13) into Eq. (2), we obtain

$$\frac{d\mathbf{x}(t)}{dt} = (A - BK)\,\mathbf{x}(t) + BF\mathbf{r}(t) - BK\mathbf{e}(t), \tag{14a}$$

$$\mathbf{y}(t) = (C - DK)\,\mathbf{x}(t) - DK\mathbf{e}(t) + DF\mathbf{r}(t). \tag{14b}$$

Here we note that the state of the feedback system is uniquely specified by $\mathbf{x}(t)$ and $\mathbf{e}(t)$, and therefore a state-variable model is obtained by combining Eq. (8) with Eq. (14). Thus,

$$\begin{bmatrix} \dot{\mathbf{x}}(t) \\ \dot{\mathbf{e}}(t) \end{bmatrix} = \begin{bmatrix} A - BK & -BK \\ 0 & A - M_3 H \end{bmatrix} \begin{bmatrix} \mathbf{x}(t) \\ \mathbf{e}(t) \end{bmatrix} + \begin{bmatrix} BF \\ 0 \end{bmatrix} \mathbf{r}(t), \tag{15a}$$

$$\mathbf{y}(t) = [C - DK \quad -DK] \begin{bmatrix} \mathbf{x}(t) \\ \mathbf{e}(t) \end{bmatrix} + DF\mathbf{r}(t), \tag{15b}$$

where the vector $[\mathbf{x}(t)\ \mathbf{e}(t)]^T$ is of order $2n \times 1$. Therefore, the characteristic polynomial of the closed-loop system is as follows:

$$\rho(s) = \begin{vmatrix} sI - A + BK & 0 \\ 0 & sI - A + M_3 H \end{vmatrix}$$

$$= |sI - A + BK| \cdot |sI - A + M_3 H| \tag{16}$$

which is the product of the closed-loop polynomial $|sI - A + BK|$ of the system (2) with the ideal state feedback law $\mathbf{u}(t) = F\mathbf{r}(t) - K\hat{\mathbf{x}}(t)$ and the characteristic polynomial $|sI - A + M_3 H|$ of the observer, Eq. (11). The state feedback matrix K is commonly specified in order to allocate the eigenvalues of $A - BK$.

From classical control theory we know that there are several techniques for choosing K and M_3. For example, one method of choosing M_3 is to note from Eq. (9) that the estimation error $\mathbf{e}(t)$ decays at a rate dependent on the eigenvalues of $A - M_3 H$. Consequently, M_3 can be chosen in such a way that the eigenvalues of $A - M_3 H$ have specified values, that is, M_3 is specified by solving a pole allocation problem.* From the control law (12) and the observer

---

* The *pole allocation problem* can be stated as follows: Given a specified $r \times 1$ measurement vector $\mathbf{z}(t)$ and a specified set of complex numbers $\mu_1, \mu_2, \ldots, \mu_n$, find an $l \times r$ feedback matrix K such that the characteristic polynomial $\rho_c(s)$ is given by $\rho_c(s) = |sI - A + BKH| = (s - \mu_1)(s - \mu_2)\cdots(s - \mu_n)$.

characteristic polynomial $\rho(s)$, M_3 can be identified with K by setting $M_3 = K^T$ in the observer equation

$$\frac{d\hat{\mathbf{x}}(t)}{dt} = A\,\hat{\mathbf{x}}(t) + K\,[\mathbf{z}(t) - H\,\hat{\mathbf{x}}(t)] + B\mathbf{u}(t).$$

Finally, we note that in practice all the eigenvalues have negative real parts, and hence the closed-loop system is stable.

4.6 DIVERGENCE

We have seen earlier that the Kalman filter theoretically produces an increasingly accurate estimate of the state parameters as additional data become available and processed by the filter. However, the performance of the filter under actual operating conditions can be seriously degraded from the theoretical performance indicated by the state covariance matrix. In particular, the magnitude of the estimation errors as measured by the determinant of the estimation error covariance matrix is a monotonically decreasing function of the number of observations. That is, divergence takes place when the covariance matrix becomes too small or optimistic. However, it should be pointed out that the actual error may become unbounded, even though the error covariance in the Kalman filter algorithm is vanishingly small. Another reason is that the gain $K(k)$ in the Kalman filter algorithm approaches zero too rapidly. As the gain becomes small, subsequent observations are ignored by the filter. Hence the estimate becomes decoupled from the observation sequence and is not affected by the growing observation error. Furthermore, we also know that, since the system model is usually an approximation to a physical situation, the model parameters and noise statistics are seldom exact. Specifically, the system model used in constructing the filter differs from the real system that generates the observations. It is clear, therefore, that an inexact filter model will degrade the filter performance, causing the filter to diverge. Consequently, the filter designer in designing a filter model must perform a tradeoff study or evaluate the effect on performance of various approximations made.

In actual applications, divergence manifests itself by the inconsistency of residuals (or innovations) in

$$\gamma_r(k+1|k) = \mathbf{z}(k+1) - H(k+1)\,\hat{\mathbf{x}}(k+1|k) \qquad (4.69)$$

with their predicted statistics. Residuals become biased (i.e., nonzero mean) and larger in magnitude than their rms values. One method to prevent divergence is to increase the covariance matrix $P(k)$. This is justified in that such an increase in the covariance matrix compensates for the model errors. Allowing the filter to overestimate its own errors slightly, so as to minimize the likelihood of divergence, is a commonly used method of generating a conservative filter design. Model errors in some sense add uncertainty to the system, which should reflect itself in some degradation in certainty [increase in $P(k)$]. In reference [59] it is shown that actual estimation errors can be bounded by adding noise to the system.

The major causes of divergence in the Kalman filter can be summarized as follows:

1. Inaccuracies in the modeling* process used to determine the message or observation model. That is, an inexact filter model will degrade the filter performance.
2. Failure of linearization.
3. Lack of complete knowledge of the physical problem.
4. Errors in the statistical modeling of noise variances and mean.
5. Wordlength and roundoff errors in the implementation of the filter algorithm on a digital computer.
6. Loss of positive definiteness or symmetry of the error covariance matrix.

Divergence in the filter can be prevented by:

1. Direct increase of the gain matrix.
2. Limiting of the error covariance.
3. Artificial increase of the system (or plant) noise variance.

The first step in the design of a Kalman filter is the development of a mathematical model of the system. This model is developed based on the knowledge of the physical system and statistical data. Therefore, once the state (i.e., message) and measurement models are determined, all the parameters need to be identified. Commonly, there are two types of modeling errors associated with Kalman filtering. The first type, which may be associated with the filter design, arises from a mismatch between the estimator model and the real process being estimated. The other type arises from the fact that optimum Kalman filtering requires an exact knowledge of the process noise covariance matrix Q, the measurement noise covariance matrix R, and the initial estimate $P(0)$ of the error covariance matrix. Clearly, in a practical problem these matrices are either unknown or known only approximately. This problem is sometimes referred to as *tuning* of the filter, since the filter designer must adjust Q, R, and $P(0)$ for satisfactory performance. Both the tuning and the validation of the estimator model can be accomplished by performing simple statistical tests on the innovations sequence.

---

*Modeling is the process of obtaining a mathematical model of the physical system under consideration.

It should be pointed out here that the innovations or residual sequence was defined as the difference between the actual output of the system and its predicted value, given the previous outputs. This sequence can be viewed as an error or residual in the sense that the predicted value was off from the actual output by that amount. When the filter is performing satisfactorily, the innovations sequence should be small, that is, filter is predicting the state accurately. If the filter is operating in a nonoptimal fashion, then the residuals should be large. Consequently, this is a very important result, which can be used as a criterion for testing filter performance. In the tuning of the filter it is assumed that some estimates of Q, R, and $P(0)$ are available, and the question is whether the filter designed with these estimates is performing in an optimal manner. (Note that the matrices Q and R can be statistically estimated when they are not known.) The tuning of the filter is clearly a hypothesis testing problem.

Finally, it should be mentioned here briefly that the round off-error problem is a difficult one. However, the filter designer can take certain steps in order to avoid it. Among the techniques recommended for minimizing the roundoff errors are: (1) to avoid using fixed-point arithmetic by using double-precision arithmetic, particularly for off-line operations, (2) to avoid propagating the error covariance matrix P in many small steps between measurements, (3) to force the matrix P to be symmetric at every recursive step, and (4) to avoid deterministic processes in the filter modeling (e.g., a random constant).

For a more detailed discussion of divergence, references [31, 37, and 59] are recommended.

4.7 THE *U–D* COVARIANCE ALGORITHM IN KALMAN FILTERS

The discrete-time Kalman filter has been highly successful in a wide variety of real-time estimation problems. Its essence lies in updating the state estimate and its covariance matrix. Of the various algorithms that have been developed to compute these updates, the so-called $U–D$ algorithm (also known as the $U–D$ covariance factorization algorithm) is perhaps the most attractive, for reasons that will be discussed shortly. In this section we will derive the $U–D$ algorithm for measurement updates and time updates. Measurement updating using the $U–D$ factorization preserves the nonnegative definite structure of the covariance matrix P, and is equivalent to the Carlson method without square-root computations.

Bierman [9] recognized that the square-root calculations required by the Carlson algorithm are often costly and proposed a square-root-free measurement update scheme. Bierman's method employs the covariance factorization

$$P = UDU^T, \qquad (4.70)$$

where U is a unit upper triangular matrix and D is a positive diagonal matrix. As with other filtering algorithms, the $P = UDU^T$ decomposition was chosen by Bierman in order to develop a numerically stable filtering algorithm, algebraically equivalent to the Kalman filter. Specifically, the U–D algorithms get their name from the fact that they involve the unit upper triangular matrix U (all diagonal elements of U are one, and all elements below the diagonal are zero) and the diagonal matrix D which uniquely factors the $n \times n$ symmetric covariance matrix P as $P = UDU^T$. Therefore, in order to start the algorithm, the initial covariance matrix P_0 must be converted into its U–D factors. The U–D algorithms have several advantages over alternative algorithms. They assure the positive definiteness of the covariance matrices and implicitly preserve their symmetry. In addition, as square-root algorithms, they reduce the dynamic range of the numbers entering into the computations. This fact, together with the greater accuracy of the square-root algorithms, reduces the computer wordlength requirements for a specified accuracy. A rule of thumb is that square-root algorithms can use half the wordlength required by conventional algorithms. Moreover, the U–D algorithms generally involve significantly less computational cost than the conventional algorithms. When U–D algorithms are used in Kalman filter applications, only the U–D factors are found, and not the covariance matrices themselves. However, any covariance matrix P can be found if desired by computing $P = UDU^T$.

The U–D data-processing algorithm can be derived by factoring Eq. (4.29). For convenience, the time index will be dropped; for example, $P(t_i^-) = P^-$, or $P(k|k) = U(k|k) D(k|k) U^T(k|k) = P^+$, and so on. Moreover, we will consider only scalar measurements. Therefore, using this simplified notation, the measurement update equations in a Kalman filter with scalar measurements are

$$\mathbf{k} = P^-\mathbf{h}(\mathbf{h}^T P^-\mathbf{h} + r)^{-1}, \tag{4.71}$$

$$\hat{\mathbf{x}}^+ = \hat{\mathbf{x}}^- + \mathbf{k}(\mathbf{y} - \mathbf{h}^T\hat{\mathbf{x}}^-), \tag{4.72}$$

$$P^+ = P^- - \mathbf{k}\mathbf{h}^T P^-. \tag{4.73}$$

Note that in Eq. (4.72) we have substituted $\mathbf{y}$ for the measurement instead of $\mathbf{z}$. As in the conventional discrete-time case, we will develop the measurement update and time update equations. This development is based on the excellent works of Bierman, Thornton, and Maybeck [9, 67, 68 and 48], respectively.

Measurement Update Given the factors U^-, D^- of P^- (satisfying $P^- = U^- D^- U^{-T}$), the U–D measurement algorithm generates the factors U^+, D^+ of P^+ (satisfying $P^+ = U^+ D^+ U^{+T}$) so that Eq. (4.73) is obeyed. In addition, it computes $\mathbf{k}$ defined in Eq. (4.71), from which $\hat{\mathbf{x}}^+$ is computed by Eq. (4.72). The U–D measurement algorithm is derived by substituting Eq. (4.71) and the U–D factors of P^- in to Eq. (4.73). This yields

$$P^+ = U^- D^- U^{-T} - U^- D^- U^{-T}\mathbf{h}\left(\frac{1}{\alpha}\right)\mathbf{h}^T U^- D^- U^{-T}$$

$$= U^-\left[D^- -\frac{1}{\alpha}(D^-U^{-T}\mathbf{h})\,(D^-U^{-T}\mathbf{h})^T\right]U^{-T}$$

$$= U^-\left[D^- -\frac{1}{\alpha}\mathbf{v}\mathbf{v}^T\right]U^{-T}, \tag{4.74}$$

where

$$\alpha = \mathbf{h}^T P^- \mathbf{h} + r, \tag{4.75}$$

$$\mathbf{f} = U^{-T}\mathbf{h}, \tag{4.76}$$

$$\mathbf{v} = D^- \mathbf{f} = D^- U^{-T}\mathbf{h}. \tag{4.77}$$

The bracketed factor in Eq. (4.74) is positive definite and hence may be factored as $\hat{U}\hat{D}\hat{U}^T$. Thus, if a unit upper triangular matrix $\hat{U}$ and a diagonal matrix $\hat{D}$ can be found that satisfy

$$\hat{U}\hat{D}\hat{U}^T = D^- -\frac{1}{\alpha}\mathbf{v}\mathbf{v}^T, \tag{4.78}$$

then $U^+ = U^-\hat{U}$ and $D^+ = \hat{D}$ will be the desired factors of P^+, since one has

$$U^+ D^+ U^{+T} = U^- \hat{U}\hat{D}\hat{U}^T U^{-T} \tag{4.79}$$

$$= U^-\left[D^- -\frac{1}{\alpha}\mathbf{v}\mathbf{v}^T\right]U^{-T}$$

$$= P^+.$$

An algorithm that generates $\hat{U}$ and $\hat{D}$ is given below, without proof, as Proposition 1. This proposition is then used in Proposition 2, which contains the $U\!-\!D$ measurement algorithm that generates U^+ and D^+.

Proposition 1. Let the unit upper triangular matrix $\hat{U}$ and the diagonal matrix $\hat{D} = \text{diag}(\hat{d}_1, ..., \hat{d}_n)$ be defined by setting

$$c_n = -\frac{1}{\alpha} \tag{4.80}$$

computing recursively for $j = n, n-1, ..., 2$

$$\hat{d}_j = d_j^- + c_j v_j^2, \tag{4.81}$$

$$c_{j-1} = c_j d_j^- /\hat{d}_j, \tag{4.82}$$

$$\hat{U}(i,j) = c_j v_j v_i /\hat{d}_j, \qquad i = 1, 2, ..., j-1, \tag{4.83}$$

and finally letting

$$\hat{d}_1 = d_1^- + c_1 v_1^2, \tag{4.84}$$

where α and $v^T = (v_1, \ldots, v_n)$ are defined in Eqs. (4.75)–(4.77) and $D^- = \operatorname{diag}(d_1^-, \ldots, d_n^-)$. Then $\hat{U}$ and $\hat{D}$ satisfy Eq. (4.78).

Proposition 2. Let $\mathbf{f}^T = (f_1, \ldots, f_n)$ and $v^T = (v_1, \ldots, v_n)$ be defined by Eqs. (4.76)–(4.77), and U_j^- be the jth column of U^-. Set

$$\alpha_1 = r + v_1 f_1, \qquad d_1^+ = d_1^- r/\alpha_1, \tag{4.85}$$

$$\mathbf{k}_1^T = (v_1, 0, \ldots, 0), \tag{4.86}$$

and then for $j = 2, 3, \ldots, n$ let

$$\alpha_j = \alpha_{j-1} + v_j f_j, \tag{4.87}$$

$$d_j^+ = d_j^- \alpha_{j-1}/\alpha_j, \tag{4.88}$$

$$\lambda_j = -f_j/\alpha_{j-1}, \tag{4.89}$$

$$U_j^+ = U_j^- + \lambda_j K_{j-1}, \tag{4.90}$$

$$K_j = K_{j-1} + v_j u_j^-. \tag{4.91}$$

Finally, let

$$K = K_n/\alpha_n, \tag{4.92}$$

$$\hat{\mathbf{x}}(t^+) = \hat{\mathbf{x}}(t^-) + K[\mathbf{z}(t) - H(t)\hat{\mathbf{x}}(t^-)].$$

If D^+ and U^+ are defined by

$$D^+ = \operatorname{diag}(d_1^+, \ldots, d_n^+),$$

$$U^+ = [U_1^+ \mid U_2^+ \mid \cdots \mid U_n^+]$$

that is, the jth column of U^+ is U_j^+, then $D^+ = \hat{D}$ and $U^+ = U^- \hat{U}$. Hence, by the argument leading up to Eq. (4.79), D^+ and U^+ are the U–D factors of P^+. In addition, K defined by Eq. (4.92) is equal to the Kalman gain $\mathbf{k}$ defined by Eq. (4.71).

To show that $K = \mathbf{k}$, we first note that

$$K_j = K_{j-1} + v_j U_j^-$$

$$= U^- v_{(j-1)} + v_j U_j^- \qquad (4.93)$$

$$= U^- v_j.$$

Therefore, making use of Eqs. (4.71),(4.75),(4.77), and (4.93), respectively, we obtain

$$\mathbf{k} = P^- h (h^T P^- h + \alpha)^{-1}$$

$$= U^- D^- U^{-T} h / \alpha$$

$$= U^- v / \alpha$$

$$= K_n / \alpha$$

$$= K.$$

As a final comment, we note that the characterization of d_j^+ given by Eq. (4.88) is preferable to the equivalent characterization given by Eq. (4.81). Since all the d's are theoretically positive, it follows from Eq. (4.82) that the c_j's will have the same sign, which is negative. Hence, the updated diagonals d_j are computed in Eq. (4.81) as differences. Such calculations are susceptible to a loss of accuracy due to cancellation, and this might even result in negative $\hat{d}$ terms being computed. The representation in Eq. (4.88) avoids this problem.

Figure 4.20 contains a flowchart for the *U–D* measurement update algorithm given by Eqs. (4.76)–(4.77) and Eqs. (4.85)–(4.92), while Figure 4.21 presents a FORTRAN mechanization algorithm that is very efficient in storage and time [9]. An efficient FORTRAN program for computing $P = UDU^T$ is given in Figure 4.22 [69]. In addition, a FORTRAN program for computing only the variances (the diagonal elements of P) from the $U–D$ factors is included in Figure 4.23.

To start a Kalman filter that uses the $U–D$ algorithms, the initial (*a priori*) covariance matrix P_0 must be factored into $P_0 = U_0 D_0 U_0^T$. This can be done by the *Cholesky factorization* (see [38, 43, 69]). However, initial covariance matrices are commonly diagonal, in which case $D_0 = P_0$ and $U_0 = I$. If the problem to which the Kalman filter is applied contains biases, correlated process noise, or considered parameters, improvements can be made to the $U–D$ algorithms which exploit the special structure induced by these effects [68].

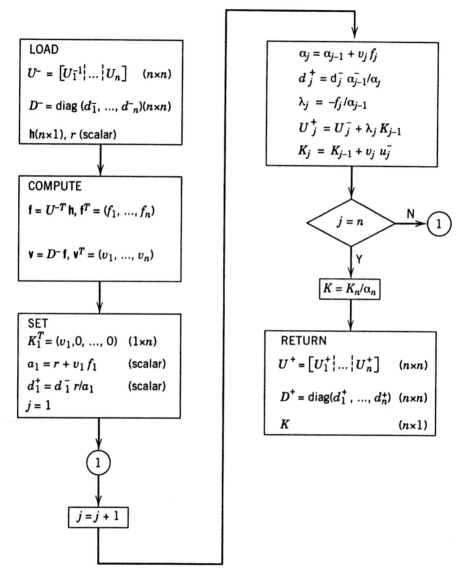

Figure 4.20 $U-D$ measurement update algorithm.

UD ALGORITHM

```
 1   C--   SUBROUTINE MEAS (U, H, R, N, Y, X,)
 2   C--   PARAMETER JN = 50
 3   C--   DIMENSION U (N, N), X (N), H (N)
 4   C--   DIMENSION B (JN)
 5   C
 6   C--   THIS SUBROUTINE PERFORMS THE U – D MEASUREMENT
 7   C--   UPDATE FOR A DISCRETE KALMAN FILTER. ALTHOUGH THIS
 8   C--   SUBROUTINE IS VERY SPACE EFFICIENT, IT DESTROYS
 9   C--   THE INPUTED MATRIX U, VECTORS H AND X, AND SCALAR Y. IT
10   C--   PRESERVES THE SCALAR R AND INTEGER N.
11   C
12   C--   THE COVARIANCE MATRIX MEASUREMENT UPDATE EQUATION IS
13   C--      P⁺ = (P⁻) − K*Hᵀ*(P⁻)
14   C--   WHERE
15   C--      K = (P⁻)*H/(Hᵀ*(P⁻)H + R).
16   C--   THIS SUBROUTINE INVOLVES THE U – D FACTORS OF P⁻ AND
17   C--   P⁺ NAMELY THE UNIT UPPER TRIANGULAR MATRICES U⁻, U⁺
18   C--   AND THE DIAGONAL MATRICES D⁻, D⁺ SUCH THAT
19   C--   P⁺ = (U⁺)* (D⁺)*(U⁺)ᵀ AND P⁻ = (U⁻)*(D⁻)*(U⁻)ᵀ. SINCE
20   C--   THE DIAGONALS OF U⁺ AND U⁻ ARE ALWAYS ONE, THE
21   C--   MATRICES D⁺, D⁻ ARE STORED ALONG THE DIAGONALS OF U⁺,
22   C--   U⁻.THE MATRIX U IN THE CALL LIST IS USED TO INPUT THE OLD
23   C--   FACTORS U⁻, D⁻ AND OUTPUT THE NEW FACTORS U⁺, D⁺
24   C
25   C--   THE NEW AND OLD ESTIMATES OF THE STATE VECTOR, X⁺ AND
26   C--   X⁻ ARE RELATED BY
27   C--      X⁺ = (X⁻) + K*(Y − Hᵀ*(X⁻)).
28   C--   THE VECTOR X IN THE CALL LIST IS USED TO INPUT X⁻ AND
29   C--   OUTPUT X⁺
30   C
31   C--   THE PARAMETER JN MUST BE EQUAL TO OR GREATER THAN N.
32   C
33   C
34   C--   INPUTS (PRE-UPDATE)
35   C
36   C--   N                    = DIMENSION OF THE STATE VECTOR
37   C--   U (N, N)             = CONTAINS U – D FACTORS OF COVARIANCE
38   C---                           MATRIX P⁻
39   C--   H(N)                 = STATE MEASUREMENT VECTOR
40   C--   R (SCALAR)           = VARIANCE OF MEASUREMENT NOISE
41   C--   Y (SCALAR)           = OBSERVATION
42   C--   X (N)                = STATE ESTIMATE X⁻
43   C
44   C--   OUTPUTS (POST-UPDATE)
45   C
46   C--   U (N, N)             = CONTAINS U–D FACTORS OF COVARIANCE
47   C--                           MATRIX P⁺
```

Figure 4.21   FORTRAN program for $U-D$ measurement update algorithm.

```
48   C--   X(N) = STATE ESTIMATE OF X⁺.
49   C
50   C--   AT THE END OF THIS PROGRAM SEVERAL QUANTITIES HAVE
51   C--   BEEN COMPUTED WHICH ARE OF INTEREST IN SOME
52   C--   APPLICATIONS. THESE QUANTITIES AND THEIR STORAGE
            LOCATIONS FOLLOW:
53   C
54   C--      A =Hᵀ*(U⁻)*(D⁻)*(U⁻)ᵀ*H+R =Hᵀ*(P⁻)*H+R,
55   C
56   C--   WHICH IS THE VARIANCE OF THE INNOVATIONS Y − Hᵀ*(X⁻).
57   C--      AINV = 1./A
58   C         Y    = Y(OBSERVATION) − Hᵀ*(X⁻) = INNOVATIONS
59   C--   KALMAN GAIN = B*AINV.
60   C
61   C--   REPLACE MEASUREMENT Y BY THE INNOVATIONS Y−Hᵀ*(X⁻).
62   C
63   C         DO 1 J = 1, N
64   C       1 Y = Y − H(J)*X(J)                    ● Eq. (4.76)
65   C
66   C--   COMPUTE F = (U⁻)ᵀ*H AND PLACE IN H.
67   C--   COMPUTE V = (D⁻)*F AND PLACE IN B.
68   C
69         IF (N. EQ. 1) GO TO 6
70         DO 3 J = N, 2, − 1
71             DO 2 K = 1, J − 1
72             H(J) = H(J) + U(K, J)*H(K)           ● Eq. (4.76)
73      2      CONTINUE
74      3  B(J) = U(J, J)*H(J)                       ● Eq. (4.77)
75      6  CONTINUE
76   C         B(1) = U(1, 1)*H(1)                    ● Eq. (4.77)
77   C
78   C--   PERFORM MEASUREMENT UPDATE OF U−D FACTORS.
79   C--   FOR EACH J, K(1),..., K(J) IS STORED IN B(1),...,B(J)
80   C--   AND V(J), ..., V(N) IS STORED IN B(J), ..., B(N). NOTICE THAT
81   C--   K(J) = V(J) IN THE JTH ITERATION.
82   C
83         A = R + B(1)*H(1)                          ● Eq. (4.85)
84         AINV = 1./A
85         U(1, 1) = U(1, 1)*R*AINV                   ● Eq. (4.85)
86         IF (N. EQ. 1) GO TO 11
87         DO 10 J = 2, N
88         AOLD = A                                   ● AOLD = A/SUB J − 1
89         A = A + B(J)*H(J)                          ● Eq. (4.87)
90         ALAM = − H(J)*AINV                         ● Eq. (4.89)
91         AINV = 1/.A
92         U(J, J) = U(J, J)*AOLD*AINV                ● Eq. (4.88)
93             DO 10 I = 1, J − 1
94             C = U(I, J)
95             U(I, J) = C + ALAM*B(I)                ● Eq. (4.90)
```

Figure 4.21 (*Continued*)

```
 96    10   B(I) = B(I) + B(J)*C                    ● Eq. (4.91)
 97    11   CONTINUE
 98   C
 99   C--  THE KALMAN GAIN K CAN BE COMPUTED AT THIS POINT BY
100   C--  K = B*AINV, THAT IS, BY
101            DO 30 J = 1, N
102   C 30    K(J) = B(J)*AINV                       ● Eq. (4.92)
103   C
104   C--  COMPUTE UPDATED STATE ESTIMATE  X⁺ = (X⁻) + K*Y = (X⁻) +
             B*AINV, WHERE Y IS NOW THE INNOVATIONS COMPUTED
105   C--  ABOVE.
106   C
107            C = Y*AINV
108            DO 15 J = 1, N
109    15   X(J) = X(J) + B(J)*C                     ● Eqs. (4.72) and (4.92)
110            RETURN
111            END
```

Figure 4.21 (*Continued*).

```
 1            SUBROUTINE PCOMPT (U, P, N)
 2            PARAMETER JN = 50
 3            DIMENSION U(N, N), P(N, N)
 4            DIMENSION V(JN)
 5   C
 6   C--  THIS SUBROUTINE COMPUTES THE MATRIX P = UDUᵀ WHERE U
 7   C--  IS UPPER UNIT TRIANGULAR AND D IS DIAGONAL. IT IS
 8   C--  ASSUMED THAT THE MATRIX D IS STORED ALONG THE
 9   C--  DIAGONAL OF U. ALL MATRICES HAVE DIMENSION (N, N).
10   C--  THE PARAMETER JN MUST BE EQUAL TO OR GREATER THAN N.
11   C
12   C--  IF (N. EQ. 1) GO TO 50
13            NM 1 = N − 1
14            IF (N.EQ. 2) GO TO 40
15            DO 30 I = 1, N − 2
16   C
17   C--  P(I, N)  (I = 1, N − 2) IS COMPUTED HERE.
18   C
19            V(N) = U(I, N)*U(N, N)
20            P(I, N) = V(N)
21            P(N, I) = V(N)
22   C
23            P(I, I) = U(I, I) + U(I, N) * V(N)
24   C
25            DO 20 J = NM1, I + 1, − 1
26   C
27   C--  P(I, I) IS COMPUTED HERE.
28   C
```

Figure 4.22 FORTRAN program for computing $P = UDU^T$.

```
29                V (J) = U (I, J) * U (J, J)
30                P (I, I) = P (I, I) + U (I, J) * V (J)
31      C
32      C--       P (I, J) (I = 1, N − 2  J = 1 + 1, N − 1) IS COMPUTED HERE.
33      C
34                P (I, J) = V (J)
35                  DO 10 K = J + 1, N
36                  P (I, J) = P (I, J) + V (K) * U (J, K)
37       10       CONTINUE
38                P (I, J) = P (I, J)
39       20       CONTINUE
40       30       CONTINUE
41      C
42       40       CONTINUE
43      C
44      C-- AT THIS POINT THERE ARE STILL FOUR ENTRIES TO BE
45      C-- COMPUTED.
46      C-- THEY FOLLOW:
47      C
48                P (NM1, N) = U (NM1, N) * U (N, N)
49                P (N, NM1) = P (NM1, N)
50                P (NM1, NM1) = U (NM1, NM1) + U (NM1, N) * P (NM1, N)
51       50   P (N, N) = U (N, N)
52                RETURN
53                END
```

Figure 4.22   (*Continued*)

```
1                SUBROUTINE PVAR (U, PV, N)
2                DIMENSION U (N, N), PV (N)
3       C
4       C-- THIS SUBROUTINE COMPUTES THE DIAGONALS OF THE CO-
5       C-- VARIANCE MATRIX P = UDUᵀ, WHERE U IS UNIT UPPER TRI-
6       C-- ANGULAR AND D IS DIAGONAL. IT IS ASSUMED THAT THE
7       C-- MATRIX D IS STORED ALONG THE DIAGONAL OF U. THE
8       C-- DIAGONALS OF P ARE PLACED IN THE OUTPUT VECTOR PV. ALL
10      C-- MATRICES HAVE DIMENSION (N, N)
11               IF (N. EQ. 1) GO TO 30
12               DO 20 I = 1, N − 1
13               PV (I) = U (I, I)
14                 DO  10 J = I + 1, N
15                 PV (I) = PV (I) + U (I, J) * U (I, J) * U (J, J)
16       10       CONTINUE
17       20 CONTINUE
18       30 CONTINUE
19               PV (N) = U (N, N)
20               RETURN
21               END
```

Figure 4.23   FORTRAN program for computing the diagonal elements of UDU^T.

Before we leave this section, the Cholesky factorization (or decomposition) algorithm will be given for the interested reader. The Cholesky decomposition algorithm can be stated as follows: if P is a symmetric, positive definite $n \times n$ matrix, it can be uniquely factored (i.e., in terms of an upper triangular and a lower triangular matrix) as [43]

$$P = S^T S$$

namely

$$
\begin{bmatrix}
P_{11} & P_{12} & \cdots & P_{1n} \\
P_{12} & P_{22} & \cdots & P_{2n} \\
\vdots & \vdots & \ddots & \vdots \\
P_{1n} & P_{n2} & \cdots & P_{nn}
\end{bmatrix}
=
\begin{bmatrix}
S_{11} & S_{21} & \cdots & S_{n1} \\
S_{21} & S_{22} & \cdots & S_{n2} \\
\vdots & \vdots & \ddots & \vdots \\
0 & 0 & \cdots & S_{nn}
\end{bmatrix}
\begin{bmatrix}
S_{11} & 0 & \cdots & 0 \\
S_{21} & S_{22} & \cdots & 0 \\
\vdots & \vdots & \ddots & \vdots \\
S_{n1} & S_{n2} & \cdots & S_{nn}
\end{bmatrix}
= S_n^T S_n,
$$

where S is an upper triangular matrix. S is commonly called the square-root matrix of P, or the Cholesky square root of P. The recursive algorithm for computing S in terms of its entries S_{ij} is as follows:

$$S_{11} = (P_{11})^{1/2},$$

$$S_{1j} = P_{ij}/S_{11}, \qquad \text{for} \quad j = 2, \ldots, n,$$

$$S_{22} = [P_{22} - (S_{12})^2]^{1/2},$$

$$S_{2j} = \frac{P_{2j} - S_{12}S_{1j}}{S_{22}} \qquad \text{for} \quad j = 3, \ldots, n,$$

$$\vdots$$

$$S_{ii} = \left[P_{ii} - \sum_{k=1}^{i-1} (S_{ki})^2 \right]^{1/2} \qquad \text{for} \quad i = 2, \ldots, n, \qquad (4.94)$$

$$S_{ij} = \frac{P_{ij} - \sum_{k=1}^{i-1} S_{ki} S_{kj}}{S_{ii}} \qquad \text{for} \quad j = i+1, \ldots, n. \qquad (4.95)$$

The Cholesky decomposition is also seen in the literature in the form

$$P = SS^T,$$

where

$$S = \begin{bmatrix} S_{11} & 0 & \cdots & 0 \\ S_{21} & S_{22} & \cdots & 0 \\ \vdots & \vdots & \ddots & \vdots \\ S_{n1} & S_{n2} & \cdots & S_{nn} \end{bmatrix}.$$

The elements of the lower triangular matrix S are computed, as before, recursively, beginning in the upper left corner as follows:

$$s_{ij} = \begin{cases} \left(q_{ii} - \sum_{k=1}^{i-1} s_{ik}^2 \right)^{1/2}, & j = i, \\ 0 & j > i, \\ \dfrac{q_{ij} - \sum_{k=1}^{j-1} s_{ik} s_{jk}}{s_{ji}}, & j = 1, 2, \ldots, i-1 \end{cases}$$

(Note that the sequence of nonzero computations is $s_{11}, s_{21}, s_{22}, s_{31}, \ldots$, and that the summations are *not* evaluated when the upper limit is less than one.)

Time Update We discussed earlier that between the jth and $(j + 1)$ st measurements, the state estimate and covariance matrix are propagated by the following equations [Eqn. (4.26), (4.27)]:

$$\hat{\mathbf{x}}_{j+1}^- = A_j \hat{\mathbf{x}}_j^+, \tag{4.96a}$$

$$P_{j+1}^- = A_j P_j^+ A_j^T + G_j Q_j G_j^T, \tag{4.96b}$$

where A represents the state-transition matrix ($A \equiv \Phi$). In the sequel, all time subscripts j will be suppressed. Therefore, the equation for time updating the covariance matrix in a Kalman filter becomes

$$P^- = AP^+ A^T + GQG^T \tag{4.97}$$

(as before, the superscripts $-$ and $+$ indicate respectively the values just before and just after the jth measurement).

Now, it is convenient to rewrite Eq. (4.97) as

$$P^- = W\hat{D}\mathbf{W}^T \tag{4.98}$$

where

$$W = [AU^+ \vdots G], \qquad \hat{D} = \left[\begin{array}{c|c} D^+ & 0 \\ \hline 0 & Q \end{array}\right], \tag{4.99}$$

and $P^+ = U^+ D^+ U^{+T}$. The dimensions of W and $\hat{D}$ are $n \times p$ and $p \times p$, respectively, where $p = n + m$. A unit upper triangular matrix U^- and a diagonal matrix D^-, both of dimension $n \times n$, will be found that are the U–D factors of the time-updated covariance P^-. That is,

$$U^- D^- U^{-T} = W\hat{D}W^T.$$

The only properties of D and W that are used are the positive definiteness of D and the linear independence of the rows of W. This last property is satisfied by W defined in Eq. (4.99), since P^- is positive definite and hence has rank n. Consequently, both of its factors, W and $\hat{D}$ in Eq. (4.98) have at least rank n.

It is convenient to express W as

$$\mathbf{W} = \begin{bmatrix} w_1^T \\ \vdots \\ w_n^T \end{bmatrix},$$

where $w_1, \ldots, w_n$ are linearly independent vectors of dimension p. The time update (or time propagation) of the U–D factors employs a generalized Gram–Schmidt orthogonalization in order to preserve numerical accuracy while at the same time attaining computational efficiency. Therefore, the keystone of the derivation is the Gram–Schmidt orthogonalization with respect to the weighted inner product defined for any p-vectors $\mathbf{v}$ and $\mathbf{w}$ [44]

$$\langle \mathbf{w}, \mathbf{v} \rangle_{\hat{D}} \triangleq \mathbf{w}^T \hat{D} \mathbf{v}. \tag{4.100}$$

More specifically, we will need n vectors (of dimension p) $\mathbf{v}_1, \ldots, \mathbf{v}_n$ generated by $\mathbf{w}_1, \ldots, \mathbf{w}_n$ and orthogonal to each other with respect to this inner product; that is, $\langle \mathbf{v}_k, \mathbf{v}_j \rangle_{\hat{D}} = 0, k \neq j$. For notational simplicity, we let $U = U^-$ and $D = D^-$ in the following arguments. Hence, matrices U, D are sought that satisfy

$$UDU^T = W\hat{D}W^T. \tag{4.101}$$

Before we develop the time-update algorithm, it is necessary to discuss the Gram–Schmidt orthogonalization. Let $\mathbf{w}_1, \ldots, \mathbf{w}_n$ be linearly independent p-vectors with $p \triangleq n$, and let D be a $p \times p$ positive definite matrix. The vectors $\mathbf{v}_1, \ldots, \mathbf{v}_n$ are defined recursively by

$$\mathbf{v}_n = \mathbf{w}_n \tag{4.102}$$

and

$$\mathbf{v}_k = \mathbf{w}_k - \sum_{j=k+1}^{n} \frac{\langle \mathbf{w}_k, \mathbf{v}_j \rangle_{\hat{D}}}{\langle \mathbf{v}_j, \mathbf{v}_j \rangle_{\hat{D}}} \mathbf{v}_j, \qquad k = n-1, \dots, 1, \tag{4.103}$$

are nonzero and orthogonal with respect to D, that is $\langle \mathbf{v}_k, \mathbf{v}_j \rangle_{\hat{D}} = 0$, $k \neq j$ ($k < j \leqslant n$). Equation (4.103) expresses the w's as a linear combination of the v's. That is,

$$\mathbf{w}_k = \mathbf{v}_k + \sum_{j=k+1}^{n} U(k, j) \mathbf{v}_j, \qquad k = n-1, \dots, 1, \tag{4.104}$$

with

$$U(k, j) = \frac{\langle \mathbf{w}_k, \mathbf{v}_j \rangle_{\hat{D}}}{\langle \mathbf{v}_j, \mathbf{v}_j \rangle_{\hat{D}}}, \qquad j = k+1, \dots, n. \tag{4.105}$$

Furthermore,

$$W = UV, \tag{4.106}$$

where U is the unit upper triangular matrix with $U(k, j)$ defined by Eq. (4.105) for $j > k$. The matrices U and D satisfy Eq. (4.101), since they both have dimension $n \times n$ and

$$W\hat{D}W^T = UV\hat{D}V^T U^T = UDU^T.$$

In order to find matrices U and D that satisfy Eq. (4.101), let

$$V = \begin{bmatrix} \mathbf{v}_1^T \\ \vdots \\ \mathbf{v}_n^T \end{bmatrix}$$

and let

$$D \triangleq VDV^T = \mathrm{diag}\,(d_1, \dots, d_n), \tag{4.107}$$

where

$$d_k = \langle \mathbf{v}_k, \mathbf{v}_k \rangle_{\hat{D}}. \tag{4.108}$$

From the above discussion, we see that the factorization of the updated covariance has been found with U and D defined by Eqs. (4.102), (4.103), (4.105), (4.107), and (4.108). This concludes the derivation of the Gram–Schmidt orthogonalization.

A more efficient computational form of the classical Gram–Schmidt orthogonalization is the so-called *modified Gram–Schmidt* (MGS) algorithm. This algorithm has been determined to be numerically superior to the conventional Gram–Schmidt method, exhibiting less susceptibility to roundoff errors, and it uses less storage. The MGS can be developed as follows: let $\mathbf{w}_1, \ldots, \mathbf{w}_n$ be linearly independent p-vectors with $p \geq n$, and let $\hat{D}$ be a positive definite $p \times p$ matrix. Consider now the recursions.

$$\mathbf{v}_k^{(n)} = \mathbf{w}_k, \qquad k = 1, \ldots, n, \tag{4.109}$$

and for $j = n, \ldots, 2$

$$d_j = \langle \mathbf{v}_j^{(j)}, \mathbf{v}_j^{(j)} \rangle_{\hat{D}}, \tag{4.110}$$

$$U(k,j) = \langle \mathbf{w}_k, \mathbf{v}_j^{(j)} \rangle_{\hat{D}} / d_j, \qquad k = 1, \ldots, j-1 \tag{4.111}$$

$$\mathbf{v}_k^{(j-1)} = \mathbf{v}_k^{(j)} - U(k,j)\mathbf{v}_j^{(j)}, \qquad k = 1, \ldots, j-1, \tag{4.112}$$

and

$$d_1 = \langle \mathbf{v}_1^{(1)}, \mathbf{v}_1^{(1)} \rangle_{\hat{D}}. \tag{4.113}$$

Then

$$\mathbf{v}_j^{(j)} = \mathbf{v}_j, \qquad j = n, \ldots, 1, \tag{4.114}$$

where $\mathbf{v}_j$ is defined in Eqs. (4.102)–(4.103) and the values $U(k,j)$ given by Eqs. (4.105) and (4.111) are equal. It is worth noting here that Eqs. (4.109)–(4.113) are merely an algebraic rearrangement of the Gram–Schmidt orthogonalization as given by Eqs. (4.102), (4.103), (4.105), (4.108). Furthermore, we note that the vectors $\mathbf{v}_k^{(k)} = \mathbf{v}_k$ are equal, since $\mathbf{v}_k^{(K)}$ is formed by starting with $\mathbf{v}_k^{(n)} = \mathbf{w}_k$ and by successively cycling $U(k,j)\mathbf{v}_j^{(j)} = (\langle \mathbf{w}_j, \mathbf{v}_j^{(j)} \rangle \hat{D}/d_j)v_j$ for $j = n, \ldots, k+1$, which yields $\mathbf{v}_k$ in Eq. (4.103). Finally, Eq. (4.111) may be written alternatively as

$$U(k,j) = \langle \mathbf{v}_k^{(j)}, \mathbf{v}_j^{(j)} \rangle_{\hat{D}} / d_j, \qquad k = 1, \ldots, j-1, \tag{4.115}$$

for $j = n, \ldots, 2$, since $\langle \mathbf{v}_k^{(j)}, \mathbf{v}_j^{(j)} \rangle_{\hat{D}} = \langle \mathbf{w}_k, \mathbf{v}_j^{(j)} \rangle_{\hat{D}}$. This last equality follows immediately from Eq. (4.109) if $j = n$. Using Eq. (4.115) in place of Eq. (4.111) reduces storage requirements and contributes to improved numerical accuracy of this algorithm.

Figure 4.24 presents in flowchart form the $U-D$ time update algorithm given by Eqs. (4.109), (4.110), (4.112), (4.113), and (4.115). Figure 4.25 gives the FORTRAN program of this algorithm, which is very efficient in storage and time [9].

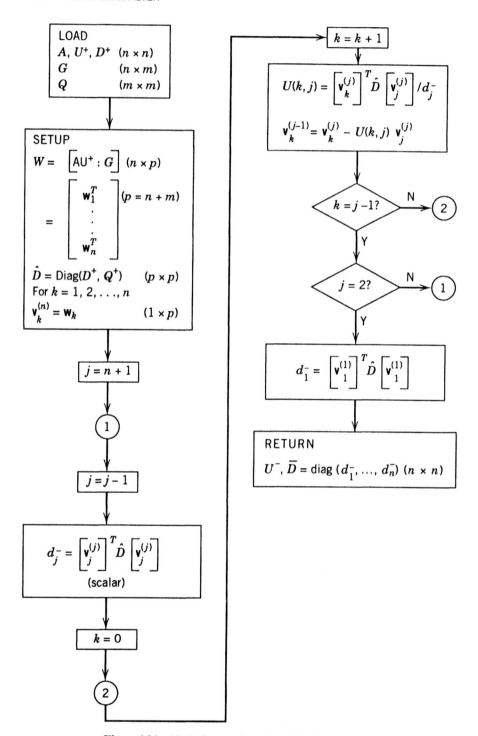

Figure 4.24 *U–D* time update algorithm flowchart.

```
 1  C--    SUBROUTINE TIME (U, A, Q, G, N, M, X)
 2  C--    PARAMETER JNPM = 50
 3  C--    DIMENSION U(N, N), A(N, N), Q(M), G(N, M), X(N)
 4  C--    DIMENSION B(JNPM), C(JNPM) D(JNPM)
 5  C
 6  C--    THIS SUBROUTINE PERFORMS THE U–D TIME UPDATE
 7  C--    FOR A DISCRETE KALMAN FILTER. ALTHOUGH THIS
 8  C--    SUBROUTINE IS VERY SPACE AND TIME EFFICIENT, IT
 9  C--    DESTROYS THE INPUTED MATRICES U AND G AND THE
10  C--    VECTOR X. IT PRESERVES THE MATRIX A, THE VECTOR Q,
    C--    AND THE INTEGERS N AND M.
11  C
12  C--    THE COVARIANCE MATRIX TIME UPDATE EQUATION IS
13  C--       $P^- = A*(P^+)*A^T + GQQG^T$.
14  C--    THIS SUBROUTINE INVOLVES THE U–D FACTORS OF $P^-$
15  C--    AND $P^+$, NAMELY, THE UNIT UPPER TRIANGULAR MATRICES
16  C--    $U^-$, $U^+$ AND THE DIAGONAL MATRICES $D^-$, $D^+$ SUCH
17  C--    THAT $P^+ = (U^+)*(D^+)*(U^+)^T$ AND $P^- = *(U^-)*(D^-)*(U^-)^T$.
18  C--    SINCE THE DIAGONALS OF $U^+$, $U^+$, $U^-$
19  C--    ARE ALWAYS ONE, THE MATRICES $D^+$ AND $D^-$
20  C--    ARE STORED ALONG THE DIAGONALS OF $U^+$, $U^-$. THE
21  C--    MATRIX U IN THE CALL LIST IS USED TO INPUT THE OLD
    C--    FACTORS $U^+$, $D^+$ AND OUTPUT THE NEW FACTORS $U^-$ $D^-$.
22  C
23  C--    THE NOISE COVARIANCE MATRIX QQ IS ASSUMED TO BE
    C--    DIAGONAL
24  C--    ITS DIAGONALS ARE CONTAINED IN THE INPUT VECTOR Q.
25  C
26  C--    THE NEW AND OLD ESTIMATES OF THE STATE VECTOR $X^-$
27  C--    AND $X^+$, ARE RELATED BY
28  C--       $X^- = A*(X^+)$.
29  C--    THE VECTOR X IN THE CALL LIST IS USED TO INPUT $X^+$ AND
30  C--    OUTPUT $X^-$.
31  C
32  C--    THE PARAMETER JNPM MUST BE EQUAL TO OR GREATER
       THAN N + M.
33  C
34  C
35  C--    INPUTS (PRE-UPDATE)
36  C
37  C--    N        = DIMENSION OF THE STATE VECTOR
38  C--    M        = DIMENSION OF THE NOISE VECTOR
39  C--    U(N, N) = CONTAINS U–D FACTORS OF COVARIANCE
    C--                 MATRIX P+.
40  C--    A(N, N) = STATE TRANSITION MATRIX
41  C--    Q(N)    = CONTAINS DIAGONALS OF NOISE
    C--                 COVARIANCE MATRIX
42  C--    G(N, M) = NOISE DISTRIBUTION MATRIX
43  C--    X(N)     = STATE ESTIMATE $X^+$
44  C
45  C--    OUTPUTS (POST-UPDATE)
46  C
```

Figure 4.25 FORTRAN program for $U–D$ time update algorithm.

```
47  C--      U(N, N,) CONTAINS U-D FACTORS OF COVARIANCE
                 MATRIX P⁻
48  C--      X(N)    = STATE ESTIMATE X⁻
49  C
50  C
51           NPM = N + M
52  C
53  C--      THE MATRIX W = [AU;G] IS WRITTEN IN [U;G], I.E., EACH
54  C--      ROW OF W HAS ITS FIRST N ELEMENTS AND LAST M
55  C-       ELEMENTS STORED IN THE CORRESPONDING ROW OF U
                 AND G RESPECTIVELY.
56  C
57  C--      COMPUTE A*U AND PLACE IN U.
58  C--      RETRIEVE DIAGONALS OF U AND STORE IN D.
59  C        COMPUTE A*X AND PLACE IN X
60  C
61           DO 1 I = 1, N
62           B(I) = X(I)
63        1  X(I) = 0.
64           IF(N.EQ.1) GO TO 6
65           DO 5 J = N, 2, -1
66  C
67  C--      USE D(1),..., D(J-1) TO TEMPORARILY STORE
68  C--      U(1, J),..., (J-1, J).
69  C--      USE D(J) TO PERMANENTLY STORE U(J, J)
70  C
71           DO 2 K = 1, J
72        2  D(K) = U(K, J)
73           DO 4 I = 1, N
74           X(J) = X(J) + A(J, I)*B(I)             ● Eq. (4.96a)
75           U(I, J) = A(I, J)
76           DO 4 K = 1, J-1
77        4  U(I, J) = U(I, J) + A(I, K)*D(K)       ● Eq. (4.99)
78           U(J, 1) = A(J, 1)                      ● Eq. (4.99)
79        5  X(1) = X(1) + A(1, J)*B(J)             ● Eq. (4.96a)
80        6  CONTINUE
81           D(1) = U(1, 1)
82           U(1, 1) = A(1, 1)                      ● Eq. (4.99)
83           X(1) = X(1) + A(1, 1)*B(1)             ● Eq. (4.96a)
84  C
85  C--      PERFORM UPDATE FOR J = N, N-1, ..., 1.
86  C--      THE NEW VALUES OF U ARE TEMPORARILY STORED IN
87  C--      THE LOWER LEFT TRIANGULAR PART OF U.
88  C
89           DO 20 J = N, 1, -1
90           S = 0.
91  C
92  C--      COMPUTE D-/SUB J AND PLACE IN U(J, J).
93  C--      STORE COMPUTATIONS OF (D-/HAT)*(V/SUB J) IN
                 VECTOR C.
```

Figure 4.25 *(Continued)*

```
 94  C--    USE B(1),..., B(N), B(N + 1),...,B(N + M) TO TEMPORARILY
            STORE
 95  C--    U(J, 1),..., U(J, N), G(J, 1),..., G(J, M).
 96  C
 97  C      DO 10 I = 1, N
 98         B(I) = U(J, I)
 99         C(I) = D(I)*B(I)
100     10  S = S + B(I)*C(I)                              ● Eq. (4.110)
101         DO 11 I = 1, M
102         NPI = N + 1
103         B(NPI) = G(J, I)
104         C(NPI) = Q(I)*B(NPI)
105     11  S = S + B(NPI)*C(NPI)                          ● Eq. (4.110)
106         U(J, J) = S
107  C
108         IF(J.EQ.1) GO to 20
109         DINV = 1./S
110  C
111         DO 19 K = 1, J − 1
112  C
113  C--    COMPUTE ELEMENT U(K, J) OF NEW U-D FACTOR AND
114  C--    PLACE IN U(J, K).
115  C
116         S = 0.
117         D0 12 I = 1, N
118     12  S = S + U(K, I)*C(I)                           ● Eq. (4.111)
119         DO 13 I = 1, M
120     13  S = S + G(K, I)*C(N + I)                        ● Eq. (4.111)
121         S = S*DINV                                     ● Eq. (4.111)
122         U(J, K) = S
123  C
124  C--    UPDATE VECTOR V/SUB K, WHICH IS STORED IN KTH
125  C--    ROW OF U AND G.
126  C
127         DO 14 I = 1, N
128     14  U(K, I) = U(K, I) − S*B(I)                     ● Eq. (4.112)
129         DO 15 I = 1, M
130     15  G(K, I) = G(K, I) − S*B(N + I)                 ● Eq. (4.112)
131     19  CONTINUE
132     20  CONTINUE
133  C
134         IF(N.EQ.1) GO TO 25
135         DO 24 J = 2, N
136         DO 24 I = 1, J − 1
137         U(I, J) = U(J, I)
138         U(J, I) = 0.
139     24  CONTINUE
140     25  RETURN
141         END
```

Figure 4.25   (*Continued*)

Table 4.3 gives a comparison of the number of multiplications required by the U–D and the conventional algorithm to compute the Kalman gain **k** for measurement updates, state estimates, and updated covariance matrices. The integers n and m are the dimensions of the state vector **x** and the noise vector **w**, respectively. The conventional algorithm is the direct implementation of Eqs. (4.71)–(4.73). Finally, we note that if the U–D algorithms are used and the covariance matrix is desired, the extra computation $P = UDU^T$ must be performed. This requires $\frac{1}{6}n^3 + \frac{1}{2}n^2 - \frac{2}{3}n$ multiplications. If only the diagonals of P (that is, the variances of the state estimates) are desired, then $n^2 - n$ multiplications are required.

4.8 THE EXTENDED KALMAN FILTER

We have seen in Section 4.3 that the conventional discrete-time Kalman filter is a recursive data-processing algorithm, which is usually implemented in software on a digital computer. At update time, it combines all available measurements, plus prior knowledge about the system and the measuring devices, to produce an estimate of the state $\mathbf{x}(t)$ in such a manner that the mean square is minimized statistically. During propagation, it advances the estimate in such a way as to again maintain optimality. Thus, the conventional Kalman filter performs the above tasks for linear systems and linear measurements in which the driving and measurement noises are assumed to be mutually uncorrelated, white, zero-mean, and Gaussian.

In nature, however, most physical problems or processes are nonlinear. Consequently, the nonlinear systems must be linearized (that is, approximated) before the linear filter theory can be applied. Specifically, the problem of combined state and parameter estimation was originally posed as a nonlinear state estimation problem using the extended Kalman filter (EKF). Since this requires a linear approximation of a nonlinear system about the current estimate, divergence may result if the initial estimate is poor. Moreover, not much is known about the convergence properties of the extended Kalman filter, and the conditions for acceptability of the solution are so to speak vague.

The problem of identification of nonlinear systems is divided into identification of (1) deterministic systems (that is, noise-free system), and (2) stochastic systems (implying the existence of plant or system and observation noise). Systems of the later kind can be solved by means of the extended Kalman filter

Table 4.3 Multiplication Counts for Conventional and U–D Algorithms

Algorithm	Measurement Update	Time Update
Conventional $P^+ = P^- - \mathbf{k}h^T P^-$	$1.5n^2 + 4.5n$	$1.5n^3 + 0.5(m+3)n^2 - 1.5\,mn$
U–D	$1.5n^2 + 5.5n$	$1.5n^3 + (m+2)n^2 + (m-0.5)(n-1)$

in which the system is linearized along a reference trajectory, then transformed into a discrete-time equivalent form.

As stated above, the EKF method assumes the validity of linearization. The extended Kalman filter is a variation of the conventional filter which relaxes the requirements that the system and measurements be linear. It is the filter generally used in practice for nonlinear applications.

In this section, we will briefly discuss continuous-time nonlinear systems with discrete, nonlinear observations, also known as the continuous–discrete extended Kalman filter (CDEKF). Consider now the continuous-time, non-linear physical system (or dynamics model) described by the n-vector differential equation [17, 23, 37, 49, 58]

$$\dot{\mathbf{x}}(t) = \mathbf{f}[\mathbf{x}(t), t] + G(t)\mathbf{w}(t) \tag{4.116}$$

and the nonlinear m-dimensional vector observation or measurement* equation

$$\mathbf{z}(t_i) = \mathbf{h}[\mathbf{x}(t_i), t_i] + \mathbf{v}(t_i) \tag{4.117}$$

with the initial condition $\mathbf{x}(0) \sim N(\bar{\mathbf{x}}(0), P(0))$. Note that the mean $\bar{\mathbf{x}}(0)$ and the covariance matrix $P(0)$ are known. As before [see Eq. (4.5)], $\mathbf{w}(t)$ is a zero-mean, white Gaussian random process with

$$\mathscr{E}\{\mathbf{w}(t)\mathbf{w}^T(t+\tau)\} = Q(t)\delta(\tau),$$

and $\mathbf{v}(t)$ is a zero-mean, white Gaussian random sequence independent [i.e., $\mathscr{E}\{\mathbf{v}(t)\mathbf{w}^T(\tau)\} = 0$] of $\mathbf{w}(t)$:

$$\mathscr{E}\{\mathbf{v}(t)\mathbf{v}^T(t+\tau)\} = R(t)\delta(\tau).$$

In the above equations, $Q(t)$ is the spectral density covariance matrix of $\mathbf{w}(t)$, and $R(t)$ is the covariance matrix of the noise $\mathbf{v}(t)$. Furthermore, the symmetric matrices $P(0)$ and $Q(t)$ must be positive semidefinite, while the symmetric matrix $R(t)$ must be positive definite. Also, the usual conditions of controllability and observability will be assumed to be satisfied here.

In many extended Kalman–Bucy filter applications, the elements of Q are adjustable "tuning" parameters. That is, the elements are designated by the filter developer to aid in the covergence of the filter. While there is significant

---

* Let $\mathbf{z}(t_i) \equiv \mathbf{z}_i$ be a vector of discrete measurements at times $t_1, t_2, .., t_n$. Then the set of measurements $\mathbf{Z}_n = \{\mathbf{z}_1, \mathbf{z}_2, ..., \mathbf{z}_n\}$ act indirectly as a sequence of initial conditions for which Eq. (4.116) must be solved.

theory for the creation of an appropriate matrix Q, it is formed in practice via engineering intuition based on experience.

Now, the Kalman optimal filter can be written as

$$\frac{d\hat{\mathbf{x}}(t)}{dt} = F(t)\hat{\mathbf{x}}(t) + \mathscr{K}(t)[\mathbf{z}(t) - H(t)\hat{\mathbf{x}}(t)]$$

$$= \mathbf{f}[\hat{\mathbf{x}}(t), t] + P(t)H^T(t)R^{-1}(t)\{\mathbf{z}(t) - \mathbf{h}[\hat{\mathbf{x}}(t), t]\}. \qquad (4.118)$$

Associated with the state vector $\mathbf{x}$ is the error covariance matrix P. This matrix is the expected value of the outer product of the state vector minus its mean: $P(t) = \mathscr{E}\{[\mathbf{x}(t) - \bar{\mathbf{x}}(t)][\mathbf{x}(t) - \bar{\mathbf{x}}(t)]^T\}$ and $\bar{\mathbf{x}}(t) = \mathscr{E}\{\mathbf{x}(t)\}$. It evolves according to the linear variance equation,

$$\frac{dP(t)}{dt} = F(t)P(t) + P(t)F^T(t)$$
$$- P(t)H^T(t)R^{-1}(t)H(t)P(t) + G(t)Q(t)G^T(t). \qquad (4.11)$$

The nonlinear extended Kalman filter requires that the matrix* F, which consists of the partial derivatives of Eq. (4.116) with respect to each state variable, be evaluated at the current state. That is,

$$F(t, \hat{\mathbf{x}}(t)) = \left.\frac{\partial \mathbf{f}[\mathbf{x}, t]}{\partial \mathbf{x}}\right|_{\mathbf{x} = \hat{\mathbf{x}}(t)} \qquad (4.119)$$

In essence, if one assumes that the conditional-mean estimate of $\mathbf{x}(t)$ is known, then $\mathbf{f}[\mathbf{x}(t), t]$ can be expanded in a Taylor series about $\mathbf{x}(t) = \hat{\mathbf{x}}(t)$, thereby obtaining a set of linear equations in $\mathbf{x}(t)$. Thus, for the homogeneous non-linear differential equation

$$\dot{\mathbf{x}}(t) = \mathbf{f}[\mathbf{x}(t), t], \qquad \mathbf{x}(t_0) = \mathbf{x}_0$$

we have

$$\dot{\mathbf{x}}(t) = \mathbf{f}[\mathbf{x}(t), t]$$
$$= \mathbf{f}[\hat{\mathbf{x}}(t), t] + \left.\frac{\partial \mathbf{f}[\mathbf{x}(t), t]}{\partial \mathbf{x}(t)}\right|_{\mathbf{x}(t) = \hat{\mathbf{x}}(t)} [\mathbf{x}(t) - \hat{\mathbf{x}}(t)] + \cdots. \qquad (4.120)$$

---

* Since $F(\mathbf{x}(t), t)$ is evaluated at the current estimate $\hat{\mathbf{x}}(t)$, Eqs. (4.116) and (4.11) are coupled.

The same procedure applies for the observation model. The extended Kalman filter recalculates the elements of the matrices F and H during each update and propagation cycle by evaluating the partial derivatives of the vector functions $\mathbf{f}$ and $\mathbf{h}$:

$$F(t, \hat{\mathbf{x}}(t)) = \frac{\partial \mathbf{f}[\mathbf{x}, t]}{\partial \mathbf{x}}\bigg|_{\mathbf{x} = \hat{\mathbf{x}}(t_i^-)} = \begin{bmatrix} \dfrac{\partial f_1(\mathbf{x})}{\partial x_1} & \cdots & \dfrac{\partial f_1(\mathbf{x})}{\partial x_n} \\ \vdots & & \vdots \\ \dfrac{\partial f_n(\mathbf{x})}{\partial x_1} & \cdots & \dfrac{\partial f_n(\mathbf{x})}{\partial x_n} \end{bmatrix}_{\mathbf{x} = \hat{\mathbf{x}}(t_i^-)} \tag{4.121}$$

$$G(t) = G[\hat{\mathbf{x}}(t), t], \tag{4.122}$$

$$H(t, \hat{\mathbf{x}}(t)) = \frac{\partial \mathbf{h}[\mathbf{x}, t_i]}{\partial \mathbf{x}}\bigg|_{\mathbf{x} = \hat{\mathbf{x}}(t_i^-)} = \begin{bmatrix} \dfrac{\partial h_1(\mathbf{x})}{\partial x_1} & \cdots & \dfrac{\partial h_1(\mathbf{x})}{\partial x_m} \\ \vdots & & \vdots \\ \dfrac{\partial h_m(\mathbf{x})}{\partial x_1} & \cdots & \dfrac{\partial h_m(\mathbf{x})}{\partial x_m} \end{bmatrix}_{\mathbf{x}(t) = \hat{\mathbf{x}}(t_i^-)} \tag{4.123}$$

Furthermore, the $n \times n$ matrix $F(\hat{\mathbf{x}}(t), t)$ and the $m \times n$ matrix $H(\hat{\mathbf{x}}(t), t)$ may be viewed as sensitivity matrices that relate small perturbations in $\mathbf{x}(t)$ to changes in $\dot{\mathbf{x}}(t)$ and $\mathbf{z}(t)$. The matrices $F(\hat{\mathbf{x}}(t), t)$ and $H(\hat{\mathbf{x}}(t), t)$ (i.e., the first derivatives of the nonlinearities) are known as the *Jacobians*. The discrete filter time propagation equations for this system form t_{i-1} to t_i are

$$\hat{\mathbf{x}}(t_i^-) = \hat{\mathbf{x}}(t_{i-1}^+) + \int_{t_{i-1}}^{t_i} \mathbf{f}[\hat{\mathbf{x}}(\tau), \tau]\, d\tau, \tag{4.124}$$

$$P(t_i^+) = \Phi[t_i, t_{i-1}; \hat{\mathbf{x}}(\tau)]\, P(t_{i-1})\Phi[t_i, t_{i-1}; \hat{\mathbf{x}}(\tau)]$$
$$+ \int_{t_{i-1}}^{t_i} \Phi[t_i, t_{i-1}; \hat{\mathbf{x}}(\tau)]\, G(\tau)Q(\tau)G^T(\tau)\Phi^T[t_i, t_{i-1}; \hat{\mathbf{x}}(\tau)]\, d\tau. \tag{4.125}$$

The function $\Phi[t_i, t_{i-1}; \hat{\mathbf{x}}(\tau)]$ represents the solution to the differential equation

$$\dot{\Phi}[t_i, t_{i-1}; \hat{\mathbf{x}}(\tau)] = F[t; \hat{\mathbf{x}}(\tau)]\, \Phi[t_i, t_{i-1}; \hat{\mathbf{x}}(\tau)] \tag{4.126}$$

with boundary condition

$$\Phi[t_i, t_{i-1}; \hat{\mathbf{x}}(\tau)] = I. \tag{4.127}$$

The measurement model and the measurement update equations for the non-linear filter at time t_i are

$$\mathbf{z}(t_i) = \mathbf{h}[\mathbf{x}(t_i), t_i] + \mathbf{v}(t_i) \tag{4.128}$$

$$K(t_i) = P(t_i^-)H^T[t_i; \hat{\mathbf{x}}(t_i^-)]\{H[t_i; \hat{\mathbf{x}}(t_i^-)] P(t_i^-)H^T$$

$$[t_i; \hat{\mathbf{x}}(t_i^-)] + R(t_i)\}^{-1}, \tag{4.129}$$

$$\hat{\mathbf{x}}(t_i^+) = \mathbf{x}(t_i^-) + K(t_i)\{\mathbf{z}(t_i) - [\hat{\mathbf{x}}(t_i^-), t_i]\} \tag{4.130}$$

$$P(t_i^+) = P(t_i^-) - K(t_i)H[t_i; \hat{\mathbf{x}}(t_i^-)]P(t_i^-) \tag{4.131}$$

for the iterative procedure with $i = 1, 2, 3, \ldots$

The recursive CDEKF algorithm is initialized by setting the state estimate $\hat{\mathbf{x}}(t) = \hat{\mathbf{x}}(0)$ and the estimation error covariance matrix $P(t) = P(0)$. For a nonlinear state model described by the state differential equation

$$\dot{\mathbf{x}}(t) = \mathbf{f}[\mathbf{x}(t), \mathbf{u}(t), t], \qquad \mathbf{x}(t_0) = \mathbf{x}_0,$$

where $\mathbf{u}(t)$ may be identified as a control function, expansion into a Taylor series yields

$$\dot{\mathbf{x}}(t) = \mathbf{f}[\mathbf{x}(t), \mathbf{u}(t), t]$$

$$= \mathbf{f}[\mathbf{x}^*(t), \mathbf{u}^*(t), t] + F(t)[\mathbf{x}(t) - \mathbf{x}^*(t)] + G(t)[\mathbf{u}(t) - \mathbf{u}^*(t)] + \cdots,$$

where

$$F(t) = \left.\frac{\partial \mathbf{f}}{\partial \mathbf{x}}\right|_* = \begin{bmatrix} \dfrac{\partial f_1}{\partial x_1} & \cdots & \dfrac{\partial f_1}{\partial x_n} \\ \vdots & & \vdots \\ \dfrac{\partial f_n}{\partial x_1} & \cdots & \dfrac{\partial f_n}{\partial x_n} \end{bmatrix}_*,$$

$$G(t) = \left.\frac{\partial \mathbf{f}}{\partial \mathbf{u}}\right|_* = \begin{bmatrix} \dfrac{\partial f_1}{\partial u_1} & \cdots & \dfrac{\partial f_1}{\partial u_r} \\ \vdots & & \vdots \\ \dfrac{\partial f_n}{\partial u_1} & \cdots & \dfrac{\partial f_n}{\partial u_r} \end{bmatrix}_*,$$

in which $|_*$ denotes that we evaluate thus the matrix obtained $\mathbf{x} = \mathbf{x}^*$ and $\mathbf{u} = \mathbf{u}^*$.

The extended Kalman filter has found extensive use in orbit determination problems, aircraft tracking, instrumentation accuracy, etc. For example, if we wish to estimate the bias error of an instrument, then the discrete-time non-linear system state equation takes the form

$$\mathbf{x}(t_{i+1}) = f[\mathbf{x}(t_i), t_i] + \mathbf{w}(t_i) \qquad (4.132a)$$

with the corresponding observation model

$$\mathbf{z}(t_i) = h[\mathbf{x}(t_i), t_i] + \mathbf{v}(t_i). \qquad (4.132b)$$

Then, the bias to be estimated, $\mathbf{b}$, becomes a state of the augmented system as follows:

$$\mathbf{b}(t_{i+1}) = \mathbf{b}(t_i) \qquad (4.133)$$

$$\mathbf{z}(t_i) = \mathbf{h}[\mathbf{x}(t_i), t_i] + \mathbf{v}(t_i) + \mathbf{b}(t_i), \qquad (4.134)$$

assuming the condition of observability is satisfied.

In simulations of physical systems, a variable-step-size Runge–Kutta–Fehlberg routine for numerical integration of the system nonlinear equations [Eqs. (4.116) and (4.117)] is commonly used. A variable step size is computationally more efficient; however, its use will depend to a great extent on the system dynamics.

When applying the extended Kalman filter to real-time applications, the requirement to recalculate the matrices F and H continually can impose a severe computational burden. This nonlinear formulation also makes it impossible to precompute the Kalman gains and covariance matrices (K and P), a technique often used to reduce filter processing time for linear Kalman filters.

Two variations of the conventional EKF used occasionally are the *iterated EKF* and *second-order* filters. Both of these are higher-order filters. The approach in the former is to iterate within the linearization step a few times, thus improving the quality of the estimates; however, this results in an increase in the number of operations. The second-order filter, as the name implies, is obtained by including second-order terms in the expansion for $\mathbf{f}[\mathbf{x}(t), t]$ and $\mathbf{h}[\mathbf{x}(t), t]$.

Finally, it should be noted that difficulties may arise in applying the extended Kalman filter, in that the amount of computation may be excessive due to the linearization required after each iteration of estimation.

Example Consider the case of an air-to-air missile attempting to intercept a passive target (e.g., an aircraft). This can be seen as a classical proportional navigation (PN) problem in which the guidance law is given by $A_L = NV_c \lambda$, where A_L is the interceptor missile's lateral acceleration, N is the PN constant (usually between 3 and 5), V_c is the closing velocity, and λ is the line-of-sight (LOS) angle. Before we proceed with the analysis of this problem, we note that the Kalman filter is a computational algorithm that processes measurements to deduce a minimum-variance, unbiased error estimate of the state of a system by using knowledge of the system measurement dynamics, assumed statistics of system noises and measurement dynamics, assumed statistics of system noises and measurement errors, and initial condition information. For nonlinear modeling problems, the extended Kalman filter (EKF) is used to extend the linearized Kalman filter design by linearizing about each estimate once it has been computed. That is, the EKF uses the linearization of the nonlinear measurement function about the *a priori* state estimate and then uses the standard linear Kalman filter equations. Therefore, as soon as a new state estimate is made, a new and better reference-state trajectory is incorporated into the estimation process. In this manner, one enhaces the validity of the assumption that deviations from the reference (or nominal) trajectory are small enough to allow linear perturbation techniques to be employed with adequate results.

Consequently, this is a problem of estimating the state of a system consisting of a measurement platform (the missile) and the target (relative position, relative velocity, and acceleration) from noisy measurements of the LOS angle. Consider now the intercept geometry as illustrated in Figure 4.26. Furthermore, let the measured angle λ be expressed in terms of the Cartesian x, y coordinate system.

From the diagram, the LOS (or bearing) angle is given by

$$\tan \lambda = y/x$$

or

$$\lambda = \tan^{-1}(y/x). \tag{1}$$

In the diagram, $\mathbf{r}_T$ represents the target position relative to the interceptor missile and is given (in three-dimensional form) by

$$\mathbf{r}_T = [r_x \quad r_y \quad r_z]^T.$$

Equation (1) indicated a nonlinear relationship between the LOS angle and the coordinates x, y. The system state vector is given by

$$\mathbf{x} = [x \quad y \quad \dot{x} \quad \dot{y} \quad \ddot{x}_T \quad \ddot{y}_T]^T, \tag{2}$$

where $\ddot{x}_T$ and $\ddot{y}_T$ are the target acceleration components. From the observation equation

$$\mathbf{z}(t) = \mathbf{h}[\mathbf{x}(t), t] + \mathbf{v}(t), \tag{3}$$

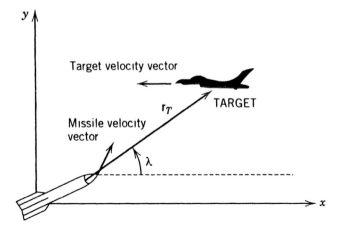

Figure 4.26 Intercept geometry.

where $\mathbf{v}(t)$ is zero-mean, Gaussian noise with covariance R, the measurement at time $t = k$ is given by the expression

$$\mathbf{z}(k) = \lambda(k) + \mathbf{v}(k0 = \tan^{-1}(y_k/x_k) + \mathbf{v}(k). \tag{4}$$

Linearization of $\mathbf{h}[\mathbf{x}(k), k]$ about the current estimate of the state is given by

$$\mathbf{H}(k) = \frac{\partial \mathbf{h}(\mathbf{x})}{\partial \mathbf{x}}\bigg|_{\mathbf{x} = \hat{\mathbf{x}}_k} = \left[\frac{-\hat{y}_k}{\hat{r}_k^2} \quad \frac{-\hat{x}_k}{\hat{r}_k^2} \quad 0 \quad 0 \quad 0 \quad 0 \right], \tag{5}$$

where

$$\hat{r}_k^2 = \hat{x}_k^2 + \hat{y}_k^2.$$

As a result of the linearization, the state of the system $\hat{\mathbf{x}}(k|k)$ and the covariance matrix $P(k|k)$ are given by the usual linear Kalman filter equations

$$\mathbf{x}(k + 1|k) = \Phi(k + 1, k)\, \mathbf{x}(k) + \mathbf{w}(k),$$

$$P(k|k - 1) = \Phi(k, k - 1)P(k - 1)\, \Phi^T(k, k - 1) + Q(k),$$

$$\hat{\mathbf{x}}(k|k) = \hat{\mathbf{x}}(k|k - 1) + K(k)[\mathbf{z}(k) - H(k)\hat{\mathbf{x}}(k|k - 1)],$$

$$P(k|k) = [I - K(k)H(k)]\, P(k|k - 1),$$

where

$$K(k) = P(k|k - 1)\, H^T(k)[H(k)\, P(k|k - 1)\, H^T(k) + R(k)]^{-1},$$

$\mathbf{w}(k)$ is the system noise vector, and $Q(k)$ is the system covariance matrix; the initial state estimate vector $\mathbf{x}(0)$ and the initial system covariance matrix $P(0)$ must be provided.

A flowchart of the extended Kalman filter is given in Figure 4.27, where we have:

Dynamics (or Message) Model:

$$\mathbf{x}(k+1) = \Phi[\mathbf{x}(k), k] + \Gamma[\mathbf{x}(k), k]\,\mathbf{w}(k), \qquad k = 0, 1, \ldots$$

Observation Model:

$$\mathbf{z}(k) = \mathbf{h}[\mathbf{x}(k), k] + \mathbf{v}(k), \qquad k = 1, 2, \ldots$$

Statistics:

$$\mathscr{E}\{\mathbf{w}(k)\,\mathbf{w}(j)^T\} = Q(k)\,\delta_{kj},$$
$$\mathscr{E}\{\mathbf{v}(k)\,\mathbf{v}(j)^T\} = R(k)\,\delta_{kj},$$

[It is assumed that $\{\mathbf{w}(k)\}$ and $\{\mathbf{v}(k)\}$ are independent.]

We have noted earlier in this chapter that when the truth model of a system is composed of linear dynamics and measurement equations, a linear Kalman filter based upon this system is an optimal estimator by almost any of the known criteria. Moreover, when a full-order filter of the truth model is implemented, the dynamics of the measurement noises used to construct the filter will normally match the truth model noises. However, when filter order reduction is performed on this linear filter, the noise strengths may have to be adjusted in order to compensate for the missing and/or combined states. Similarly, when a linearized extended full-order Kalman filter is implemented to simulate a nonlinear system, the noise strengths may need to be adjusted in both the full-order and reduced-over filters to compensate for the approximations used in implementing linearized equations to model the nonlinear system. As mentioned in Section 4.6, this filter noise strength adjustment, or tuning, as it is commonly called, will now be examined more closely.

The closer the full-order filter approximates the truth model, the less noise strength $Q(t)$ and measurement covariance matrix $R(t)$ are adjusted (i.e., normally increased) in order to compensate for nonlinearities in the true dynamics. In implementing full-order filters, there are normally three reasons why the dynamics noise strength $Q(t)$ on some of the states must be increased. These reasons are as follows:

1. States that have noise associated with them in the truth model need increased noise in the filter model to track the dynamics of the true state

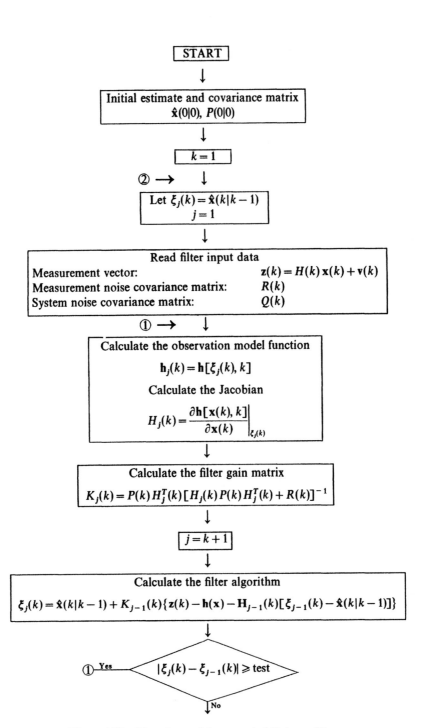

Figure 4.27   Flowchart of the extended Kalman filter.

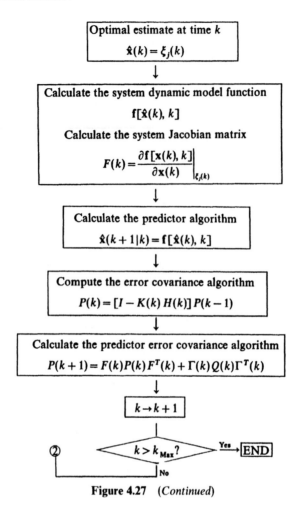

Figure 4.27   (*Continued*)

adequately. Adding noise increases the uncertainty of the filter-assumed state's dynamics to compensate for nonlinear behavior or to keep the Kalman filter gains in that channel from going to zero. Usually, the states in the filter needing this type of noise adjustment are those that are not part of the measurement equation. For example, in an inertial navigation system, some of the states that are not part of the measurement equation, need a small amount of added noise to compensate for nonlinear dynamics. If the states are not directly part of the measurement equation and the dynamics equations do not totally describe the true states behavior, the extended Kalman filter has difficulty estimating these states.

2. The dynamics noise needs to be added into a filter model when the filter variance of a state goes negative. This numerical difficulty is a normal

occurrence when the range of numbers the filter estimates is large. The problem of negative variances arises due to the limited numerical precision of computers which must multiply high order matrices together. To alleviate this problem, a small amount of added noise in the filter will keep the variance of the state positive and will not degrade the filter's state estimate.

3. Dynamics noise is added because of filter order reduction. States which are eliminated do not appear directly in the measurement equations, but impact the states which are part of the measurement equations. Therefore, in order to compensate for the eliminated states, the noise they affect is increased. The noise increase is small since the eliminated states all have small magnitudes, but is necessary to ensure a well-tuned filter.

From the above discussion, it is clear that the measurement covariance matrix $R(t_i)$ must be adjusted upward in the full-order filters to increase the uncertainty in the measurement equation due to using linearized equations to model nonlinear systems. It should be pointed out that the EKF has access only to measurements that are a combination of states and not to the states themselves. Also, increasing the measurement noise covariance matrix is due to filter order reduction. That is, when states are eliminated in the filter model that are part of the measurement equation it is necessary to increase $R(t_i)$. Finally, it should be noted that these procedures are not universal to all Kalman filter tuning. They are basic reasons why Kalman filters need to be tuned and the engineer involved in Kalman filter tuning needs to be familiar with them.

4.9 SHAPING FILTERS AND COLORED NOISE

Up to now we have discussed systems whose noise processes are white, uncorrelated noise sequences. However, there are situations in the real world where white Gaussian noise may not be adequate to model the system. In this section we will discuss the basic problem of a system (or plant) driven by *colored*, or *time-correlated*, noise.

The concept of a *shaping filter* has been used for many years in the analysis of physical systems, because it allows a system with a random input to be replaced by an *augmented system* (the original system plus the shaping filter) excited only by white noise. The shaping filter greatly simplifies the mathematics and has found wide application in the construction of optimum estimators via both the Wiener and Kalman techniques.

The shaping-filter approach is generally applied in problems where the mean-square values of outputs (e.g., mean-square errors in estimation or control) are of prime importance. In such cases, only second-order statistics are important, so that complex input processes can sometimes be represented by simple shaping filters. In Section 2.7 the Wiener filter was discussed, noting that if one is looking for the best *linear* estimation, the filter will have an impulse response $h(t)$ such that the convolution integral holds [see Eq. (2.45)]:

$$\hat{x}(t) = \int_{-\infty}^{\infty} h(t)\, x(t-\tau)\, d\tau.$$

Consider now the input $x(t)$ and output $y(t)$ of a linear system characterized by the impulse response function $g(t, \tau)$, which are related by the equation

$$y(t) = \int_{-\infty}^{t} g(t, \tau)x(\tau)\, d\tau \qquad (4.135)$$

and from which the corresponding autocorrelation function can be derived. This autocorrelation function is given by

$$\varphi_{yy}(t_1, t_2) = \int_{-\infty}^{t_1} g(t_1, \tau_1) \int_{-\infty}^{t_2} g(t_2, \tau_2)\, \varphi_{xx}(\tau_1, \tau_2)\, d\tau_2\, d\tau_1. \qquad (4.136)$$

From the above discussion, it is evident that the second-order statistics of $y(t)$ (i.e., its mean-square value) will not be changed when the input $x(t)$ is replaced by another having the same autocorrelation function.

In Figure 4.28 $x(t)$ is the output of a shaping filter whose impulse response function is $h(t)$, excited by white noise $w(t)$ (generally nonstationary) with correlation function

$$\varphi_{ww}(t_1, t_2) = q(t_1)\, \delta(t_2 - t_1). \qquad (4.137)$$

[Note that if A is the noise amplitude, then $q(t)$ can be computed from the relation $q(t) = A^2/(t_2 - t_1)$ for $t_1 < t < t_2$ and 0 otherwise.]

A noise sequence $w(t)$ whose power spectral density $\Phi_w(\omega)$ is not constant is said to be *colored*. In the continuous case, the power spectral density can be expressed as

$$\Phi_w(s) = H(s) H^T(-s), \qquad (4.138)$$

where $s = j\omega$ and $H(s)$ is the transfer function. According to the *spectral factorization theorem*, if $\Phi_w(z)$ is rational and if $|\Phi_w(z)| \neq 0$ for almost every z,

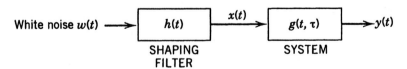

Figure 4.28 Augmented system with white-noise input.

then there is a square, rational asymptotically stable spectral factor $H(z)$ with zeros inside or on the unit circle such that [1]

$$\Phi_w(z) = H(z) H^T(z^{-1}),\qquad(4.139)$$

evaluated at $z = e^{j\omega}$. Now, if a linear system $H(z)$ is driven by white noise $w'(t)$ with unit power spectral density, then the output of the system has power spectral density $\Phi_w(z)$. Therefore, the system that produces colored noise $w(t)$ with a given spectral density from white noise can be represented by a shaping filter as shown in Figure 4.29.

Based on the above development, assume now that the system and observation models can be expressed by the usual forms [48]:

$$\dot{\mathbf{x}}(t) = F(t)\,\mathbf{x}(t) + G(t)\mathbf{w}'(t),\qquad(4.140)$$

$$\mathbf{z}(t) = H(t)\,\mathbf{x}(t) + \mathbf{v}(t),\qquad(4.141)$$

where $\mathbf{w}'(t)$ is a nonwhite (i.e., time-correlated), Gaussian noise process, while $\mathbf{v}(t)$ is a zero-mean white-noise process with $\mathbf{v}(t) = (0,\ R)$ and uncorrelated with $\mathbf{x}(0) = (\mathbf{x}_0, P_0)$. Furthermore, assume that the noise $\mathbf{w}'(t)$ is generated by a linear shaping filter of the form

$$\dot{\mathbf{x}}_f(t) = F_f(t)\,\mathbf{x}_f(t) + G_f(t)\mathbf{w}'(t),\qquad(4.142)$$
$$\mathbf{w}'(t) = H_f(t)\,\mathbf{x}_f(t),\qquad(4.143)$$

where the subscript f denotes the shaping filter. An *augmented state vector* can now be written in the form

$$\mathbf{x}_a(t) = \left[\begin{array}{c} \mathbf{x}(t) \\ \hline \mathbf{x}_f(t) \end{array}\right],\qquad(4.144)$$

so that the augmented system state equation can now be expressed in terms of Eqs. (4.140) and (4.142) as follows:

$$\left[\begin{array}{c} \dot{\mathbf{x}}(t) \\ \hline \dot{\mathbf{x}}_f(t) \end{array}\right] = \left[\begin{array}{c|c} F(t) & G(t) H_f(t) \\ \hline 0 & F_f(t) \end{array}\right] \left[\begin{array}{c} \mathbf{x}(t) \\ \hline \mathbf{x}_f(t) \end{array}\right] + \left[\begin{array}{c} 0 \\ \hline G_f(t) \end{array}\right] \mathbf{w}(t).\qquad(4.145)$$

White noise $w'(t)$ ⟶ | $H(z)$ | ⟶ Colored noise $w(t)$ . . . to system

SHAPING FILTER

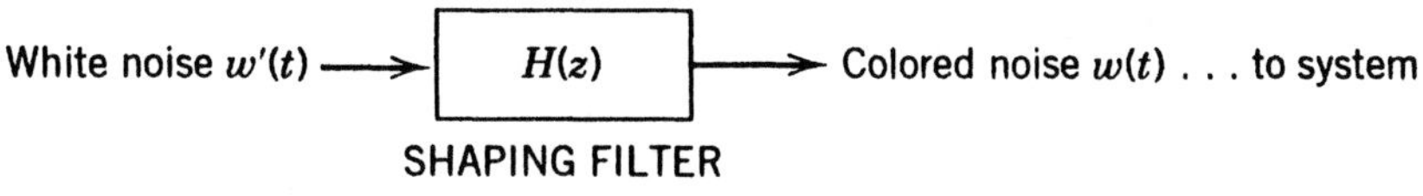

Figure 4.29   Representation of the shaping filter.

The system output (observation) equation takes the augmented form

$$\mathbf{z}(t) = [H(t) \quad 0]\begin{bmatrix} \mathbf{x}(t) \\ ----- \\ \mathbf{x}_f(t) \end{bmatrix} + \mathbf{v}(t). \tag{4.146}$$

The above equations represent the augmented linear system model driven only by white noise. Specifically, the shaping filter's output time-correlated noise $\mathbf{w}'(t)$ [Eq. (4.143)] drives the system, (4.140). Similarly, an analogous augmented system and observation model can be developed for the discrete-time case. For more details on colored noise and shaping filters the reader is referred to references [12, 48, and 58].

4.10 CONCLUDING REMARKS

This chapter has covered various aspects of estimation theory. In particular, the subject of estimation discussed in this chapter is essential for the analysis and control of stochastic systems. Estimation theory can be defined as a process of selecting the appropriate value of an uncertain quantity, based on the available information. Specifically, the need for sequential estimation arises when the measurement data are obtained *on line* as the process evolves and the measurements arrive one at a time.

Based on the Kalman–Bucy filter, state estimation presents only a limited aspect of the problem, since it yields optimal estimates only for linear systems and Gaussian random processes. Approximation techniques can usually be extended to such application contexts with observational and computational noise. The stochastic processes involved are often modeled as Gaussian ones to simplify the mathematical analysis of the corresponding estimation problems. If the process is not Gaussian, the estimator is still minimum in the mean-square sense, but not necessarily most probable. In the case of nonlinear systems, linearization techniques have been developed to handle the Kalman–Bucy filter. It should be pointed out, however, that this approach has limited value for small deviations from the point of linearization, and no guarantee of the so-called *approximate optimality is made*.

As mentioned in Chapter 1, the Kalman–Bucy filter, an algorithm which generates estimates of variables of the system being controlled by processing available sensor measurements, is one of the most widely used estimation algorithms. This is done in real time by using a model of the system and using the difference between the model predictions and the measurements. Then, in conjunction with a closed-loop algorithm, the control signals are fine tuned in order to bring the estimates into agreement with nominal performance, which is stored in the computer's memory. Recent research indicates that estimation theory is applicable not only to aerospace problems, but to the automotive industry, chemical industry, etc.

PROBLEMS

4.1 Write the equation $\ddot{x} + x = u$ in the equivalent state-space notation.

4.2 The spring–mass system shown below can be described by the second-order differential equation

$$m\ddot{x}_0 + cx_0 = x_i,$$

where m is the mass, c is the spring constant, x_0 is the output (i.e., displacement) and $x_i (x_i = F)$ is the input force. Set this equation in state-space notation (i.e., in the form $dx/dt = Fx + Gu$) and draw a flow diagram for programming and solving this equation in a digital computer. (Use $x_0 = x_1$ and $\dot{x}_0 = x_2$; see Example 3, Section 4.4.)

4.3 Given the RLC network shown in the diagram below, described by the equation

$$LC\ddot{x}_0 + CR\dot{x}_0 + x_0 = x_i.$$

With the state variables given as $x_0 = x_1$, $\dot{x}_0 = x_2 = \dot{x}_1$, set the above equation in state-space notation and draw the flow diagram for programming it in a digital computer. (*Note:* For programming, change the constants to $1/LC$ and R/L and use the same flow diagram as in Problem 4.2.)

4.4 Write the following third-order differential equation in state-space notation:

$$a_3\dot{x}_a + a_2\ddot{x}_a + a_1\dot{x}_a + a_0x_a = b_0x_e.$$

Also, draw the flow diagram for programming this equation in a digital computer. (*Note:* The flow diagram will be the same as that of Problem 4.3.)

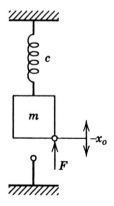

Figure Problem 4.2

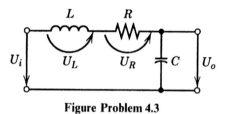

Figure Problem 4.3

4.5   Given the third-order differential equation

$$x_a + \dddot{x}_a - 4\dot{x}_a - 9x_a = 0.5x_e,$$

write this equation in state-space notation and draw the programming flow diagram.

4.6   Find:

(a) A vector matrix equation for the system

$$\ddot{y}_1 + 3\dot{y}_1 + 2y_2 = u_1,$$

$$\ddot{y}_2 + y_1 + \dot{y}_2 = u_2$$

by defining internal states x_i and placing the equations in the canonical form

$$\dot{\mathbf{x}} = F\mathbf{x} + G\mathbf{u},$$

$$\mathbf{y} = C\mathbf{x}.$$

(b) A vector matrix equation for the *RLC network*. Identify the states of the system, and place the state equations in canonical form, using u as the input and y as the output.

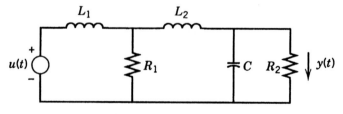

Figure Problem 4.6

4.7 Consider the system with scalar input $u(t)$ and the scalar output $y(t)$ related by differential equation

$$\frac{d^2 y(t)}{dt^2} + y(t) = \frac{du(t)}{dt}.$$

Determine b_1 and b_2 such that the state-variable representation for the system takes the form

$$\dot{x}_1(t) = -x_2(t) + b_1 u(t),$$
$$\dot{x}_2(t) = x_1(t) + b_2 u(t),$$
$$y(t) = x_1(t) + x_2(t).$$

4.8 Starting with the homogeneous (i.e., *unforced*) matrix differential equation

$$\frac{d\mathbf{x}}{dt} = F(t)\mathbf{x}(t),$$

show that the solution to this equation is given by

$$\mathbf{x}(t) = \Phi(t, t_0)\mathbf{x}(t_0).$$

[*Hint:* Expand $\mathbf{x}(t)$ into a Taylor series about some time t_0 in the form

$$\mathbf{x}(t) = \mathbf{x}(t_0) + \dot{\mathbf{x}}(t_0)(t - t_0) + \ddot{\mathbf{x}}(t_0)[1/2(t - t_0)^2/2!] + \cdots$$

and noting that the state transition matrix obeys Eq. (4.38).]

4.9 Let the scalars δv and δp represent velocity and position errors, respectively, for a one-dimensional inertial navigation system. The dynamics of δv and δp are described by the error equations

$$\delta\dot{v} = \delta a,$$
$$\delta\dot{p} = \delta v,$$

where δa is an accelerometer error whose model can be described by the equation

$$\delta a = \beta + \eta a(t) + \zeta a^2(t)$$

where $\beta = $ bias error,
$\quad\quad\eta = $ scale-factor error,
$\quad\quad\zeta = $ scale-factor nonlinearity term,
$\quad\quad a(t) = $ known time function representing the nominal acceleration profile as seen by the accelerometer.

The state vector **x** is defined as

$$\mathbf{x}^T = [\delta v \quad \delta p],$$

and the input vector **u** is defined as

$$\mathbf{u}^T = [\beta \quad \eta \quad \zeta].$$

(a) Determine the matrices F and G which permit the error propagation to be described by $\dot{\mathbf{x}} = F\mathbf{x} + G\mathbf{u}$.

(b) Show that the state transition matrix is given by

$$\Phi(t, t_0) = \begin{bmatrix} 1 & 0 \\ (t - t_0) & 1 \end{bmatrix}$$

using Eq. (4.3), that is, $\dot{\Phi}(t, t_0) = F\Phi(t, t_0)$.

(c) Outline the procedure to determine the sensitivity matrix $S(t, t_0) = \int_{t_0}^{t} \Phi(t, t_0) G(\tau) d\tau$. [Note that $S(t, t_0)$ obeys the differential equation $\dot{S}(t, t_0) = FS(t, t_0) + G(t)$ with $S(t, t_0) = 0$. Show a flowchart of this process.]

(d) Determine the errors $\delta v(t)$ and $\delta p(t)$ due to the initial errors $\delta v(t_0)$, $\delta p(t_0)$.

(e) Repeat part (a) for the case of a bias error which grows linearly with time, that is, $\beta = Bt$, where $B = $ constant.

4.10 For a simple, single-axis inertial navigation system for an aircraft (or ship), the system error equations can be written from the accompanying error diagram as follows:

$$\mathbf{x}^T(t) = [\delta v \quad \delta p \quad \delta \theta \quad e_a \quad e_g \quad e_p \quad e_v],$$

$$\delta \dot{v} = -g(\delta\theta + e_a),$$

$$\delta \dot{p} = \delta v,$$

$$\delta \dot{\theta} = (1/R)\delta v + e_g,$$

$$\dot{e}_a = -\beta_a e_a + u_a,$$

$$\dot{e}_g = -\beta_g e_g + u_g,$$

$$\dot{e}_p = -\beta_p e_p + u_p,$$

$$\dot{e}_v = -\beta_v e_v + u_v,$$

where $\delta v = $ velocity error,
    $\delta p = $ position error,

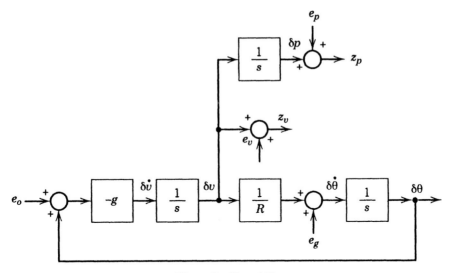

Figure Problem 4.10

$\delta\theta$ = platform tilt error,
g = acceleration of gravity,
R = radius of the earth,
e_a = accelerometer error,
e_g = gyroscope drift rate error,
e_p = position measurement error,
e_v = velocity measurement error due to Doppler radar (or ocean
     currents),
z_p = position measurement (e.g., using Loran),
z_v = velocity measurement (e.g., using Doppler radar or EM log),
u = white-noise terms,
$\beta = 1/\tau$ (τ = correlation time),
$1/s$ = integrator.

Write the above equations in the canonical form $\dot{\mathbf{x}}(t) = F\mathbf{x}(t) + G\mathbf{u}(t)$.

4.11   The gyroscope is the primary component of an inertial navigation system. However, gyroscopes exhibit a phenomenon called *drift rate*. In the present problem, we will assume that this drift rate can be modeled as a first-order Markov random process, characterized by an exponential autocorrelation function of the form

$$\psi_{gg}(\tau) = \sigma_g^2 e^{-\beta\tau}, \tag{1}$$

where β is the inverse correlation time of the drift rate and σ_g^2 is the associated variance. The state equation corresponding to Eq. (1) is

$$\frac{de_g}{dt} = -\beta e_g + w_g, \tag{2}$$

where w_g is a white driving noise with covariance Q_g:

$$\mathscr{E}\{w_g(t)w_g(t+\tau)\} = Q_g\delta(\tau), \tag{3}$$

where $Q_g = 2\beta_g\sigma_g^2$. Assuming that the system can be expressed in state-space notation as

$$\dot{x}_1 = x_2,$$

$$\dot{x}_2 = -\beta x_2 + w(t),$$

determine the error covariance matrix $P(t)$ by solving the equation

$$\dot{P} = FP + PF^T + GQG^T.$$

Hint: Place the above system in state-space notation as follows:

$$\begin{bmatrix} \dot{x}_1 \\ \dot{x}_2 \end{bmatrix} = \begin{bmatrix} 0 & 1 \\ 0 & -\beta \end{bmatrix} \begin{bmatrix} x_1 \\ x_2 \end{bmatrix} + \begin{bmatrix} 0 & 0 \\ 0 & 1 \end{bmatrix} \begin{bmatrix} w_1(t) \\ w_2(t) \end{bmatrix}.$$

4.12 The equations of a process are

$$\dot{x}_1(t) = 0,$$

$$\dot{x}_2(t) = x_1(t),$$

while the observation equation is given by

$$z(t) = x_2(t) + v(t),$$

where

$$x(0) = 0, \qquad Q(0) = 0, \qquad R(t) = r, \qquad H(t) = [0 \quad 1],$$

$$F(t) = \begin{bmatrix} 0 & 0 \\ 1 & 0 \end{bmatrix}, \qquad G(t) = \begin{bmatrix} 0 & 0 \\ 0 & 1 \end{bmatrix}, \qquad P(0) = \begin{bmatrix} p_{11}(0) & 0 \\ 0 & p_{22}(0) \end{bmatrix}.$$

This system could represent the straight-line motion of a vehicle, such as an automobile, moving with constant speed. Determine the error covariance matrix $P(t)$ for this optimal estimator using the equation

$$\dot{P}(t) = F(t)P(t) + P(t)F^T(t) - P(t)H^T(t)R^{-1}(t)H(t)P(t) + G(t)Q(t)G^T(t).$$

4.13 An unspecified system is being driven by a noise source ω_c which is definitely not white. In diagram form, this system is illustrated as follows:

$$\omega_c \rightarrow \boxed{\text{SYSTEM}} \rightarrow$$

It has been determined experimentally that this noise source can be modeled by the following expression:

$$\ddot{\omega}_c + 2\beta\dot{\omega}_c + \beta^2\omega_c = \omega, \tag{1}$$

where β is a known constant and ω is a white noise.

(a) Draw a block diagram for this expression where ω_c is the output and ω is the input:

$$\omega \rightarrow \boxed{?} \rightarrow \omega_c$$

(b) Write the state equation for the subsystem (i.e., shaping filter) in the form

$$\dot{\mathbf{x}} = F\mathbf{x} + G\mathbf{u}. \tag{2}$$

Find F and G.

(c) To drive this filter with the appropriate white noise, we need σ_ω^2. Do we have enough information to obtain σ_ω^2? What do we need?

(d) Assume that $\sigma_\omega^2 = q$. Find $\sigma_{\omega c}^2$. (Use $\dot{P} = FP + PF^T + GQG^T$ with $P = 0$ being the steady state.)

(e) Make an estimate of the time constant associated with the noise.

4.14 In the system $\dot{\mathbf{x}}(t) = F\mathbf{x}(t) + G\mathbf{u}(t)$, if the matrix F is given by

$$F = \begin{bmatrix} 0 & 1 & 0 \\ 0 & 0 & 1 \\ 0 & 0 & 0 \end{bmatrix},$$

determine the state transition matrix $\Phi(t, t_0)$ using only up to second-order terms.

4.15 Second-order vector equations of the form

$$\ddot{\mathbf{x}}(t) + Q\mathbf{x}(t) = \mathbf{0}, \qquad Q = Q^T$$

arise frequently in physical applications. The transformation

$$x_1 = x,$$
$$x_2 = \dot{x}$$

yields

$$\begin{bmatrix} \dot{x}_1(t) \\ \dot{x}_2(t) \end{bmatrix} = \begin{bmatrix} 0 & I \\ -Q & 0 \end{bmatrix} \begin{bmatrix} x_1(t) \\ x_2(t) \end{bmatrix}.$$

Show that $\det \Phi(t, t_0) \equiv 1$ for this set of equations.

4.16 Suppose that A is a nonsingular $n \times n$ matrix. Show that $\int_0^t e^{A\tau} d\tau = A^{-1} (e^{A\tau} - I) = (e^{A\tau} - I) A^{-1}$.

4.17 (a) Show that if $A(t)$ is continuous and commutes with its integral, that is,

$$A(t) \int_{t_0}^t A(\sigma) d\sigma = \int_{t_0}^t A(\sigma) d\sigma \, A(t), \qquad t_0 < t < t_1,$$

then the state transition matrix for $\dot{x}(t) = A(t)x(t)$ is given by

$$\Phi(t, t_0) = \exp\left(\int_{t_0}^t A(\sigma) d\sigma \right).$$

Also, indicate the point at which the assumption of commutativity is used.

(b) Show that in general $e^{A+B} \neq e^A e^B$. Under what condition does equality hold? (See Section 4.4.)

4.18 In Section 4.4 it was shown that $(sI - F)^{-1}$ is the Laplace transform of e^{Ft}.

(a) Find the solution of

$$\dot{x}(t) = F x(t), \qquad x(0) = x_0$$

using Laplace transforms.

(b) Do the same for the system

$$\dot{x}(t) = F x(t) + f(t), \qquad x(0) = x_0.$$

Identify the natural response and the forced response in this case.

(c) Consider the single-input, single-output (SISO) system

$$\mathbf{u}(t) \rightarrow \boxed{\begin{array}{l} \dot{\mathbf{x}}(t) = F\mathbf{x}(t) + g\mathbf{u}(t) \\ \mathbf{y}(t) = c\mathbf{x}(t) \end{array}} \rightarrow \mathbf{y}(t),$$

and assume that $x(0) = 0$. Find the transfer function

$$\hat{h}(s) = \hat{y}(s)/\hat{u}(s)$$

from input to output. The output $y(t)$ [for $x(0) = 0$] may be expressed as the convolution of the impulse response with $u(t)$. Do this, and identify the impulse response.

(d) Generalize part (c) for the case where $\mathbf{u}(t)$ and $\mathbf{y}(t)$ are vectors.

4.19 (a) Consider the plant

$$\dot{\mathbf{x}}(t) = F(t)\mathbf{x}(t) + G(t)\mathbf{u}(t), \qquad \mathbf{x}(0) = \mathbf{x}_0.$$

If $\mathbf{u}(t)$ is constrained to lie in the *convex* set

$$\{\mathbf{u}(t): \|\mathbf{u}(t)\| < 1\}$$

for all $t \in [0, T]$, then the set of values which x can take on at T is called the *set of reachable states*. Show that this set is convex in general. [A set S is convex if $\mathbf{x}_1, \mathbf{x}_2 \in S$ implies that $(1 - \alpha)\mathbf{x}_1 + \alpha\mathbf{x}_2 \in S$, $0 < \alpha < 1$.]

(b) Now consider the system

$$\dot{\mathbf{x}}(t) = \begin{bmatrix} -1 & -1 \\ 0 & -2 \end{bmatrix} \mathbf{x}(t) + \begin{bmatrix} 1 \\ 1 \end{bmatrix} \mathbf{u}(t),$$

and let $\mathbf{x}(0) = \mathbf{0}$. Constrain $|u(t)| < 1$, and find the set of reachable states for $T = 1$ and $T = 2$. What happens if $\mathbf{x}(0) \neq \mathbf{0}$?

4.20 Calculate e^{Ft} for the following matrices:

$$F_1 = \begin{bmatrix} 0 & 0 & 0 \\ 0 & 0 & 0 \\ 0 & 0 & 0 \end{bmatrix}, \qquad F_2 = \begin{bmatrix} 0 & 1 & 1 \\ 0 & 1 & 1 \\ 0 & 0 & 0 \end{bmatrix},$$

$$F_3 = \begin{bmatrix} \lambda & \alpha \\ \beta & \lambda \end{bmatrix},$$

where $\alpha\beta \equiv -1$.

4.21   Let

$$F(t) = \begin{bmatrix} -1 + \cos t & 0 \\ 0 & -2 + \cos t \end{bmatrix}.$$

(a) Calculate the state transition matrix $\Phi(t,0)$ for the system, $\dot{x}\ (t) = F(t)x(t)$.

(b) Since this system is periodic with period 2π, the state transition matrix can be written in the form $\Phi(t,0) = P(t,0)\, e^{Rt}$. What are $P(t,0)$ and R?

(c) Are all solutions of this system bounded? Do they decay to zero as $t \to \infty$?

4.22   Linearize the system

$$\dot{x}_1 = 2x_2 x_3,$$

$$\dot{x}_2 = x_3,$$

$$\dot{x}_3 = -x_2 + u^2 + 2u$$

about the trajectory

$$x_1{}^* = \cos^2 t, \qquad x_3{}^* = -\sin t,$$
$$x_2{}^* = \cos t, \qquad u^* = 0.$$

4.23   Assume that the matrix

$$W(0, T) = \int_0^T e^{-Ft} GG^T e^{-Ft} dt$$

is invertible. Show that the input

$$u(t) = -G^T e^{-Ft} W^{-1}(0, T) x_0$$

drives the state of the system

$$\dot{x}(t) = F x(t) + G u(t), \qquad x(0) = x_0$$

to 0 in T seconds.

4.24   Let

$$F = \begin{bmatrix} \lambda & 0 & 0 \\ 0 & 0 & 1 \\ 0 & -1 & 0 \end{bmatrix}, \qquad \lambda \neq 0.$$

(a) Determine whether the system

$$\dot{\mathbf{x}}(t) = F\mathbf{x}(t) + \begin{bmatrix} 0 \\ \sin t \\ 0 \end{bmatrix}$$

has any periodic solutions with period 2π. If so, are there more than one?

(b) Does the system

$$\dot{\mathbf{x}}(t) = F\mathbf{x}(t) + \begin{bmatrix} 1 \\ 0 \\ \cos 2t \end{bmatrix}$$

have solutions of period π?

4.25 Given the matrix

$$F(t) = \begin{bmatrix} \cos t & 1 & 0 \\ -1 & \cos t & 0 \\ 0 & 0 & \lambda \end{bmatrix}, \quad \text{where} \quad \lambda < 0,$$

consider the system, $\dot{\mathbf{x}}(t) = F(t)\mathbf{x}(t)$.

(a) Calculate the state transition matrix $\Phi(t,0)$ for this system.

(b) Is this system periodic? What is the least period?

(c) Does the system

$$\dot{\mathbf{x}}(t) = F(t)\mathbf{x}(t) + \begin{bmatrix} \cos t \\ 0 \\ \sin t \end{bmatrix}$$

have any periodic (2π) solutions? More than one?

4.26 Verify that the matrix F commutes with e^{Ft}.

4.27 Consider the system

$$\dot{\mathbf{x}}(t) = \begin{bmatrix} -3 & 4 \\ -2 & 3 \end{bmatrix} \mathbf{x}(t) + \begin{bmatrix} 1 \\ 1 \end{bmatrix} \mathbf{u}(t),$$

$$\mathbf{y}(t) = [-1 \quad 2] \mathbf{x}(t).$$

Is this system controllable? Observable? Is it stable? If not, can you stabilize it with output feedback, that is, with $\mathbf{u}(t) = k\mathbf{y}(t)$ for some constant k? What can you conclude about the modes of this system?

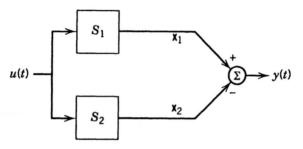

Figure Problem 4.28

4.28 Given the two systems

$$S_1: \quad \dot{x}_1(t) = \alpha_1 x_1(t) + u(t),$$
$$S_2: \quad \dot{x}_2(t) = \alpha_2 x_2(t) + u(t).$$

Under what conditions on α_1 and α_2 is the parallel system shown below controllable and/or observable?

4.29 For the network assume that $i_L = v_c = 0$ at $t = 0$.

(a) Does there exist a voltage $v(t)$, $0 < t < \pi$, such that $i_L = v_c = 1$ at $t = \pi$? If so, find one.

(b) Suppose that we want to maintain $v_c = i_L = 1$ for all $t \geqslant \pi$. Can a suitable $v(t)$, $t \geqslant \pi$, be found? If so, what is it?

(c) Assume now that the system $\dot{x}(t) = Fx(t) + Gu(t)$, is controllable. Do there exist controls in general which make $x(t)$ follow arbitrary trajectories? If not, under what conditions (involving F and G) will this be true? (Assume that the arbitrary trajectory is differentiable and that the initial condition x_0 may be chosen to fit it.)

4.30 Given the system

$$\dot{x}(t) = \begin{bmatrix} t & 0 \\ -t & \sin t \end{bmatrix} u(t),$$

$$y(t) = [t \quad 1] x(t).$$

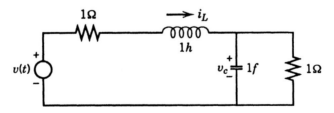

Figure Problem 4.29

(a) Determine whether this system is controllable on $[0, T]$ for all $T > 0$.

(b) Is the system observable on $[0, T]$ for all T?

4.31 Consider the simple first-order, discrete-time Kalman filter whose message and observation models are given by

$$x(k+1) = x(k) + w(k),$$

$$z(k) = x(k) + v(k)$$

with initial conditions

$$\hat{x}(0) = 0, \qquad P(0) = 100, \qquad H(k) = 1, \qquad Q(k) = 1, \qquad R(k) = 2.$$

(a) Find the state estimate $\hat{x}(k)$ if the received observation sequence is

$$Z(k) = \{1, 2, -1, 4, 2, 4, \ldots\}.$$

Tabulate $\hat{x}(k)$, $K(k)$, $P(k|k-1)$, and $P(k)$ for $k = 0, 1, 2, 3, 4, 5, 6, \infty$.

(b) What is the steady-state ($k = \infty$)—that is, the limiting—form for this filter? Write only the limiting form for the estimate

$$\hat{x}(k) = \hat{x}(k-1) + K(k)[z(k) - H(k)\hat{x}(k-1)].$$

4.32 In Problem 4.10 a single-axis inertial navigation system was presented. We will now investigate another important aspect of inertial systems, namely, that of gravity modeling. It is well known that a simple model of an inertial system horizontal channel is the so-called Schuler loop. In its purest form, this model is an undamped second-order feedback loop.

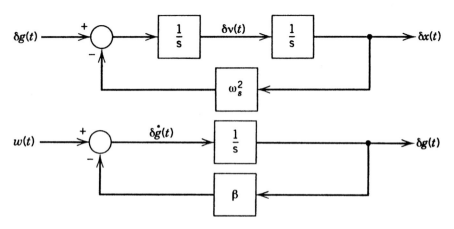

Figure Problem 4.32

The feedback signal represents, to first order, the corrective influence of the gravity model, given the position errors. Consider now the model for gravity error, δg, which corresponds to the vertical deflection, represented as the output of a linear filter driven by white, Gaussian noise. The Schuler loop can be represented diagramatically as follows: where $\omega_s^2 = g/R_e$ (g is the acceleration of gravity, R_e is the radius of the earth, and ω_s is the Schuler frequency). From the above diagram, we can write the canonical equation $dx/dt = Fx + Gw$ as

$$\begin{bmatrix} \delta\dot{x} \\ \delta\dot{V} \end{bmatrix} = \begin{bmatrix} 0 & 1 \\ -\omega_s^2 & 0 \end{bmatrix} \begin{bmatrix} \delta x \\ \delta V \end{bmatrix} + \begin{bmatrix} 0 \\ 1 \end{bmatrix} \delta g \tag{1}$$

with the state transition matrix given as

$$\Phi(t_1, t_2) = \begin{bmatrix} \cos\omega_s(t_1 - t_2) & (1/\omega_s)\sin\omega_s(t_1 - t_2) \\ -\omega_s\sin\omega_s(t_1 - t_2) & \cos\omega_s(t_1 - t_2) \end{bmatrix}. \tag{2}$$

The block diagram below shows the basic relationship between the noise $w(t)$ and the disturbance $\delta g(t)$.

In this diagram, β is the inverse correlation time given by $\beta = 1/\tau = V/d$ where d is the correlation distance (typically, 20 nm) and V is the velocity. Assuming that

$$\mathscr{E}\{w(t_1)w(t_2)\} = 2\beta\sigma^2\delta(t_2 - t_1) \tag{3}$$

with $Q = 2\beta\sigma^2$, the augmented state equation can now be written in the form

$$\begin{bmatrix} \delta\dot{x} \\ \delta\dot{V} \\ \hline \delta\dot{g} \end{bmatrix} = \begin{bmatrix} 0 & 1 & 0 \\ -\omega_s^2 & 0 & 1 \\ \hline 0 & 0 & -\beta \end{bmatrix} \begin{bmatrix} \delta x \\ \delta V \\ \hline \delta g \end{bmatrix} + \begin{bmatrix} 0 \\ 0 \\ \hline 1 \end{bmatrix} w(t). \tag{4}$$

Determine the elements (i.e., $P_{11}, P_{12}, ..., P_{33}$) of the covariance matrix

$$dP(t)/dt = FP(t) + P(t)F^T + GQG^T. \tag{5}$$

For this problem assume that

$$\mathscr{E}\{\delta g\{t)^2\} = \sigma^2$$

and

$$P(0) = \begin{bmatrix} 0 & 0 & 0 \\ 0 & 0 & 0 \\ \hline 0 & 0 & \sigma^2 \end{bmatrix}.$$

CHAPTER 5

LINEAR REGULATORS

5.1 INTRODUCTION

Optimal control of linear deterministic systems with quadratic performance criteria and bounded controls has been studied extensively. Historically, the theory and design of optimal control systems have a common mathematical foundation in the calculus of variations. The calculus of variations is the branch of mathematics which is concerned with the finding of trajectories that maximize or minimize a given functional. Optimization techniques have been investigated extensively in the United States by Athans, Bellman, Bryson, Kalman, Leitmann, Miele, and Wiener among others, and in Russia by Boltyanskii, Butkovskii, Fel'dbaum, Gamkrelidge, Krasovskii, Letov, Mishchenko, and Pontryagin. Modern optimal control techniques such as the minimum principle of Pontryagin and the dynamic programming of Bellman are derived from the calculus of variations.

For more than a quarter of century now, much attention has been focused on optimizing the behavior of systems, such as maximizing the range of a rocket, determining optimal flight paths, and minimizing the error in estimation of the position of a vehicle. Finding the control which attains the desired objective while maximizing (or minimizing) a defined criterion constitutes the fundamental problem of optimization theory. For example, the Bolza formulation in the calculus of variations leads into the proof of the Pontryagin minimum principle. Moreover, many problems in modern system theory may be simply stated as extreme-value problems. By adding a sufficient number of variables, almost all solvable problems in optimal control can be solved by the calculus of variations.

In solving optimal control problems, the control engineer often sets up a mathematical performance criterion or index and then tries to find solutions that optimize that particular measure. In recent years, several methods have been advanced which help find solutions to certain class of problems. All these methods are related to the calculus of variations. Time-optimal, fuel-optimal, and energy-optimal problems are a class of optimal control problems that the control engineer is often called on to solve. In missile guidance, for instance, the approach normally taken in designing the missile guidance system, whether it is for air-to-air or air-to-ground application, is to model those missile functions that respond to attitude commands. Also, a missile may follow a preassigned trajectory specified by the mission planner, but be diverted in flight by the guidance system's update commands. In this section we will briefly discuss the optimal linear–quadratic–regulator (LQR) problem. In particular, we will discuss the deterministic linear–quadratic state regulator, develop the optimal feedback control law in a framework which is readily applicable by the control engineer, and discuss its solution in terms of the solution to the matrix Riccati differential equation.

Deterministic models possess the following characteristics: (1) there are no plant (e.g., aircraft) disturbances; (2) plant and output variables can be measured exactly, and controller dynamics are known exactly. Although several different approaches to the problem have been developed, the feedback law generally arises in conjunction with an operator Riccati differential or integral equation. Consequently, numerical methods for approximating the gain rely on approximating the solution to the associated Riccati equation. As we shall see later in Section 5.5, a full optimization study involves finding the optimum control law in the presence of stochastic disturbances. For linear problems with quadratic performance criteria, the *separation theorem** may be invoked, which decouples the full stochastic control problem into two separate parts: (1) the control portion of the decoupled problem solves for the optimum deterministic controller with a quadratic performance measure, assuming that we have exact and complete knowledge of all the state variables of the plant (system), and (2) the remaining portion of the problem is that of a stochastic estimator which uses the noisy and incomplete measurements of the states of the system to give the least-square-error estimates of the system states. These estimates are then used as if they were known exactly by the optimum controller. The separation theorem assures that the composite system of controller and estimator will be the jointly optimum stochastic controller. In general, this performance index may take a variety of forms containing constraints or penalties on the control energy expended and on the deviations of the states from the desired values. The general regulator problem has been treated by many authors [2, 28, and 58].

---

*Simply stated, the separation theorem obtained its name from the ability of such systems to perform state estimatation and optimal control separately.

5.2 THE ROLE OF THE CALCULUS OF VARIATIONS IN OPTIMAL CONTROL

As stated in the introduction (Section 5.1), modern optimal control theory has its roots in the calculus of variations. Specifically, modern optimal control theory techniques such as the minimum principle of Pontryagin and dynamic programming of Bellman are based on and/or inspired by the classical calculus of variations.

In this section, we will discuss briefly the basic concepts of the calculus of variations necessary for understanding and solving the type of problems encountered in optimal control theory. In particular, the calculus of variations will be discussed by means of the Euler–Lagrange equation and associated transversality conditions. In the presentation that follows, we will forgo much of the mathematical rigor. For the reader who wishes to obtain a deeper insight into the calculus of variations, references [11 and 34] are recommended.

The basic problem in the calculus of variations is to determine a function such that a certain definite integral involving that function and certain of its derivatives takes on a maximum or minimum value. Furthermore, the elementary part of the theory is concerned with a *necessary* condition (generally in the form of a differential equation with boundary conditions) which is required must satisfy. One of the earliest instances of a variational problem is the problem of the brachistochrone (i.e., the curve of quickest descent), first formulated and solved by John Bernoulli (1696).

We begin our discussion by noting that the simplest single-stage process with equality constraints is to maximize or minimize a cost function (or performance index) of the form

$$J = \theta[\mathbf{x}, \mathbf{u}] \tag{5.1a}$$

subject to the equality constraint

$$\mathbf{f}(\mathbf{x}, \mathbf{u}) = \mathbf{0}, \tag{5.1b}$$

where $\mathbf{x}$ is an n-dimensional state vector

$$\mathbf{x}^T = [x_1, x_2, \ldots, x_n],$$

$\mathbf{u}$ is an m-vector

$$\mathbf{u}^T = [u_1, u_2, \ldots, u_m],$$

and $\mathbf{f}$ is an n-vector function

$$\mathbf{f}^T(\mathbf{x}, \mathbf{u}) = [f_1(\mathbf{x}, \mathbf{u}), f_2(\mathbf{x}, \mathbf{u}), \ldots, f_n(\mathbf{x}, \mathbf{u})].$$

The *Lagrangian* function is formed by adjoining the cost function to the given constraint [i.e., adjoining Eq. (5.1a) to (5.1b)] via the *Lagrange multiplier* technique:

$$L(\mathbf{x}, \mathbf{u}, \lambda) = \theta(\mathbf{x}, \mathbf{u}) + \lambda^T \mathbf{f}(\mathbf{x}, \mathbf{u}), \tag{5.2}$$

where λ is the Lagrange multiplier (also known as the *adjoint vector*)

$$\lambda^T = [\lambda_1, \lambda_2, \ldots, \lambda_n].$$

In order for L to be a maximum or minimum, this requires that

$$\frac{\partial L}{\partial \mathbf{x}} = \frac{\partial \theta}{\partial \mathbf{x}} + \frac{\partial}{\partial \mathbf{x}} \mathbf{f}^T(\mathbf{x}, \mathbf{u}) \lambda = 0,$$

$$\frac{\partial L}{\partial \mathbf{u}} = \frac{\partial \theta}{\partial \mathbf{u}} + \frac{\partial}{\partial \mathbf{u}} \mathbf{f}^T(\mathbf{x}, \mathbf{u}) \lambda = 0,$$

where

$$\left(\frac{\partial L}{\partial \mathbf{u}}\right)^T = \left[\frac{\partial L}{\partial u_1}, \frac{\partial L}{\partial u_2}, \ldots, \frac{\partial L}{\partial u_m} \right].$$

It should be emphasized here that in order for J to be an *extremum* (either a maximum or a minimum), not only must

$$\frac{\partial L}{\partial \mathbf{x}} = 0, \qquad \frac{\partial L}{\partial \mathbf{u}} = 0,$$

but also the second variation of L must be: (1) greater than zero for a minimum, or (2) less than zero for a maximum. Note that this discussion is applicable to necessary and sufficient conditions for a *local* extremum.

The type of problem we encounter in the calculus of variations can be more generally characterized as follows: we are given a function F of three variables

$$F = F(y, y', x),$$

where we take F independent of x (however, it is not necessary to make this restriction), and $y' = dy/dx$. Furthermore, we are given the definite integral [34]

$$J = \int_\alpha^b F(y, y', x)\, dx \tag{5.3}$$

with the boundary conditions

$$f(a) = \alpha, \qquad f(b) = \beta.$$

The problem is to find a function

$$y = f(x) \tag{5.4}$$

which will make the integral J an extremum, or at least give it a *stationary* value. (An extremal which satisfies the appropriate end conditions is often called a *stationary function* of the variational problem, and is said to make the relevant integral *stationary*, even if it does not actually make the integral maximal or minimal relative to all slightly varied admissible functions. The reader is cautioned that the usage of the terms *extremal* and *stationary function* varies within the literature.)

Consider now the function $y = f(x)$, which by hypothesis gives a stationary value to the definite integral (5.3). In order to prove that we do have a stationary value, we must evaluate the same integral for a slightly modified function which we shall designate $y = \overline{f(x)}$, and show that the rate of change of the integral due to the change in the function becomes zero. The modified function can be written in the form [11]

$$\overline{f(x)} = f(x) + \varepsilon\phi(x), \tag{5.5}$$

where $\phi(x)$ is an arbitrary new function which satisfies the same general conditions as $\overline{f(x)}$. Hence, $\phi(x)$ must be continuous and differentiable. Referring to Figure 5.1, and making use of the variable parameter ε, we can modify the function $f(x)$ by arbitrarily small amounts. To this end, let ε decrease toward zero.

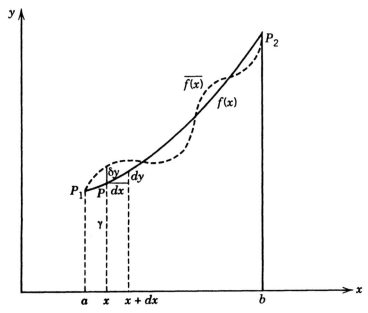

Figure 5.1 Plot of $f(x)$ and $\overline{f(x)}$.

We can now compare the values of the modified function $\overline{f(x)}$ with the values of the original function $f(x)$ at a certain definite point x of the independent variable, by forming the difference between $\overline{f(x)}$ and $f(x)$. This difference is called the *variation* of the function $f(x)$, and we will denote it by δy:

$$\delta y = \overline{f(x)} - f(x) = \varepsilon \phi(x), \tag{5.6}$$

or

$$\overline{f(x)} = f(x) + \varepsilon \phi(x).$$

The variation function δy is an infinitesimal change, since the parameter ε decreases toward zero. Hence, $\phi(x)$ can be chosen arbitrarily, as long as the general continuity conditions are satisfied.

In deriving the Euler–Lagrange equation, we will change our notation in order to conform with the standard notation used in optimization theory. We begin by considering the problem of selecting a continuously differentiable function $x: [t_0, t_f] \to R$ to minimize the cost functional [46]

$$J(x) = \int_{t_0}^{t_f} \phi\left[x(t), \frac{dx(t)}{dt}, t\right] dt \tag{5.7}$$

with respect to the set of all real-valued, continuously differentiable functions in the interval $[t_0, t_f]$. Furthermore, following the above discussion, let $x^*(t)$ be an *admissible* optimal state trajectory for all $t \in [t_0, t_f]$, and form the function [similar to Eq. (5.6)]

$$x(t) = x^*(t) + \varepsilon \eta(t), \tag{5.8a}$$

where $\eta(t)$ is a continuously differentiable function [i.e., a variation in $x(t)$], ε is a small parameter, and $\eta_i(t_1) = \eta_i(t_2) = 0$ by definition. Differentiating Eq. (5.8a) with respect to t, we obtain

$$\dot{x}(t) = \dot{x}^*(t) + \varepsilon \dot{\eta}(t). \tag{5.8b}$$

Substituting Eqs. (5.8a) and (5.8b) into Eq. (5.7), we obtain [46]

$$J(\varepsilon) = \int_{t_0}^{t_f} \phi[x^*(t) + \varepsilon\eta(t), \dot{x}^*(t) + \varepsilon\dot{\eta}(t), t] dt. \tag{5.9}$$

Now, if $x(t)$ is the minimizing function, then by definition $J(\varepsilon)$ will always be greater than the minimum possible value, except when $\varepsilon = 0$. Furthermore, $J(\varepsilon)$

will have a stationary point at $\varepsilon = 0$. Thus,

$$\frac{dJ(\varepsilon)}{d\varepsilon}\bigg|_{\varepsilon=0} = 0. \tag{5.10}$$

Carrying out the operation of Eq. (5.10), we have

$$\frac{\partial J(\varepsilon)}{\partial \varepsilon}\bigg|_{t=0} = \int_{t_0}^{t_f} \left[\eta(t)\frac{\partial \phi(x^*, \dot{x}^*, t)}{\partial x} + \dot{\eta}(t)\frac{\partial \phi(x^*, \dot{x}^*, t)}{\partial \dot{x}^*} \right] dt = 0, \tag{5.11}$$

or

$$0 = \int_{t_0}^{t_f} \eta(t)\frac{\partial \phi(x^*, \dot{x}^*, t)}{\partial x^*} dt + \int_{t_0}^{t_f} \dot{\eta}(t)\frac{\partial \phi(x^*, \dot{x}^*, t)}{\partial \dot{x}^*} dt. \tag{5.12}$$

After simplification, Eq. (5.12) becomes

$$0 = \int_{t_0}^{t_f} \eta(t)\left(\frac{\partial \phi}{\partial x^*} - \frac{d}{dt}\frac{\partial \phi}{\partial \dot{x}^*} \right) dt + \frac{\partial \phi}{\partial \dot{x}^*}\eta(t)\bigg|_{t_0}^{t_f}. \tag{5.13}$$

Clearly, if

$$\frac{\partial \phi}{\partial x^*} - \frac{d}{dt}\frac{\partial \phi}{\partial \dot{x}^*} = 0, \tag{5.14}$$

$$\frac{\partial \phi}{\partial \dot{x}^*}\eta(t) = 0 \qquad \text{for} \quad t = t_0, t_f \tag{5.15}$$

along the optimal trajectory, then Eq. (5.13) is satisfied. Equation (5.14) is the well-known *Euler–Lagrange* equation, developed independently by Euler and Lagrange. Here we note that

$$\lim_{r \to 0} J(x) = J(x^*), \qquad \lim_{r \to 0} x(t) = x^*(t), \tag{5.16}$$

so that Eq. (5.13) becomes simply

$$\frac{\partial \phi}{\partial x} - \frac{d}{dt}\left(\frac{\partial \phi}{\partial \dot{x}} \right) = 0.$$

If several independent variables exist, then Eq. (5.7) takes the form

$$J(x) = \int_{t_0}^{t_f} \phi[x_1, ..., x_n; \dot{x}_1, ..., \dot{x}_n; t] dt.$$ (5.17)

Again, assuming that ϕ is twice continuously differentiable with respect to each of its arguments, we proceed in the same manner as before. Define n arbitrary functions $\eta_1(t) = \eta_2(t) = 0$, and consider variation about the possible optimum solution $x_i(t)$ and $\dot{x}_i(t)$ by considering what happens to the integral when

$$x_i(t) = x^*(t) + \varepsilon \eta_i(t), \qquad i = 1, 2, ..., n,$$ (5.18a)

and

$$\dot{x}_i(t) = \dot{x}^*(t) + \varepsilon \dot{\eta}_i(t), \qquad i = 1, 2, ..., n,$$ (5.18b)

are substituted in the integral (5.17). The Euler–Lagrange equation for this case, namely,

$$\frac{\partial \phi}{\partial x_i} - \frac{d}{dt}\left(\frac{\partial \phi}{\partial \dot{x}_i}\right) = 0, \qquad i = 1, ..., n,$$ (5.19)

represents a system of simultaneous differential equations that must be satisfied at every point of an optimal trajectory.

We now examine the case of unspecified or variable end points. From the above discussion, if $\eta_i(t)$ is arbitrary, then for each i, the first and second terms on the right side of Eq. (5.13) must vanish separately. The vanishing of the first term will lead to the Euler–Lagrange equation (5.14) as before, but the first term, for those variables with an unspecified end point at t_f, will lead to the so-called *transversality* conditions

$$\left.\frac{\partial \phi}{\partial \dot{x}_i}\right|_{i=t_f} = 0$$ (5.20)

[note that Eq. (5.20) applies for those $x_i(t)$ whose end points $x_i(t_f)$ are unspecified]. When the end point is not free, the transversality conditions become more involved.

The above discussion will now be extended to the problem of determining an *admissible* control function $\mathbf{u}$ in order to minimize the cost function [12]

$$J = \theta[\mathbf{x}(t), t]\Big|_{t=t_0}^{t=t_f} - \int_{t_0}^{t_f} \phi[\mathbf{x}(t), \mathbf{u}(t), t] dt,$$ (5.21)

where θ and ϕ possess continuous partial derivatives in $\mathbf{x}$ and $\mathbf{u}$. The integrand in Eq. (5.21) is sometimes referred to as the *Lagrangian*. Futhermore, the cost function, Eq. (5.12), is to be minimized subject to the n-dimensional dynamic constraint

$$\dot{\mathbf{x}}(t) = \mathbf{f}[\mathbf{x}(t), \mathbf{u}(t), t], \qquad \mathbf{x}(t_0) = \mathbf{x}_0, \tag{5.22}$$

where $\mathbf{x}(t)$ is an n-dimensional state vector

$$\mathbf{x}^T(t) = [x_1(t), x_2(t), \ldots, x_n(t)]$$

and $\mathbf{u}(t)$ is an m-dimensional control vector

$$\mathbf{u}^T(t) = [u_1(t), u_2(t), \ldots, u_m(t)].$$

Therefore, the cost function expressed by Eq. (5.21) is equivalent, under certain smoothness assumptions, to three other useful cost functions:

1. The *Mayer* type

$$J(x) = \theta[\mathbf{x}(t_f), t_f],$$

   which contains only a terminal penalty.
2. The *Lagrange* type, which contains only the integral term.
3. The *Bolza* type, containing both terms of Eq. (5.21) (i.e., both terminal and integral costs).

The constraint (5.22) represents a model of a large class of physical systems, where $\mathbf{x}(t)$ is an n-vector trajectory, and $\mathbf{u}(t)$ is an m-vector control function. Using the Lagrange multiplier method, we can adjoin the system differential equality to the cost function, yielding [12]

$$J = \theta[\mathbf{x}(t), t]\Big|_{t=t_0}^{t=t_f} + \int_{t_0}^{t_f} (\phi[\mathbf{x}(t), \mathbf{u}(t), t] + \lambda^T(t)\{\mathbf{f}[\mathbf{x}(t), \mathbf{u}(t), t] - \dot{\mathbf{x}}\})dt, \tag{5.23}$$

where λ is the Lagrange multiplier. Defining the *Hamiltonian* as

$$\mathscr{H}[\mathbf{x}(t), \mathbf{u}(t), \lambda(t), t] = \phi[\mathbf{x}(t), \mathbf{u}(t), t] + \lambda^T(t)\mathbf{f}[\mathbf{x}(t), \mathbf{u}(t), t], \tag{5.24}$$

the cost function becomes

$$J = \theta[\mathbf{x}(t), t]\Big|_{t=t_0}^{t=t_f} + \int_{t_0}^{t_f} \{\mathscr{H}[\mathbf{x}(t), \mathbf{u}(t), \lambda(t), t] - \lambda^T(t)\dot{\mathbf{x}}\}dt \tag{5.25a}$$

with $\lambda(t)$ satisfying the differential equation

$$\dot{\lambda}^T(t) = -\frac{\partial \mathcal{H}\left[\mathbf{x}(t), \mathbf{u}(t), \lambda(t), t\right]}{\partial \mathbf{x}}, \qquad t_0 \leqslant t \leqslant t_f. \qquad (5.25b)$$

Integrating Eq. (5.25) by parts, we obtain

$$J = \{\theta[\mathbf{x}(t), t] - \lambda^T(t)\mathbf{x}(t)\}\Big|_{t=t_0}^{t=t_f} + \int_{t_0}^{t_f} \{\mathcal{H}[\mathbf{x}(t), \mathbf{u}(t), \lambda(t), t] + \dot{\lambda}^T\mathbf{x}(t)\} dt. \quad (5.26)$$

Taking the first variation in the state $\mathbf{x}(t)$ and control $\mathbf{u}(t)$ yields

$$\delta J = \delta \mathbf{x}^T \left(\frac{\partial \theta}{\partial \mathbf{x}} - \lambda\right)\Big|_{t=t_0}^{t=t_f} + \int_{t_0}^{t_f}\left[\delta \mathbf{x}^T\left(\frac{\partial \mathcal{H}}{\partial \mathbf{x}} + \dot{\lambda}\right) - \delta \mathbf{u}^T \frac{\partial \mathcal{H}}{\partial \mathbf{u}}\right] dt. \qquad (5.27)$$

A necessary condition for a minimum is obviously that the first variation in J vanishes for arbitrary variations $\delta \mathbf{x}$ and $\delta \mathbf{u}$. Therefore, we have as the necessary condition for a minimum the well-known relations

$$\delta \mathbf{x}^T\left(\frac{\partial \theta}{\partial \mathbf{x}} - \lambda\right) = 0 \qquad \text{for} \quad t = t_0, t_f, \qquad (5.28)$$

$$\dot{\lambda} = -\frac{\partial \mathcal{H}}{\partial \mathbf{x}}, \qquad \dot{\mathbf{x}} = \mathbf{f}(\mathbf{x}, \mathbf{u}, t) = \frac{\partial \mathcal{H}}{\partial \lambda}, \qquad (5.29)$$

$$\frac{\partial \mathcal{H}}{\partial \mathbf{u}} = \mathbf{0}. \qquad (5.30)$$

Optimization of the integral cost function subject to a dynamic constraint is a *two-point boundary value problem* (TPBVP), as constants of integration for the state (5.22) and its adjoint

$$\lambda^T(t_f) = \frac{\partial \phi[\mathbf{x}(t), t]}{\partial \mathbf{x}}\Big|_{t=t_f}$$

are specified at opposite ends (i.e., boundaries) of the time interval. For nonlinear systems, there is no general way to solve the TPBVP. As we shall see later in this chapter, this is the reason why most of the solved optimal control problems involve a linear plant. However, there are numerical procedures which permit the solving of some nonlinear TPBVPs through iteration.

Example The above theory will now be illustrated by means of a specific example. Consider a two-dimensional air-to-air engagement, where an interceptor missile is attempting to intercept a moving target using lateral thrust control. Futhermore, we want the target to be intercepted in minimum time.

To this end, let θ be the direction of the interceptor axis (i.e., thrust vector **a**) measured from the horizontal x-axis, and α the angle measured from the thrust to the velocity vector, as illustrated in Figure 5.2. This figure depicts the dynamic configuration of the interceptor missile. The thrust acceleration will be assumed to be constant and fixed along the longitudinal axis of the missile, while the lateral thrusters, **u**, are either off or on with maximum thrust. For instance, when the thrusters are on, they provide a rotational torque which produces a constant angular acceleration of the missile in pitch. For simplicity, assume that the interceptor missile can be characterized by a six-component state vector and associated boundary conditions as follows:

System equations:	Boundary conditions:	
$\dot{x} = v,$	$x(0) = 0, \quad x(T) = x_f,$	(1a)
$\dot{y} = w,$	$y(0) = 0, \quad y(T) = y_f,$	(1b)
$\dot{v} = a \cos \theta,$	$v(0) = 0,$	(1c)
$\dot{w} = a \sin \theta,$	$w(0) = w_0,$	(1d)
$\dot{\theta} = q,$	$\theta(0) = \pi/2,$	(1e)
$\dot{q} = u, \quad |u| = u_0 \text{ or } 0,$	$q(0) = 0,$	(1f)

where T is the final time, and a and u_0 are specified constants. The cost function (or performance index) is [58]

$$J = \int_{t_0}^{T} dt. \tag{2}$$

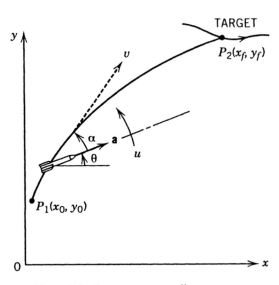

Figure 5.2 Interceptor coordinate system.

The system Hamiltonian is

$$H = 1 + \lambda_x v + \lambda_y w + \lambda_v a \cos \theta + \lambda_w a \sin \theta + \lambda_\theta q + \lambda_q u. \tag{3}$$

(Note that in this case H is constant, since it is not explicitly a function of time.) Following the development from references [2 and 12], it can be shown that the costate variables are *influence coefficients* on the cost function [see also next section, Eqs. (5.40), (5.41), and (5.43)]. For example, λ_x represents the effect or influence on the cost function of a change in x (i.e., $\lambda_x = \partial J / \partial x$). Therefore, the costate equations for this problem are

$$\dot{\lambda}_x = -\frac{\partial H}{\partial x} = 0, \tag{4a}$$

$$\dot{\lambda}_y = -\frac{\partial H}{\partial y} = 0, \tag{4b}$$

$$\dot{\lambda}_v = -\frac{\partial H}{\partial v} = -\lambda_x, \qquad \lambda_v(T) = 0, \tag{4c}$$

$$\dot{\lambda}_w = -\frac{\partial H}{\partial w} = -\lambda_y, \qquad \lambda_w(T) = 0, \tag{4d}$$

$$\dot{\lambda}_\theta = -\frac{\partial H}{\partial \theta} = a(\lambda_v \sin \theta - \lambda_w \cos \theta), \qquad \lambda_\theta(T) = 0, \tag{4e}$$

$$\dot{\lambda}_q = -\frac{\partial H}{\partial q} = -\lambda_\theta, \qquad \lambda_q(T) = 0. \tag{4f}$$

The final time is determined from $H(T) = 0$. Applying the boundary conditions to the costates results in the following equation:

$$1 + \lambda_x v_f + \lambda_y w_f = 0. \tag{5}$$

In order to minimize the cost function, it is necessary first to minimize the Hamiltonian with respect to u, the control, at each instant over the interval of integration. However, the Hamiltonian in this case has only one term dependent on u, namely, $\lambda_q u$.

Typically, the lateral thrusters can provide three values for the angular acceleration: $+u_0$, $-u_0$, and 0. This is known in the literature as *bang–bang–off* control. In determining the pitch control history of the interceptor missile that will produce the minimum-time trajectory for a specified intercept point, the bang–bang–off control history in effect turns the thrusters on to cause the missile to pitch away from the vertical, then reverses them to bring the pitch rate back to zero, and then turn them off. Therefore, since u is bounded at $\pm u_0$, $\lambda_q u$ will be minimized by $+u_0$ if $\lambda_q < 0$ and by $-u_0$ if $\lambda_q > 0$. As a result, the optimal control takes the form [62]

$$u = -u_0 \, \text{sgn} \, (\lambda_q).$$

(6)

Integrating the costate equations (4a)–(4f) backward in time from the final position results in the following equations:

$$\lambda_x = \text{constant},$$

(7a)

$$\lambda_y = \text{constant},$$

(7b)

$$\lambda_v = \lambda_x(T-t) + c_1, \qquad \lambda_v(T) = 0 \quad \Rightarrow \quad c_1 = 0,$$

(7c)

$$\lambda_w = \lambda_y(T-t) + c_2, \qquad \lambda_w(T) = 0 \quad \Rightarrow \quad c_2 = 0.$$

(7d)

Substituting Eqs. (4a)–(4d) into Eq. (4e) yields

$$\dot{\lambda}_\theta = a[\lambda_x(T-t)\sin\theta - \lambda_y(T-t)\cos\theta]$$

$$= a(T-t)(\lambda_x\sin\theta - \lambda_y\cos\theta).$$

(8)

Now, introduce two constants: a constant angle θ_0 and a constant ξ, as follows:

$$\theta_0 = \tan^{-1}\left(\frac{-\lambda_y}{-\lambda_x}\right),$$

$$\xi = (\lambda_x{}^2 + \lambda_y{}^2)^{1/2}.$$

From the boundary conditions given in Eqs. (1a)–(1f), the constant ξ can be expressed as a function of $\lambda_q(0), \theta_0$, and $T($ the time of flight), yielding

$$\xi = \frac{|\lambda_q(0)|u_0 - 1}{w_0 + aT} \csc\theta_0.$$

(9)

Using the above constants, Eq. (4e) becomes

$$\dot{\lambda}_\theta = -a\xi(T-t)\sin(\theta - \theta_0).$$

(10)

Differentiating Eq. (4f), we have

$$\ddot{\lambda}_q = a\xi(T-t)\sin(\theta - \theta_0)$$

(11)

with boundary conditions

$$\lambda_q(T) = \dot{\lambda}_q(T) = 0.$$

Finally, the control law can be determined from the expression

$$u = \ddot{\theta} = -u_0 \sin(\lambda_q), \qquad \theta(0) = \pi/2, \quad \dot{\theta} = 0, \tag{12}$$

where λ_q is given by

$$\ddot{\lambda}_q = a\zeta(T-t)\sin(\theta - \theta_0),$$

$$\lambda_q(T) = 0, \quad \dot{\lambda}_q(T) = 0.$$

5.3 THE CONTINUOUS-TIME LINEAR-QUADRATIC REGULATOR

Consider the continuous-time linear deterministic system characterized by

$$\dot{\mathbf{x}}(t) = A(t)\mathbf{x}(t) + B(t)\mathbf{u}(t), \qquad \mathbf{x}(t_0) = \mathbf{x}_0, \tag{5.31a}$$

$$\mathbf{y}(t) = C(t)\mathbf{x}(t), \tag{5.31b}$$

where $\mathbf{x}(t)$ is an n-dimensional state vector, $\mathbf{u}(t)$ is the r-dimensional plant control input vector $(0 < m \leqslant r \leqslant n)$, and $\mathbf{y}(t)$ is an m-dimensional output vector $(m \leqslant n)$; and $A(t)$, $B(t)$, and $C(t)$ are $n \times n$, $n \times r$, $m \times n$ matrices, respectively. This plant is a continuous-time, linear dynamical system, possessing the following properties:

1. A time set $\{t\}$ such that $\{t\} = (-\infty, \infty) = E_1$ for $T > t_0$, where t_0 is the initial time and T is the final or terminal time.
2. A set of states $\{\mathbf{x}\} = X = E_n$, called the state space, where E_n is the n-dimensional Euclidean vector space.
3. A set of inputs or controls $\{\mathbf{u}\} = U = E_r$, called the input space.
4. A function space Ω whose elements are bounded, measurable functions, which map E_1 into U.
5. A set of outputs $\{y\} = Y = E_m$, called the output space.

If the control $u(\cdot)$ is a given element of Ω, let $\mathbf{x}_u = \varphi(t; t_0, x_0, u(\cdot))$ denote the solution of the system Eq. (5.31a) starting from x_0 at time t_0 [that is, $\mathbf{x}(t_0) = \mathbf{x}_0$], and generated by the control $\mathbf{u}(\cdot)$. Furthermore, let $\mathbf{y}_u(t) = C(t)\mathbf{x}_u(t)$ to be corresponding output trajectory. Then, the optimal linear regulator problem is to determine the control $\mathbf{u}(\cdot)$, which, as we shall see below, minimizes a quadratic performance index (also known as cost functional). More specifically, an explicit solution to Eq. (5.31a), given an initial state x_0 at t_0 and $t \geqslant t_0$, is as follows [52]:

$$\mathbf{x}(t) = \Phi(t, t_0)\mathbf{x}_0 + \int_{t_0}^{t} \Phi(t, \tau) B(\tau)\mathbf{u}(\tau)d\tau, \tag{5.32}$$

where $\Phi(t, \tau)$ is the state transition matrix associated with $A(t)$. Similarly, the solution to Eq. (5.31b), given an initial state $\mathbf{x}_0$ and input $\mathbf{u}(t)$ on $[t_0, \infty)$, is [2]

$$\mathbf{y}(t) = C(t)\Phi(t, t_0)\mathbf{x}_0 + \int_{t_0}^{t} H(t, \tau)\mathbf{u}(\tau)d\tau, \tag{5.33}$$

where

$$H(t, \tau) = \begin{cases} C(t)\Phi(t, \tau)B(\tau) & \text{for} \quad t \geqslant \tau, \\ 0 & \text{for} \quad t < \tau \end{cases} \tag{5.34}$$

is the impulse response matrix. If $A(t)$ is a constant matrix, then the state transition matrix has the explicit form

$$\Phi(t, \tau) = e^{A(t-\tau)}. \tag{5.35}$$

Therefore, and as mentioned above, the optimal linear regulator problem for a linear dynamic system entails the determination of the optimal control $\mathbf{u}^*(t), t \in [t_0, T]$, which minimizes the quadratic performance index [2]

$$J(\mathbf{x}, t_0, T, \mathbf{u}(\cdot)) = \tfrac{1}{2}\mathbf{x}^T(T)Sx(T) + \frac{1}{2}\int_{t_0}^{T} [\mathbf{x}^T(t)Q(t)\mathbf{x}(t) + \mathbf{u}^T(t)R(t)\mathbf{u}(t)]dt, \tag{5.36}$$

where S is a real symmetric positive semidefinite (nonzero) $n \times n$ matrix; the "terminal state" $\mathbf{x}_u(T) \in X$ is unconstrained, and the terminal time T may be either fixed *a priori* or unspecified $(T > t_0)$. The superscript T denotes matrix transposition. $Q(t)$ is a real symmetric $n \times n$ positive semidefinite matrix, and $R(t)$ is a real symmetric $r \times r$ positive definite matrix. S and $Q(t)$ are not both identically zero. Since $R^{-1}(t)$ is positive definite, it possesses a unique positive definite square root, $R^{-1/2}(t)$. Similarly, the positive semidefiniteness of $Q(t)$ implies the existence of the unique positive semidefiniteness square root $Q^{-1/2}(t)$.

The notation $S \geqslant 0$ will be used to indicate that the matrix S is positive semidefinite, that is, $\mathbf{x}^T Sx \geqslant 0$. Similarly, the notation $S > 0$ will be used to indicate that S is positive definite. In order to minimize the performance index J, it is necessary that J be finite, which means that it will become infinite if uncontrollable. The weighting matrices $Q(t)$ and $R(t)$ are selected by the control-system designer to place bounds on the trajectory and control, respectively, while the matrix S and the terminal penalty cost $\mathbf{x}^T(T)Sx(T)$ are included in order to insure that $\mathbf{x}(t)$ stays close to zero near the terminal time. From a design point of view, the control system designer may design his system so that the term $\mathbf{x}^T Q(t)\mathbf{x}(t)$* is chosen to penalize deviations of the regulated state $\mathbf{x}(t)$ from the desired equilibrium condition $\mathbf{x}(t) = \mathbf{0}$, whereas the term $\mathbf{u}^T R(t)\mathbf{u}(t)$ discourages the use of excessively large control effort.

It must be pointed out, however, that the choice of the matrices $Q(t), R(t),$ and S to minimize the quadratic performance index J is somewhat arbitrary, and must be determined by experimentation. If $R(t) = 0,$ for example, we do not penalize the system for its control-energy expenditure. The optimal control, in this case, will try to bring the state to zero as fast as possible. Normally, the state vector $\mathbf{x}(t)$ represents components of position, velocity, and any other modeling parameters.

In missile interceptor problems, one selects S so that only position errors are weighted. Therefore, the cost functional (5.36) is made up of a quadratic form in the terminal state plus an integral of quadratic forms in the state and control. For example, in the case of an interceptor missile and a target (or evader), the control designer can select the performance index to be

$$J = \frac{1}{2}(\text{miss distance})^2 + \frac{1}{2}\int_{t_0}^{T} u^2 \, dt,$$

where u is the interceptor missile control. Typically, the missile control is the normal acceleration. In pursuit–evasion problems, the interceptor (pursuer) seeks to minimize the time to intercept (or capture), while the target (evader) seeks to maximize the time to capture. In tactical missile intercept (pursuit) applications, it is important that the control designer develop guidance laws, which will provide an effective means to combat high-g maneuvering targets. Thus, the missile–target engagement can be formulated as an optimal control problem, after of course a suitable performance index has been selected.

Optimal guidance laws for accelerating targets depend on adequate estimates of key states, such as line-of-sight (LOS) rate and target accelerations. Consequently, when optimum estimates of these parameters are available, superior performance can be demonstrated. The guidance laws thus developed can be implemented with current state-of-the-art digital hardware. Generally, in optimal deterministic control problems, the optimal control law becomes a function of all state variables. With the above quadratic performance index and linearized missile dynamics, linear–quadratic optimal control theory can be applied to obtain feedback control laws. These control laws are normally modified forms of proportional navigation. Consequently, the exact form of the control depends on the model used for the interceptor missile autopilot and the assumptions made about the target.

---

*Note that the quadratic term $\mathbf{x}^T Q \mathbf{x}$ is of dimension $(1 \times n)(n \times n)(n \times 1) = 1$. Furthermore, if selected values of $\mathbf{x}$ cause $\mathbf{x}^T Q \mathbf{x}$ to be zero, then it is positive or negative semidefinite. Finally, it is useful to note the following:

$$\mathbf{x}^T Q \mathbf{x} = tr(\mathbf{x}^T Q \mathbf{x}) = tr(\mathbf{x}\mathbf{x}^T Q) = tr(Q \mathbf{x}\mathbf{x}^T).$$

Example The above discussion will now be extended to a specific homing missile interceptor application (Figure 5.3a). Consider the linear dynamic system described by

$$\dot{\mathbf{x}}(t) = A(t)\mathbf{x}(t) + B(t)\mathbf{u}(t),$$

where $\mathbf{x}(t)$ is the n-dimensional state vector and $\mathbf{u}(t)$ is an l-dimensional control vector. We wish to determine $\mathbf{u}(t)$ over the interval $0 < t < T$, so as to minimize the quadratic performance index (or *cost function*)

$$J = \mathbf{x}^T(T)S_0\mathbf{x}(T) + \int_0^T (\mathbf{x}^T Q\mathbf{x} + \mathbf{u}^T R\mathbf{u})\,dt, \tag{5.37}$$

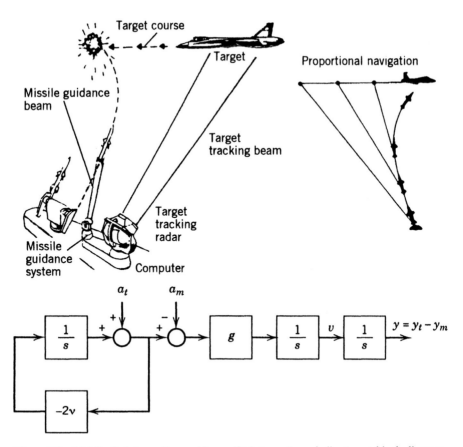

Figure 5.3 (*a*) Missile interception guidance. (*b*) Interceptor-missile–target block diagram.

where the matrices S_0, Q, and R are $n \times n, n \times n$, and $n \times m$, respectively. This is known as the noise-free regulator problem. The optimal control history can be determined as a linear combination of the states of the system. Thus, the control law takes the form

$$\mathbf{u}(t) = -C(t)\mathbf{x}(t),$$

where $C(t)$ is an $m \times n$ control gain matrix. The optimal gains are found by solving a nonlinear matrix Riccati equation backwards in time, that is, starting at the terminal time T. Furthermore, let us define the reverse (or "adjoint") time as $\tau = T - t$. Then the matrix Riccati equation to be solved for a matrix S is [12]

$$\frac{dS}{dt} = SA + A^T S - SBR^{-1}B^T S + Q, \qquad S|_{\tau=0} = S_0, \qquad (5.38)$$

where the optimal control gain matrix is determined from

$$C(t) = -R^{-1}(t)B^T(t)S(t). \qquad (5.39)$$

For this simplified homing problem, the dynamical equations are given by

$$\dot{y} = v,$$

$$\dot{v} = ga_t - ga_m,$$

$$\dot{a}_t = -2va_t,$$

or in state-space notation

$$\frac{d}{dt}\begin{bmatrix} y \\ v \\ a_t \end{bmatrix} = \begin{bmatrix} 0 & 1 & 0 \\ 0 & 0 & g \\ 0 & 0 & -2v \end{bmatrix}\begin{bmatrix} y \\ v \\ a_t \end{bmatrix} + \begin{bmatrix} 0 \\ -g \\ 0 \end{bmatrix}a_m,$$

where y is the miss distance $(y = y_t - y_m)$, g is the acceleration of gravity (9.81 m/s²), a_t is the target acceleration, a_m is the interceptor missile acceleration control signal which is to be determined, v is the frequency, and $2v$ is the target maneuver bandwidth. The target is assumed to perform evasive maneuvers with a correlation time $1/2v$ seconds.

Figure 5.3b illustrates in block diagram form this simplified interceptor missile homing problem. The performance index for this problem is chosen as

$$J = y^2(T) + \gamma \int_0^T a_m^2 dt$$

where $y(T)$ is the miss distance at the intercept time T, and γ is a weighting factor of the acceleration constraints relative to the miss distance constraint. The performance index may be written in canonical form by defining

$$S_0 = \begin{bmatrix} 1 & 0 & 0 \\ 0 & 0 & 0 \\ 0 & 0 & 0 \end{bmatrix}, \qquad Q = \begin{bmatrix} 0 & 0 & 0 \\ 0 & 0 & 0 \\ 0 & 0 & 0 \end{bmatrix}, \qquad R = \gamma.$$

The resulting optimum control law is found to be a proportional navigation scheme that shuts itself off as $t_{go} \to 0$.* An important special case for the target dynamic model is $v = 0$, which corresponds to a constant target maneuver. The Riccati equation can then be solved numerically using a Runge–Kutta integration scheme.

Many of the present-day short-range tactical air-to-air missiles employ proportional navigation (see also Section 4.8) as the guidance law. In fact, Bryson and Ho [12] used optimal control theory to show that proportional navigation is the optimal control law that minimizes the miss distance. The modern short-range air-to-air missile engagement is the most demanding tactical weapon scenario from the point of view of the guidance law, due to short engagement times (typically in the order of 3–4 sec) and drastic changes in the kinematics of the scenario.

Example As a variation of the above example, let us determine the guidance law for a simple short-range air-to-air tactical missile. Furthermore, let the system be described by the equation

$$\dot{\mathbf{x}} = A\mathbf{x} + B\mathbf{u},$$

where $\mathbf{x}$ is the state vector and $\mathbf{u}$ is the control vector. The cost functional to be minimized is given by

$$J(\mathbf{u}) = \mathbf{x}^T(T)S_T\mathbf{x}(T) + \frac{1}{2}\int_{t_0}^{T} \mathbf{u}^T R\mathbf{u}\, dt$$

with

$$S_T = \begin{bmatrix} I & 0 \\ 0 & 0 \end{bmatrix}, \qquad R = \begin{bmatrix} b & 0 & 0 \\ 0 & b & 0 \\ 0 & 0 & b \end{bmatrix},$$

where I is the identity matrix, b is a weighting factor, and T is the terminal or intercept time. The missile and target position, velocity, and acceleration

---

* t_{go} refers to the remaining flight-time the interceptor missile has to hit the target.

vectors, relative to some fixed inertial reference frame, will be given by (r_m, v_m, a_m) and (r_t, v_t, a_t), respectively. Here we will assume that the missile x-axis is pointing along the missile's longitudinal direction, the y-axis normal to the right (both x and y are in the horizontal plane), and the z-axis pointing down. The state vector and control will be defined as follows:

$$\mathbf{x}^T = [x_1 \quad x_2 \quad x_3 \quad x_4 \quad x_5 \quad x_6],$$

$$\mathbf{u} = \begin{bmatrix} -a_{mx} \\ -a_{my} \\ -a_{mz} \end{bmatrix}$$

where $x_1, x_2, x_3 =$ target–missile relative position in the x, y, z directions, respectively $x_1 = r_{tx} - r_{mx}, x_2 = r_{ty} - r_{my}, x_3 = r_{tz} - r_{mz}$,
$x_4, x_5, x_6 =$ target–missile relative velocity in the x, y, z directions, respectively $x_4 = \dot{x}_1 = v_{tx} - v_{mx}, x_5 = \dot{x}_2 = v_{ty} - v_{my}, x_6 = \dot{x}_3 = v_{tz} - v_{mz}$.

From the above definitions, a linear model describing this particular engagement can now be written in the following form:

$$\dot{x}_1 = x_4,$$

$$\dot{x}_2 = x_5,$$

$$\dot{x}_3 = x_6,$$

$$\dot{x}_4 = a_{tx} - a_{mx},$$

$$\dot{x}_5 = a_{ty} - a_{my},$$

$$\dot{x}_6 = a_{tz} - a_{mz}.$$

The Hamiltonian (see following discussion) for this case is given by

$$H = \tfrac{1}{2}\mathbf{u}^T R\mathbf{u} + \mathbf{p}^T(A\mathbf{x} + B\mathbf{u}),$$

where $\mathbf{p}$ is the costate vector $[\mathbf{p}(T) = S_T\mathbf{x}(T)]$. The solution for the control vector is

$$\mathbf{u}(t) = \left(-\frac{3t_{go}}{3b} + t_{go}^3 \right)[\mathbf{I} \quad t_{go}\mathbf{I}]\mathbf{x}(t),$$

where $t_{go} = T - t$. If the target acceleration $a_t = 0$, we can define t_{go} as

$$t_{go} = -\frac{r}{v_c},$$

where v_c is the missile's closing velocity. Using small-angle approximations, the guidance law reduces to

$$u_1 = -a_{mx} = 0,$$

$$u_2 = -a_{my} = 3v_c\dot{\lambda}_r,$$

$$u_3 = -a_{mz} = -3v_c\dot{\lambda}_e,$$

where λ_r is the relative angle from the line of sight (LOS) in the horizontal plane, and λ_e is the elevation relative angle from the LOS. Based on the assumed model, this is the proportional navigation guidance law that minimizes the terminal miss distance. Finally, we note that if the weighting factor b in the matrix R is small, it means that we are willing to expend whatever acceleration is required to minimize the terminal miss distance. (Of course, the real missile must be capable of producing and sustaining such accelerations.) If on the other hand b is large, we shall in effect limit the magnitude of acceleration available to achieve small miss distances. In other words, we are free to choose how much we wish to "pay" for terminal accuracy.

The optimization problem posed above can be solved via the Pontryagin minimum principle (see Section 5.6). The system Hamiltonian can be written as [2, 58]

$$H(\mathbf{x}(t), \mathbf{p}(t), \mathbf{u}(t), t) = \tfrac{1}{2}\mathbf{x}^T(t)Q(t)\mathbf{x}(t) + \tfrac{1}{2}\mathbf{u}^T(t)R(t)\mathbf{u}(t)$$
$$+ \mathbf{p}^T(t)[A(t)\mathbf{x}(t) + B(t)\mathbf{u}(t)], \tag{5.40}$$

where $\mathbf{p}(t)$ is the $n \times 1$ costate (or adjoint) vector. For an optimum control, we rewrite Eq. (5.31a) to designate the optimality condition as

$$\dot{\mathbf{x}}^*(t) = A(t)\mathbf{x}^*(t) + B(t)\mathbf{u}^*(t).$$

Then the minimum principle requires that

$$\dot{\mathbf{p}}^*(t) = -\frac{\partial H}{\partial \mathbf{x}(t)} = -Q(t)\mathbf{x}^*(t) - A^T(t)\mathbf{p}^*(t) \tag{5.41}$$

and

$$0 = \frac{\partial H}{\partial \mathbf{u}(t)} = R(t)\mathbf{u}^*(t) + B^T(t)\mathbf{p}^*(t) \tag{5.42}$$

with the terminal condition

$$\mathbf{p}(T) = Sx^*(T). \tag{5.43}$$

Equation (5.42) is known as the optimality condition, and is the control that minimizes H. The optimal control $\mathbf{u}^*(t)$ can be found in most textbooks on modern control theory [2, 12, 58]. Solving Eq. (5.42), the optimal control is given by

$$\mathbf{u}^*(t) = -R^{-1}(t)B^T(t)\mathbf{p}^*(t). \tag{5.44}$$

The optimal control given by Eq. (5.44) is a feedback form of control and is referred to as the linear–quadratic regulator (LQR).

As indicated earlier, the assumption that $R(t)$ is positive definite for all $t \in [t_0, T]$ guarantees the existence of $R^{-1}(t)$ for all $t \in [t_0, T]$. Substituting Eq. (5.44) into Eq. (5.31a) results in the reduced canonical equation

$$\dot{x}^*(t) = A(t)\mathbf{x}^*(t) - B(t)R^{-1}(t)B^T(t)\mathbf{p}^*(t). \tag{5.45}$$

Combining now the canonical equations (5.41) and (5.45) yields the canonical system

$$\begin{bmatrix} \dot{x}^*(t) \\ \dot{p}^*(t) \end{bmatrix} = \begin{bmatrix} A(t) & -B(t)R^{-1}(t)B^T(t) \\ -Q(t) & -A^T(t) \end{bmatrix} \begin{bmatrix} \mathbf{x}^*(t) \\ \mathbf{p}^*(t) \end{bmatrix} = K(t)\begin{bmatrix} \mathbf{x}^*(t) \\ \mathbf{p}^*(t) \end{bmatrix} \tag{5.46}$$

subject to the boundary conditions

$$\mathbf{x}^*(t_0) = \mathbf{x}_0,$$

$$\mathbf{p}^*(T) = S\mathbf{x}^*(T). \tag{5.47}$$

Equation (5.46) represents a system of $2n$ time-varying homogeneous differential equations. The $2n \times 2n$ canonical matrix $K(t)$ has eigenvalues that are located symmetrically about the imaginary axis. Figure 5.4 illustrates the location of the eigenvalues in the complex plane.

We wish now to obtain a closed-loop control system for Eq. (5.44). In order to do this, we note that the n-dimensional state and costate vectors $\mathbf{x}(t), \mathbf{p}(t)$ can be related by the linear transformation

$$\mathbf{p}^*(t) = P(t)\mathbf{x}^*(t) \tag{5.48}$$

for all $t \in [t_0, T]$. Here $P(t)$, also known as a gain matrix, is an $n \times n$ matrix which is the unique solution to the matrix Riccati differential equation

$$\dot{P}(t) = -P(t)A(t) - A^T(t)P(t) + P(t)B(t)R^{-1}(t)B^T(t)P(t) - Q(t) \tag{5.49}$$

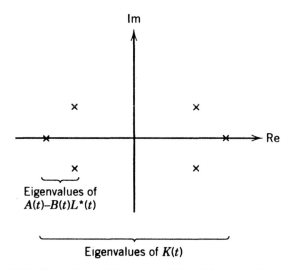

Figure 5.4   Eigenvalues of the optimal closed-loop regulator system.

subject to the boundary condition at the terminal time T

$$P(T) = S. \tag{5.50}$$

It should be noted that $P(t)$ is independent of the state. In general, $P(t)$, which has $n(n+1)/2$ nonzero elements, will be time-varying even if $A(t), B(t), Q(t)$, and $R(t)$ are constant.

Substituting now Eq. (5.48) into Eq. (5.44), the optimal control $\mathbf{u}(t)$ assumes the form

$$\mathbf{u}^*(\mathbf{x}(t), t) = -R^{-1}(t)B^T(t)P(t)\mathbf{x}^*(t) \triangleq -L^*(t)\mathbf{x}(t), \tag{5.51}$$

where $L^*(t)$ is an $m \times n$ time-varying optimal feedback gain matrix and applies for all $\mathbf{x}(t)$. Figure 5.5 depicts in block diagram form the optimal state regulator.

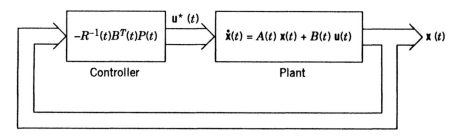

Figure 5.5   The optimal state regulator.

In order to show that $P(t)$ is indeed the unique solution to the matrix Riccati differential equation, we will assume that the terminal time T is free and proceed as follows: First, differentiate Eq. (5.48). This yields

$$\dot{\mathbf{p}}^*(t) = \dot{P}(t)\mathbf{x}^*(t) + P(t)\dot{\mathbf{x}}^*(t).$$

Next, substituting Eq. (5.51) into Eq. (5.36a), we obtain

$$\dot{\mathbf{x}}^*(t) = A(t)\mathbf{x}^*(t) - B(t)R^{-1}(t)B^T(t)\mathbf{p}^*(t)$$

$$= [A(t) - B(t)R^{-1}(t)B^T(t)P(t)]\mathbf{x}^*(t).$$

Similarly,

$$\dot{\mathbf{p}}^*(t) = [-Q(t) - A^T(t)P(t)]\mathbf{x}^*(t).$$

Substituting these equations for $\dot{\mathbf{x}}^*(t)$ and $\dot{\mathbf{p}}^*(t)$ above, we obtain

$$[-Q(t) - A^T(t)P(t)]\mathbf{x}^*(t) = [\dot{P}(t) + P(t)A(t) - P(t)B(t)R^{-1}(t)B^T(t)P(t)]\mathbf{x}^*(t),$$

which yields

$$\dot{P}(t) = -P(t)A(t) - A^T(t)P(t) + P(t)B(t)R^{-1}(t)B^T(t)P(t) - Q(t)$$

with

$$\mathbf{p}^*(T) = P(T)\mathbf{x}^*(T) = S\mathbf{x}^*(T)$$

and $P(T) = S$ as in Eq. (5.50).

Consider next the case when the final time T is zero. That is, $\mathbf{x}(T) = 0$, where 0 is the zero vector. $P(T)$ does not exist, since it would contain infinite-valued elements. For this reason, certain modifications must be made to the above results. Equation (5.49) can be solved off line by integrating backward in time from the terminal time $t = T$ to the initial time $t = t_0$. Having accomplished this, the time-varying gains $L^*(t)$ are then stored on tape in the feedback controller, to be played back upon command in real time. [However, it must be pointed out that Eq. (5.49) is a nonlinear equation and care must be exercised in determining the existence of a unique solution]. Now, substituting Eq. (5.51) into Eq. (5.31a), and assuming $A(t)$ and $B(t)$ are a controllable pair, the optimal closed-loop system is given by the linear differential equation

$$\dot{\mathbf{x}}^*(t) = [A(t) - B(t)L^*(t)]\mathbf{x}^*(t), \tag{5.52}$$

$$\mathbf{x}^*(t_0) = \mathbf{x}_0,$$

where $\mathbf{x}^*(t)$ is the optimal state generating the above optimal system. Equation (5.52) is asymptotically stable in the large. That is, all of the eigenvalues of $A(t) - B(t)L^*(t)$ are in the left half of the complex plane. Reference [41] shows that if the closed-loop system resulting from application of an optimal control law is asymptotically stable, then the matrix Riccati equation whose solution is used to define the optimal control law is computationally stable. That is, as an error propagates, it is attenuated, and there is no possibility of buildup of the error. In block diagram form, the optimal feedback control system is shown in Figure 5.6.

We will now apply Eqs. (5.40)–(5.51) to a specific case. Suppose we are given the linear time-invariant system

$$\dot{\mathbf{x}}(t) = A\mathbf{x}(t) + B\mathbf{u}(t),$$

$$\mathbf{x}(0) = \xi.$$

Futhermore, suppose we are given a function of time $\mathbf{z}(t)$, a state $\boldsymbol{\theta}$, and a final time T such that

$$\mathbf{z}(0) = \xi, \qquad \mathbf{z}(T) = \boldsymbol{\theta}.$$

The function $\mathbf{z}(t)$ can be thought of as defining a desired path from ξ to $\boldsymbol{\theta}$. Our objective here is to force the system from ξ to $\boldsymbol{\theta}$ and make the state $\mathbf{x}(t)$ be near $\mathbf{z}(t)$. We begin our analysis by defining the error between $\mathbf{z}(t)$ and $\mathbf{x}(t)$ as

$$\mathbf{e}(t) = \mathbf{z}(t) - \mathbf{x}(t).$$

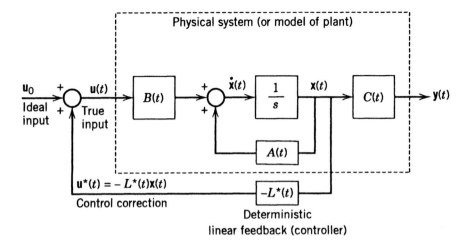

Figure 5.6 Closed-loop realization of the optimal linear state regulator.

Next, we select the cost functional (or performance index) as

$$J(\mathbf{u}) = \frac{1}{2} \int_0^T [\mathbf{e}^T(t)Q\mathbf{e}(t) + \mathbf{u}^T(t)R\mathbf{u}(t)]\, dt,$$

where Q and R are positive definite, constant, symmetric matrices. The final time T will be assumed to be known. From Eq. (5.40), the Hamiltonian in this case is given by

$$H = \tfrac{1}{2}\mathbf{e}^T Q\mathbf{e} + \tfrac{1}{2}\mathbf{u}^T R\mathbf{u} + \mathbf{p}^T[A\mathbf{x} + B\mathbf{u}],$$

while the optimality conditions are (the argument t has been omitted for convenience)

$$\dot{\mathbf{p}} = -\frac{\partial H}{\partial \mathbf{x}} = -Q\mathbf{z} + Q\mathbf{x} + A^T \mathbf{p} = Q(\mathbf{z} - \mathbf{x}) - A^T \mathbf{p},$$

$$\frac{\partial H}{\partial \mathbf{p}} = A\mathbf{x} + B\mathbf{u},$$

$$\frac{\partial H}{\partial \mathbf{u}} = R\mathbf{u} + B^T \mathbf{p} = 0, \qquad \text{or} \quad \mathbf{u} = -R^{-1}B^T \mathbf{p},$$

$$\frac{\partial^2 H}{\partial \mathbf{u}^2} = R.$$

The above results can be put in matrix form as follows:

$$\begin{bmatrix} \dot{\mathbf{x}} \\ \dot{\mathbf{p}} \end{bmatrix} = \begin{bmatrix} A & -BR^{-1}B^T \\ -Q & -A^T \end{bmatrix} \begin{bmatrix} \mathbf{x} \\ \mathbf{p} \end{bmatrix} + \begin{bmatrix} 0 \\ Q\mathbf{z} \end{bmatrix}.$$

Now let

$$\Psi = \begin{bmatrix} \psi_{11}(t) & \psi_{12}(t) \\ \psi_{21}(t) & \psi_{22}(t) \end{bmatrix} = e^{Ct},$$

where

$$C = \begin{bmatrix} A & -BR^{-1}B^T \\ -Q & -A^T \end{bmatrix}.$$

Then

$$\mathbf{x}(T) = \mathbf{0} = [\psi_{11}(T) \quad \psi_{12}(T)] \begin{bmatrix} \xi \\ \mathbf{p}(0) \end{bmatrix} + \int_0^T \psi_{12}(T-r)Q\mathbf{z}(\tau)d\tau.$$

If $\psi_{12}(T)$ is nonsingular,

$$\mathbf{p}(0) = \Psi^{-1}(T) \left[-\int_0^T \psi_{12}(T-r)Q\mathbf{z}(\tau)dr - \psi_{11}(\tau)\xi + \theta \right],$$

so that

$$\mathbf{u}(t) = -BR^{-1}B^T\mathbf{p}$$

$$= -BR^{-1}B^T \left\{ [\psi_{21}(t)\xi + \psi_{22}(t)\mathbf{p}(0)] + \int_0^T \psi_{22}(t-\tau)Q\mathbf{z}(\tau)d\tau \right\}.$$

Therefore, in order to evaluate $\mathbf{u}(t)$, $\mathbf{z}(t)$ must be known for all $t \in [0, T]$.

The proper choice of a cost functional is one of the more difficult tasks in formulating an optimal control problem. Optimal control designers are confronted with the performance problem from the outset—that is, the definition of the constraints under which the control law must be implemented. If the performance index is not chosen carefully, the controlled plant may not respond as well as expected.

Finally, it should be noted that in some instances certain elements of the S-matrix may be large enough to cause computational difficulties. If this happens, it is desirable to obtain an inverse matrix Riccati differential equation. In order to do this, let

$$P(t)P^{-1}(t) = I. \tag{5.53}$$

Differentiating Eq. (5.53), we obtain

$$\dot{P}(t)P^{-1}(t) + P(t)\dot{P}^{-1}(t) = 0. \tag{5.54}$$

The inverse matrix Riccati equation then becomes

$$\dot{P}^{-1}(t) = A(t)P^{-1}(t) + P^{-1}(t)A^T(t) - B(t)R^{-1}(t)B^T(t)$$
$$+ P^{-1}(t)Q(t)P^{-1}(t) \tag{5.55}$$

with

$$P^{-1}(T) = S^{-1}. \tag{5.56}$$

Therefore, it is now possible to solve the matrix Riccati equation with $S^{-1} = [0]$, the null matrix, which requires that every component of the state vector approach the origin as the time approaches the terminal time. The gains $L^*(t)$, or some elements of this matrix, become infinite at the terminal time in this case.

In order to illustrate the applicability of the linear–quadratic-regulator problem, several representative examples have been selected so that the reader can have a better appreciation of the theory presented, and perhaps apply these techniques to his particular problem.

Example 1   Find the optimal control and trajectory which minimize

$$J = \int_0^1 u^2 \, dt$$

for the system (plant)

$$\dot{x} = -x + u$$

with the initial state $x(0) = 1$ and terminal state $x(1) = 0$.

The Hamiltonian of this simple system is

$$H(x, u, p, t) = u^2 + p(-x + u). \tag{1}$$

The optimality condition yields

$$\frac{\partial H}{\partial u} = 0 = 2u + p \quad \Rightarrow \quad u^* = -\frac{p}{2}. \tag{2}$$

The optimal value of H becomes, after substituting Eq. (2) into Eq. (1),

$$H^*(x, p, u^*, t) = \frac{p^2}{4} - px - \frac{p^2}{2} = -px - \frac{p^2}{4}.$$

Now

$$\dot{x} = \frac{\partial H^*}{\partial p} = -x - \frac{p}{2}; \qquad \dot{x} + x = -\frac{p}{2} \;\Rightarrow\; (D+1)x = -\frac{p}{2} \tag{3}$$

$$\dot{p} = \frac{\partial H^*}{\partial x} = p; \qquad \dot{p} - p = 0 \;\Rightarrow\; (D-1)p = 0. \tag{4}$$

From Eq. (3) we have

$$p = -2(D+1)x. \tag{5}$$

Therefore,

$$(D+1)(D-1)x = 0,$$

$$x(t) = K_1 e^t + K_2 e^{-t}.$$

Using the boundary conditions, solve for K_1 and K_2. One therefore obtains

$$x(t) = -0.15652 e^t + 1.1562 e^{-t},$$

$$u^*(t) = -0.31304 e^t.$$

The state feedback (Figure 5.7) is given by

$$u^* = -Kx,$$

$$K = -\frac{u^*}{x} = \frac{0.31304 e^t}{-0.15652 e^t + 1.1562 e^{-t}}.$$

The state and optimal control are plotted in Figure 5.8.

Example 2 (See also example in Section 4.1) Consider the second-order (double-integral plant) system

$$\dot{x}_1 = x_2, \qquad x_1(0) = -1, \tag{1}$$

$$\dot{x}_2 = u, \qquad x_2(0) = 0,$$

and the cost functional

$$J = \frac{1}{2} x_1^2(T) + \frac{1}{2} \int_0^T u^2 \, dt. \tag{2}$$

Determine the optimal control u^* for this system.

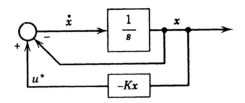

Figure 5.7 State-feedback block diagram. The state and optimal control are plotted in Figure 5.8.

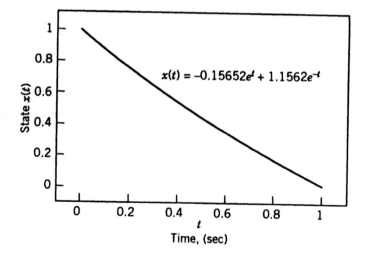

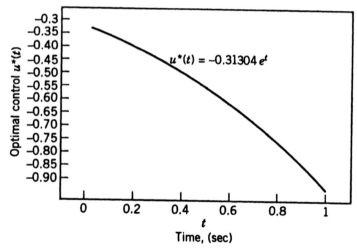

Figure 5.8   State and optimal control plots.

The Hamiltonian of the problem is

$$H = \tfrac{1}{2}u^2 + \lambda_1 x_2 + \lambda_2 u. \tag{3}$$

We have

$$\dot{\lambda}_1 = -\frac{\partial H}{\partial x_1} = 0, \qquad \lambda_1(T) = x_1(T), \tag{4}$$

$$\dot{\lambda}_2 = -\frac{\partial H}{\partial x_2} = -\lambda_1, \qquad \lambda_2(T) = 0. \tag{5}$$

The optimality condition is

$$0 = \frac{\partial H}{\partial u} = u + \lambda_2 \quad \Rightarrow \quad u^* = -\lambda_2. \tag{6}$$

Now,

$$\begin{bmatrix} \dot{x}_1 \\ \dot{x}_2 \\ \dot{\lambda}_1 \\ \dot{\lambda}_2 \end{bmatrix} = \begin{bmatrix} 0 & 1 & 0 & 0 \\ 0 & 0 & 0 & -1 \\ 0 & 0 & 0 & 0 \\ 0 & 0 & -1 & 0 \end{bmatrix} \begin{bmatrix} x_1 \\ x_2 \\ \lambda_1 \\ \lambda_2 \end{bmatrix}, \qquad \begin{cases} x_1(0) = -1, \\ x_2(0) = 0, \\ \lambda_1(T) = x_1(T), \\ \lambda_2(T) = 0. \end{cases} \tag{7}$$

Next, we set

$$\lambda(t) = K(t)\mathbf{x}(t), \tag{8}$$

where

$$K(t) = \begin{bmatrix} 1 & 0 \\ 0 & 0 \end{bmatrix}. \tag{9}$$

Note that here we have used $\lambda(t)$ for $\mathbf{p}(t)$, and $K(t)$ for S. The matrix Riccati equation for this system can be obtained from

$$\dot{\mathbf{x}} = A\mathbf{x} + B\mathbf{u}. \tag{10}$$

The cost functional becomes

$$J = \mathbf{x}^T(T)K\mathbf{x}(T) + \frac{1}{2}\int_0^T (\mathbf{x}^T Q\mathbf{x} + \mathbf{u}^T R\mathbf{u})\, dt. \tag{11}$$

Therefore, the matrix Riccati equation takes the form

$$\dot{K} = -KA - A^T K + KBR^{-1}B^T K - Q, \tag{12}$$

and the optimal control by

$$u^* = -R^{-1}B^T K\mathbf{x}. \tag{13}$$

From the system (1) we have the relations

$$A = \begin{bmatrix} 0 & 1 \\ 0 & 0 \end{bmatrix}, \qquad B = \begin{bmatrix} 0 \\ 1 \end{bmatrix}, \qquad \begin{bmatrix} 0 & 0 \\ 0 & 0 \end{bmatrix}, \qquad R = 1, \qquad K(T) = \begin{bmatrix} 1 & 0 \\ 0 & 0 \end{bmatrix}.$$

Since $u^* = -\lambda_2$, we obtain after some algebra

$$u^*(t) = \frac{3(T-t)}{3+T^3}. \tag{14}$$

Next, we have $\dot{x}_2(t) = u(t)$. Integrating, we obtain an expression for the velocity as follows:

$$x_2(t) = \left(\frac{3T}{3+T^3}\right)t - \left(\frac{3}{3+T^3}\right)\frac{t^2}{2}. \tag{15}$$

The position is obtained by integrating $\dot{x}_1(t) = x_2$:

$$x_1(t) = \left(\frac{3T}{3+T^3}\right)\frac{t^2}{2} - \left(\frac{3}{3+T^3}\right)\frac{t^3}{6} - 1. \tag{16}$$

The double-integral plant can be represented diagramatically as in Figure 5.9. In a typical interceptor guidance problem, the parameters x_1, x_2, and u will represent the interceptor missile parameters:

x_1 = displacement,
x_2 = velocity,
u = $\dot{x}_2$ = acceleration.

Figures 5.10 and 5.11 show the histories of the position (trajectory), velocity, and optimal control respectively.

The double-integral plant is very important in the study of optimal control systems.

Example 3   Minimize the performance index

$$J(u) = \int_0^T [x(t)^2 + u(t)^2]\, dt$$

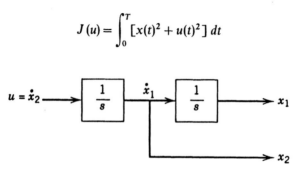

Figure 5.9   Simple double–integral plant.

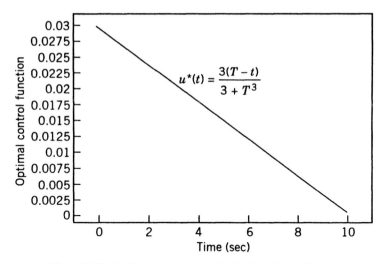

Figure 5.10 Optimal control of the double–integral plant.

subject to the constraints

(a) $\dot{x} = x + u$;
(b) $x(0) = x_0$, $x(T) = 0$;
(c) no control constraints.

(a) Forming the Hamiltonian, we obtain

$$H(x, u, p, t) = x^2 + u^2 + p(x + u).$$

The optimality condition yields

$$0 = \frac{\partial H(x, u, p, t)}{\partial u} = p + 2u = 0 \quad \Rightarrow \quad u^* = -\frac{p}{2}.$$

The optimal value for H becomes

$$H^*(x, u, p, t) = x^2 + \left(-\frac{p}{2}\right)^2 + p\left(x - \frac{p}{2}\right)$$

$$= px + x^2 - \frac{p^2}{4}.$$

Now,

$$\dot{x} = \frac{\partial H^*(x, p, t)}{\partial p} = x - \frac{p}{2}$$

$$\dot{p} = \frac{\partial H^*(x, p, t)}{\partial x} = -p - 2x.$$

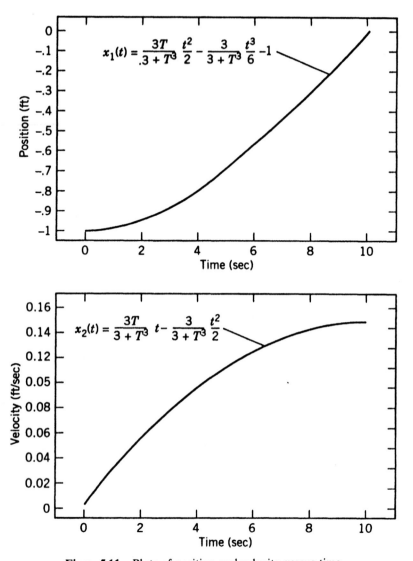

Figure 5.11 Plots of position and velocity versus time.

The state is obtained as

$$x(t) = K_1 e^{\sqrt{2}t} + K_2 e^{-\sqrt{2}t}.$$

(b) For the conditions $x(0)$ and $x(T) = 0$, we obtain

$$x(t) = \frac{x_0}{e^{-\sqrt{2}T} - e^{\sqrt{2}T}} \left[e^{\sqrt{2}(t-T)} - e^{-\sqrt{2}(t-T)} \right],$$

$$p(t) = \frac{2x_0}{e^{-\sqrt{2}t} - e^{\sqrt{2}t}} [(1 - \sqrt{2})e^{\sqrt{2}t} - (1 + \sqrt{2})e^{-\sqrt{2}t}],$$

$$u^*(t) = \frac{-x_0}{e^{-\sqrt{2}T} - e^{\sqrt{2}T}} [(1 - \sqrt{2})e^{\sqrt{2}t} - (1 + \sqrt{2})e^{-\sqrt{2}t}].$$

(c) The optimal control is a feedback control in that $u(t)$ is completely determined by the initial state x_0. The plots for the state $x(t)$ and optimal control $u^*(t)$ are shown in Figure 5.12. (*Note*: The graphs are for $T = 1$ and $x_0 > 0$.)

5.4 THE DISCRETE-TIME LINEAR–QUADRATIC REGULATOR

Up to now we have considered the continuous-time linear regulator problem. In this section we will consider the discrete-time linear plant described by the n-dimensional equation

$$\mathbf{x}(k + 1) = A(k)\mathbf{x}(k) + B(k)\mathbf{u}(k), \qquad k = 0, 1, 2, \ldots, N, \qquad (5.57)$$

$$\mathbf{x}(0) = \mathbf{x}_0$$

at time k. This equation is the discrete-time counterpart of Eq. (5.31a). The problem here is to find, as before, the optimal control $\mathbf{u}^*(\mathbf{x}(k), k)$ that

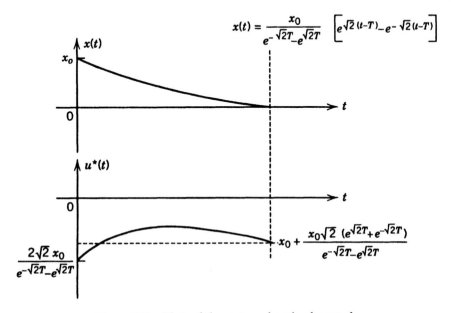

Figure 5.12 Plots of the state and optimal control.

minimizes the quadratic performance index

$$J = \tfrac{1}{2} \mathbf{x}^T(N)S\mathbf{x}(N) + \tfrac{1}{2} \sum_{k=0}^{N-1} [\mathbf{x}^T(k)Q(k)\mathbf{x}(k) + \mathbf{u}^T(k)R(k)\mathbf{u}(k)], \quad (5.58)$$

where S and $Q(k)$ are real symmetric positive semidefinite matrices, $R(k)$ is a real symmetric positive definite matrix, and N is a fixed integer $(N > 0)$. The corresponding Hamiltonian is

$$H(k) = \tfrac{1}{2}\mathbf{x}^T(k)Q(k)\mathbf{x}(k) + \tfrac{1}{2}\mathbf{u}^T(k)R(k)\mathbf{u}(k) + \mathbf{p}^T[A(k)\mathbf{x}(k) + B(k)\mathbf{u}(k)],$$

where the matrices $Q(k)$ and $R(k)$ are functions of the stage k. As before, the weighting matrices $Q(k)$ and $R(k)$ represent the individual component weights on the state and control, respectively, over the sampling period. For most problems of interest, the sampling interval is constant.

Without loss of generality and for the sake of notational simplification, assume that the matrices $A, B, Q,$ and R are constant. Making use of Eq. (5.50), define

$$J_{NN}(\mathbf{x}(N)) = \tfrac{1}{2}\mathbf{x}^T(N)S\mathbf{x}(N) = J^*{}_{NN}\mathbf{x}(N)) \triangleq \tfrac{1}{2}\mathbf{x}^T(N)P(0)\mathbf{x}(N), \quad (5.59)$$

where $P(0) \triangleq S$. The performance index over the final interval $N-1$ can be expressed as

$$J_{N-1,N}(\mathbf{x}(N-1), \mathbf{u}(N-1)) = \tfrac{1}{2}\mathbf{x}^T(N)P(0)\mathbf{x}(N) + \tfrac{1}{2}\mathbf{x}^T(N-1)Q\mathbf{x}(N-1)$$
$$+ \tfrac{1}{2}\mathbf{u}^T(N-1)R\mathbf{u}(N-1), \quad (5.60)$$

and the minimum performance index by

$$J^*_{N-1,N}(\mathbf{x}(N-1)) \triangleq \min_{\mathbf{u}(N-1)} \{J(\mathbf{x}(N-1), \mathbf{u}(N-1))\}. \quad (5.61)$$

From the above, we note that the state $\mathbf{x}(N)$ is related to the control $\mathbf{u}(N-1)$ by the equation

$$J^*_{N-1,N}(\mathbf{x}(N-1)) = \min_{\mathbf{u}(N-1)} \{\tfrac{1}{2}\mathbf{x}^T(N-1)Q\mathbf{x}(N-1) + \tfrac{1}{2}\mathbf{u}^T(N-1)R\mathbf{u}(N-1)$$
$$+ \tfrac{1}{2}[A\mathbf{x}(N-1) + B\mathbf{u}(N-1)]^T P(0)[A\mathbf{x}(N-1)$$
$$+ B\mathbf{u}(N-1)]\}. \quad (5.62)$$

Assuming now that the admissible controls are not bounded, we can minimize $J_{N-1,N}$ with respect to $\mathbf{u}(N-1)$ to obtain a stationary point:

$$\frac{\partial J_{N-1,N}}{\partial \mathbf{u}(N-1)} = \mathbf{0}. \quad (5.63)$$

Performing the operation indicated by Eq. (5.63), we obtain

$$R\mathbf{u}(N-1) + B^T P(0)[A\mathbf{x}(N-1) + B\mathbf{u}(N-1)] = [0] \tag{5.64}$$

where $[0]$ is the null matrix. In Eq. (5.64), the control that satisfies this equation may yield a minimum, a maximum, or neither. For this reason, we must look for the second partial derivatives. Taking the second partial derivatives of Eq. (5.64) yields

$$\frac{\partial^2 J_{N-1,N}}{\partial \mathbf{u}^2(N-1)} = R + B^T P(0)B. \tag{5.65}$$

Since $J_{N-1,N}$ is a quadratic function of $\mathbf{u}(N-1)$, the control that satisfies Eq. (5.65) yields the absolute or global minimum $J_{N-1,N}$. Solving Eq. (5.65), the optimal control becomes

$$\mathbf{u}(N-1) = -[R + B^T P(0)B]^{-1} B^T P(0)A\mathbf{x}(N-1) \triangleq F(N-1)\mathbf{x}(N-1). \tag{5.65a}$$

Substituting Eq. (5.65a) into Eq. (5.60) gives $J^*_{N-1,N}$ in the form

$$J^*_{N-1,N}(\mathbf{x}(N-1)) = \tfrac{1}{2}\mathbf{x}^T(N-1)\{[A + BF(N-1)]^T P(0)[A + BF(N-1)]$$
$$+ F^T(N-1)RF(N-1) + Q\}\mathbf{x}(N-1). \tag{5.66}$$

From the above, the results can be summarized for the general, time-varying kth stage case as follows:

Optimal Control Law:

$$\mathbf{u}^*(N-K) = -[R(N-K) + B^T(N-K)P(K-1)B(N-K)]^{-1} B^T$$

$$(N-K)P(K-1)A(N-K)\mathbf{x}(N-K) \tag{5.67}$$

$$\triangleq F(N-K)\mathbf{x}(N-K), \qquad K = 1, 2, \ldots, N.$$

Optimal Cost Functional:

$$J^*_{N-K,N}(\mathbf{x}(N-K)) = \tfrac{1}{2}\mathbf{x}^T(N-K)\{[A(N-K) + B(N-K)F(N-K)]^T P(K-1)$$

$$[A(N-k) + B(N-K)F(N-K)] + F^T(N-K)$$

$$R(N-K)F(N-K) + Q(N-K)\}\mathbf{x}(N-K)$$
$$\triangleq \tfrac{1}{2}\mathbf{x}^T(N-K)P(K)\mathbf{x}(N-K). \tag{5.68}$$

The equation for the optimal control, Eq. (5.67), implies that the optimal control at each stage is a linear combination of the states. Since the calculations are usually done in a digital computer, it is conceivable that the discrete version of the optimal regulator problem would be computationally more efficient as well as more accurate. Moreover, in solving the discrete-time linear regulator problem in a digital computer, the control designer may reduce the number of arithmetic operations by making certain substitutions. For example, one may let the matrix $F(N-K)$ be a constant matrix as $N \to \infty$, which means that for a large number of stages, the optimal control can be implemented by feedback of the states.

5.5 OPTIMAL LINEAR–QUADRATIC GAUSSIAN REGULATORS

5.5.1 Introduction

In order to implement the deterministic LQR discussed in Section 5.3, it is necessary to measure exactly all the states of the system under design. This is obviously not possible in the real world. What one can measure is outputs, by means of sensors in the system. These sensors (e.g., a missile seeker) have noise associated with them, which means that the measurements are not perfect. For example, although the deterministic approach admits errors in modeling via feedback, it does not take into account the many disturbances, such as wind gusts acting on an aircraft (the plant) or target glint noise present in the angle tracking signal of an interceptor missile. Consequently, in the presence of uncertainties and plant disturbances, we must invoke the stochastic estimation theory, or Kalman–Bucy filter. In addition, real systems will always have some type of noises or biases affecting them, thereby corrupting the state equations. Therefore, when noise is a significant part of the system, a linear–quadratic Gaussian (LQG) regulator must be used. Feedback controllers built to stabilize systems must be able to keep the overall system stable in the presence of external disturbances, modeling errors, and changes in the operating environment.

In this section, we will consider the continuous-time LQG regulator formulation for the design of controllers when uncertainties in the state, input, and measurement matrices are present. Many engineering applications require that the controller be *robust*.* Robustness implies that the controller provides adequate (i.e., stable closed-loop) performance in the face of uncertainties and over a wide range of operating conditions and system parameters. Therefore, a very important requirement of, say, a flight control system design is that it be

---

* In Kalman filter, *robustness* is the ability of the filter to cope with adverse environments and input conditions.

robust. Moreover, robustness becomes a concern when controllers are designed using a reduced-order model. A robust controller will provide stable closed-loop performance even when states of the real-world system have been ignored in the design model. In particular, the LQR has desirable robustness properties in its guaranteed gain and phase margins of at least –6 dB to ∞ and of ± 60°, respectively.

In the 1960s, modern optimal control theory showed promise in application to flight control problems. One disadvantage in such an application is that the resulting controllers require full-state feedback. Since, as was stated above, the measurements of all states are generally not available, it becomes necessary to include a filter or observer (the Kalman filter is an observer) in the controlled system to estimate the states. That is, we need a method to reconstruct the state equations and produce and estimate them, using noisy measurements and allowing for process noise entering the plant.

5.5.2 Properties of the LQG Regulator

We will now present the equations used in designing optimal linear–quadratic Gaussian controllers for systems that are modeled as linear, time-invariant systems driven by zero-mean, Gaussian white noise, subject to quadratic costs for defining optimality criteria. Consider the continuous-time stochastic linear system model governed by the known linear differential equation [2, 21, 28]

$$\dot{\mathbf{x}}(t) = A\mathbf{x}(t) + B\mathbf{u}(t) + G\xi(t), \tag{5.69}$$

$$\mathbf{y}(t) = C\mathbf{x}(t) + \mathbf{n}(t), \tag{5.70}$$

where

$A = n \times n$ plant matrix,

$B = n \times m$ control input matrix,

$G = n \times p$ disturbance input matrix,

$C = r \times n$ observation matrix,

$\mathbf{x}(t) = n \times 1$ state vector $[x(t) \in R^n]$

$\mathbf{y}(t) = r \times 1$ observation vector $[y(t) \in R^r]$

$\mathbf{u}(t) = m \times 1$ control vector $[u(t) \in R^m]$,

$\xi(t) = p \times 1$ random process noise vector,

$\mathbf{n}(t) = r \times 1$ observation noise.

The matrices $A, B, G,$ and C are time-invariant. The processes $\xi(t)$ and $\mathbf{n}(t)$ are assumed to be zero-mean, uncorrelated (that is, uncorrelated with each other and $\mathbf{x}_0$), Gaussian, white noises, with statistics

$$\mathscr{E}\{\mathbf{x}(t_0)\} = \mathbf{0}, \tag{5.71a}$$

$$\mathscr{E}\{\xi(t)\} = \mathscr{E}\{\mathbf{n}(t)\} = 0 \qquad \forall t, \tag{5.71b}$$

$$\mathscr{E}\{\xi(t)\xi^T(\xi)\} = Q_0(t)\,\delta(t-\tau) \tag{5.72a}$$

$$\mathscr{E}\{\mathbf{n}(t)\mathbf{n}^T(\tau)\} = R_f(t)\,\delta(t-\tau) \quad \left.\right\} \; \forall t,\tau, \tag{5.72b}$$

$$\mathscr{E}\{\xi(t)\mathbf{n}^T(\tau)\} = 0 \tag{5.73}$$

where $Q_0(t)$ and $R_f(t)$ are symmetric, positive semidefinite, and positive definite matrices, [i.e., $Q(t) = Q^T(t) \geqslant 0$, $R(t) = R^T(t) > 0$] and $\delta(t)$ is the Dirac delta function. If the system (5.69), (5.70) is controllable and observable, that is,

$$\text{rank } [B, AB, ..., A^n B] = n,$$

$$\text{rank } [C^T, C^T A^T, ..., C^T A^{nT}] = n,$$

and if a steady state is reached, then the optimal control system is time-invariant. Furthermore, the following assumptions about this system hold:

1. The matrix A is the infinitesimal generator of a strongly continuous semigroup T_t on a real Hilbert space H.
2. $B \in L\{R^n, H\}$.
3. $C \in L\{H, R^n\}$.
4. $G \in L\{Y, H\}$, where Y is a real Hilbert space. The input and output spaces are assumed to be real, since the LQG theory is developed for that type of space. (It should be noted that most physical systems are well modeled on real spaces.)

Next, we wish to produce an estimate, $\hat{\mathbf{x}}(T)$, of the state $\mathbf{x}(T)$, at times $T > t_0$, using only the noisy measurement data $\{\mathbf{y}(t): t_0 < t < T\}$. This can be done by forming the state error vector

$$\mathbf{e}(t) = \mathbf{x}(t) - \hat{\mathbf{x}}(t) \tag{5.74}$$

and minimizing the mean-square error

$$\mathbf{e}(T) = \mathscr{E}\{\|\mathbf{x}(t) - \hat{\mathbf{x}}(t)\|^2\}$$

$$= \mathscr{E}\{\mathbf{e}^T(t)\mathbf{e}(t)\}. \tag{5.75}$$

For the time-invariant case, we must assume that $\xi(t)$ and $\mathbf{n}(t)$ are wide-sense stationary, so that the matrices $Q_0(t)$ and $R_f(t)$ become constant matrices. Also, we must assume that the observation of the output begins at $t_0 = -\infty$. If the observation time is long compared to the dominant time constants of the system, this assumption is reasonably valid. Furthermore, we will assume that our estimator takes the form of an observer, given by

$$\hat{\mathbf{x}}(t) = A\,\hat{\mathbf{x}}(t) + B\mathbf{u}(t) + K_f\,[\mathbf{y}(t) - C\hat{\mathbf{x}}(t)]. \qquad (5.76)$$

The Kalman filter gain matrix K_f which minimizes Eq. (5.75) is given by

$$K_f = V_e C^T R_f^{-1}, \qquad (5.77)$$

where V_e is the variance of the error [which, under assumptions, is constant, since $e(t)$ is also stationary] and is found by solving the algebraic variance Riccati equation

$$0 = AV_e + V_e A^T + Q_f - V_e C^T R_f^{-1} C V_e \qquad (5.78)$$

with

$$Q_f = GQ_0 G^T. \qquad (5.79)$$

Notice that if $G = I$, that is, each state has its own distinct process noise, then $Q_f = Q_0$.

The algebraic variance (filter) Riccati equation shown in Eq. (5.78) has several solutions. The "correct" solution is unique and positive definite. A sufficient condition for V_e to exist to $t_0 \to -\infty$ is that the pair $[A, C]$ is completely observable. This condition may be relaxed to detectability, in which case it is necessary and sufficient, and V_e may be positive semidefinite. Assuming that V_e exists, the error dynamics of the filter are as follows [41]:

$$\dot{\mathbf{e}}(t) = \dot{\mathbf{x}}(t) - \dot{\hat{\mathbf{x}}}(t)$$

$$= [A\mathbf{x}(t) + B\mathbf{u}(t) + G\xi(t)] - [A\hat{\mathbf{x}}(t) + B\mathbf{u}(t) + K_f\{\mathbf{y}(t) - C\hat{\mathbf{x}}(t)\}]$$

$$= (A - K_f C)\mathbf{e}(t) + [G \quad -K_f]\begin{bmatrix} \xi(t) \\ \mathbf{n}(t) \end{bmatrix}. \qquad (5.80)$$

Therefore, the poles of $[A - K_f C]$ are the poles of the filter. Obviously, these poles must be stable, or the filter will fail to estimate the states (i.e., the error must become small, not infinity). A sufficient condition for the filter to be asymptotically stable is that the pair $[A, G]$ is completely controllable.

Next, we will investigate the LQG compensator and determine its closed-loop stability characteristics. Again, let the system be described by the following equations:

$$\dot{\mathbf{x}}(t) = A\mathbf{x}(t) + B\mathbf{u}(t) + G\boldsymbol{\xi}(t), \tag{5.81}$$

$$\mathbf{y}(t) = C\mathbf{x}(t) + \mathbf{n}(t), \tag{5.82}$$

$$\mathbf{z}(t) = H\mathbf{x}(t), \tag{5.83}$$

where the statistics are the same as before. Here $\mathbf{z}(t)$ is defined as the system response equation with H being a matrix such that $H^T H = Q_c$. We now wish to find a control law of the form

$$\mathbf{u}(t) = \mathbf{f}\,[\mathbf{y}(\tau), \tau \leqslant t] \tag{5.84}$$

to minimize the criterion

$$J = \mathscr{E}\left\{ \lim_{T\to\infty} \frac{1}{T} \int_0^T [\mathbf{z}^T(t)\mathbf{z}(t) + \mathbf{u}^T(t)R_c\mathbf{u}(t)]\, dt \right\}. \tag{5.85}$$

In order to do this, we must have $[A, B]$ and $[A, G]$ stabilizable as well as $[A, C]$ and $[A, H]$ detectable. Furthermore, R_c must be symmetric positive definite.

The control law that minimizes Eq. (5.85) is given by [1]

$$\mathbf{u}(t) = -K_c\hat{\mathbf{x}}(t), \tag{5.86}$$

where K_c is the regulator gain matrix given by

$$K_c = R_c^{-1}B^T P, \tag{5.87}$$

and $\hat{\mathbf{x}}(t)$ is the current estimate of the state $\mathbf{x}(t)$, based on measurements of $\mathbf{y}(\tau)$, $\tau \leqslant t$. In Eq. (5.87), P is a constant, symmetric positive semidefinite matrix which is the solution to the algebraic Riccati equation [compare with Eq. (5.78)]

$$A^T P + PA - PBR_c^{-1}B^T P + Q_c = 0, \tag{5.88}$$

where $Q_c = H^T H$. The estimate is defined in the usual manner by the Kalman filter

$$\dot{\hat{\mathbf{x}}}(t) = A\hat{\mathbf{x}}(t) + B\mathbf{u}(t) + K_f[\mathbf{y}(t) - C\hat{\mathbf{x}}(t)], \tag{5.89}$$

where the Kalman filter gain matrix is given by

$$K_f = V_e C^T R_f^{-1}. \tag{5.90}$$

Also, the following equations are valid:

$$AV_e + V_e A^T + Q_f - V_e C^T R_f^{-1} C V_e = 0, \tag{5.91}$$

$$Q_f = G Q_0 G^T. \tag{5.92}$$

Under these conditions, the regulator poles given by

$$\det(sI - A + BK_c) = 0 \tag{5.93}$$

and the filter poles given by

$$\det(sI - A + K_f C) = 0 \tag{5.94}$$

are guaranteed to be stable.

We will now investigate the expression for the LQG compensator, that is, the dynamic output feedback compensator made up of the regulator and filter equations. Reference [1] gives an excellent account of the LQG compensator. Substituting Eq. (5.86) into Eq. (5.89), we obtain

$$\hat{x}(t) = A\hat{x}(t) - BK_c\hat{x}(t) + K_f y(t) - K_f C\hat{x}(t)$$

$$= (A - BK_c - K_f C)\hat{x}(t) + K_f y(t). \tag{5.95}$$

Taking the Laplace transforms and rearranging yields

$$x(s) = (sI - A + BK_c + K_f C)^{-1} K_f y(s). \tag{5.96}$$

Substituting Eq. (5.96) into the Laplace transform of Eq. (5.86) yields

$$u(s) = -K_c(sI - A + BK_c + K_f C)^{-1} K_f y(s). \tag{5.97}$$

This is the expression for the LQG compensator. Omitting for the moment the noises in the plant, we can write the plant transfer function as

$$y(s) = C(sI - A)^{-1} Bu(s). \tag{5.98}$$

In block diagram form, the compensator and plant are illustrated in Figure 5.13. The poles of the compensator are given by

$$\det(sI - A + BK_c + K_f C) = 0 \tag{5.99}$$

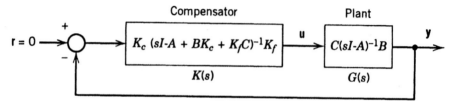

Figure 5.13   Block diagram of the LQG compensator.

and are not always stable. However, it can be shown that the closed-loop system is indeed guaranteed to be stable, which is the crucial requirement.

Closed-loop stability can be determined by looking at the eigenvalues of the closed-loop system. Combining Eqs. (5.81) and Eq. (5.86) gives [26]

$$\dot{x}(t) = Ax(t) - BK_c x(t) + G\xi(t), \tag{5.100}$$

and the combination of Eq. (5.82) and Eq. (5.96) gives

$$\dot{\hat{x}}(t) = [A - BK_c - K_f C]\hat{x}(t) + K_f Cx(t) + K_f n(t). \tag{5.101}$$

Equations (5.100) and (5.101) can be combined as follows:

$$\begin{bmatrix} \dot{x}(t) \\ \dot{\hat{x}}(t) \end{bmatrix} = \begin{bmatrix} A & -BK_c \\ K_f C & A - BK_c - K_f C \end{bmatrix} \begin{bmatrix} x(t) \\ \hat{x}(t) \end{bmatrix} + \begin{bmatrix} G & 0 \\ 0 & K_f \end{bmatrix} \begin{bmatrix} \xi(t) \\ n(t) \end{bmatrix}. \tag{5.102}$$

The closed-loop LQG poles are given by

$$\det \begin{bmatrix} sI - A & BK_c \\ K_f C & sI - A + BK_c + K_f C \end{bmatrix} = 0. \tag{5.103}$$

However, the stability of the poles is not always obvious from Eq. (5.103). Making some substitutions and rearranging, we can write Eq. (5.102) as [26]

$$\begin{bmatrix} \dot{x}(t) \\ \dot{e}(t) \end{bmatrix} = \begin{bmatrix} A - BK_c & BK_c \\ 0 & A - K_f C \end{bmatrix} \begin{bmatrix} x(t) \\ e(t) \end{bmatrix} + \begin{bmatrix} G & 0 \\ G & -K_f \end{bmatrix} \begin{bmatrix} \xi(t) \\ n(t) \end{bmatrix}, \tag{5.104}$$

so that the closed-loop poles of the LQG system are given by

$$\det \begin{bmatrix} sI - A + BK_c & -BK_c \\ 0 & sI - A + K_f C \end{bmatrix} = 0. \tag{5.105}$$

Using Schur's formula, Eq. (5.105) can be written as

$$\det(sI - A + BK_c) \det(sI + A + K_F C) = 0. \tag{5.106}$$

Therefore, the closed-loop poles of the overall LQG system are simply the poles of the regulator and the poles of the filter, which we have already seen are guaranteed to be stable. Table 5.1 summarizes the four set of poles (λ_i[ ] is the ith eigenvalue of the matrix).

In the recent years, a method known as linear–quadratic-Gaussian loop-transfer recovery (LQG LTR) has been developed in order to improve the stability of finite dimensional systems. In particular, this method was developed to demonstrate that arbitrary LQG designs do not possess the guaranteed robustness properties associated with the linear–quadratic regulators. The goal in the LQG LTR technique is to change the loop transfer function of the closed-loop system with an LQG controller at a specified loop breaking point in such a way that, asymptotically, the loop transfer function approaches a full-state LQ transfer function. This can also be accomplished by the duality principle, that is, a Kalman filter loop transfer function. The LQG LTR method uses both the frequency and time-domain concepts: the performance and robustness requirements in this method are specified in the frequency domain, but the majority of the calculations are done in the time domain.

In order to illustrate these ideas, let us consider the closed-loop system depicted in Figure 5.14a, which uses an LQG estimate-state feedback controller to stabilize the plant. It illustrates the structure for a full-state feedback controller and an observer-based controller. Let $\Phi = (sI - A)^{-1}$. Therefore, the loop transfer functions G_1 through G_4 resulting from opening the loop at the four break points are as follows [28, 66]:

Point 1: $G_1 = C\Phi BK_c(\Phi^{-1} + BK_c + K_f C)^{-1}K_f$

Point 2: $G_2 = (C\Phi BK_c + C)(\Phi^{-1} + BK_c)^{-1}K_f = C\Phi K_f$. This is the full-state Kalman filter transfer function.

Point 3: $G_3 = K_c(\Phi^{-1} + K_f C)^{-1}(K_f C\Phi B + B)K_f = K_c\Phi B$. This is the loop transfer function of the LQG regulator.

Point 4: $G_4 = K_c(\Phi^{-1} + BK_c + K_f C)^{-1}K_f C\Phi B$.

Table 5.1 Closed-Loop Poles for the Various Configurations

Configuration	Type of Poles	Remarks
Regulator	$\lambda_i[A - BK_c]$	Always stable
Filter	$\lambda_i[A - K_f C]$	Always stable
Compensator	$\lambda_i[A - BK_c - K_f C]$	Not always stable
Closed loop	$\lambda_i[A - BK_c],$	Always stable
	$\lambda_i[A - K_f C]$	

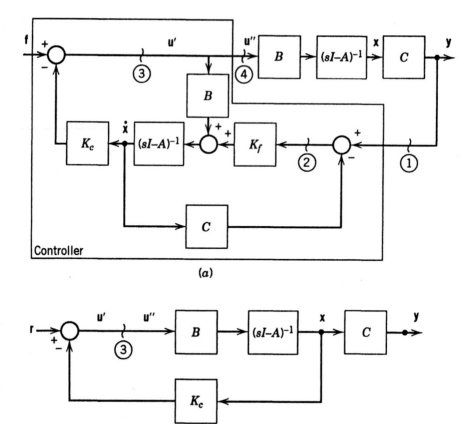

Figure 5.14 (*a*) Observer-based feedback configuration; (*b*) Full-state feedback controller.

It should be remembered that in calculating loop transfer functions, all external inputs are assumed to be zero.

From a physical point of view, points 1 and 4 are the most significant, since at these points the controller interfaces with the real-world system. Therefore, it is desirable to be able to cause the loop transfer functions at points 1 and/or 4 to be the same as a full-state loop transfer function. The value of the LQG LTR method is that it provides a way to make the loop transfer functions at points 1 and 4 look more like full-state loop transfer functions with the loop broken at either point 1 or point 4. Moreover, this method can be used to cause the loop transfer function to satisfy asymptotically the expression

$$C\Phi BK_c(\Phi^{-1} + BK_c + K_fC)^{-1}K_f = C\Phi K_f$$

at point 1, or

$$K_c(\Phi^{-1} + BK_c + K_fC)^{-1}K_fC\Phi B = K_c\Phi B$$

at point 4. In general, however, the loop transfer functions cannot be equated simultaneously at points 1 and 4. In this case, the control designer must first decide where to model the system uncertainties and then decide which method to use.

The loop transfer function at point 3 in Figure 5.14b is given by $\mathbf{u}' = K_c \Phi B \mathbf{u}''$, while the corresponding loop transfer function at point 4 is given by $\mathbf{u}' = K_c \hat{\mathbf{x}}$. Therefore, the loop transfer function at point 4 for the system in Figure 5.14a identical to the full-state LQ loop transfer function at point 3 in Figure 5.14b if

$$K_f[I + C(sI - A)^{-1}K_f]^{-1} = B[C(sI - A)^{-1}B]^{-1}$$

for all s. Reference [22] shows that if the corresponding return-difference mappings are asymptotically equal when the control loops are broken at point 4 (the point of entry of the control inputs), then the robustness properties of the observer-based controller will asymptotically approach those of all the full-state feedback controller.

Optimal regulators find use in missile guidance, aircraft performance, flight controllers, and satellite stabilization systems. Future advanced tactical fighter aircraft will be required to implement specialized control modes for execution of a particular mission. One of these control modes might be pitch pointing during, say, loft weapon delivery, in which the pitch attitude is the commanded variable while the flight-path angle is maintained constant (note that in this case, the corresponding variation in the angle of attack must likewise be compensated). This problem can be formulated as an optimal linear–quadratic regulator and synthesized by a linear state feedback control which minimizes a function of the error between the actual and ideal system responses. Another application is the thrust magnitude control capability of an air-to-air missile. Such a capability can, for example, be provided by allowing a variable fuel flow rate in a ramjet or liquid rocket engine.

5.6 PONTRYAGIN'S MINIMUM PRINCIPLE

Pontryagin's minimum principle is presented here in abridged form. The minimum principle is useful to the engineer because of the universality of its formulation and application to numerous physical problems. Reference [55] gives a rigorous proof of the minimum principle, while reference [2] provides a heuristic approach to it. (Note that in the Russian literature it is referred to as the *maximum principle* because of a different sign convention used in defining the variational Hamiltonian $\mathscr{H}$; see also Section 5.7). The minimum principle was postulated by L. S. Pontryagin in 1956, and later proved by Pontryagin and his co-workers, notably V. G. Boltyanskii and R. V. Gamkrelidge, in a number of papers.

Consider now the dynamical system of the form

$$\dot{\mathbf{x}}(t) = \mathbf{f}(\mathbf{x}(t), \mathbf{u}(t), t) \qquad t \in [t_1, t_2], \tag{5.107}$$

where $\mathbf{x}(t)$ and $\mathbf{f}$ are n-vectors and $\mathbf{u}(t)$ is an m-vector, with $0 < m \leqslant n$. Furthermore, our set of *admissible controls* is the set of all bounded piecewise continuous functions $\mathbf{u}(t)$ in the time interval $[t_1, t_2]$ such that

$$\mathbf{u}(t) \in \Omega \qquad \text{for all} \quad t \in [t_1, t_2] \tag{5.108}$$

with Ω a given subset of R^m and

$$\mathbf{u}(t-) = \mathbf{u}(t) \qquad \text{for all} \quad t \in [t_1, t_2].$$

Now, we wish to find an admissible control $\mathbf{u}(t)$ such that the system (5.107) is taken from some initial state $\mathbf{x}(t_0)$ to some final state $\mathbf{x}(t_f)$, such that the system performance is optimum in some specific sense. Namely, we seek a control $\mathbf{u}^*(t) \in \Omega$ which causes the system (5.107) to follow an admissible trajectory that minimizes the performance index in the time interval $[t_0, t_f]$ given by

$$J(\mathbf{u}) = h(\mathbf{x}(t_f), t_f) + \int_{t_a}^{t_f} L[\mathbf{x}(t), \mathbf{u}(t), t] \, dt \tag{5.109}$$

with the following boundary conditions: $\mathbf{x}(t_0)$ fixed, $h(\mathbf{x}(t_f), t_f) = 0$, and t_f free. For our problem, we define a function $\mathscr{H}(\mathbf{x}(t), \mathbf{u}(t), \mathbf{p}(t), t)$ of the n-vector $\mathbf{x}(t)$, the n-vector (costate) $\mathbf{p}(t)$, the m-vector $\mathbf{u}(t)$ and the time t by setting

$$\mathscr{H}(\mathbf{x}(t), \mathbf{u}(t), \mathbf{p}(t), t) \triangleq L[\mathbf{x}(t), \mathbf{u}(t), t] + \mathbf{p}^T(t)[\mathbf{f}(\mathbf{x}(t), \mathbf{u}(t), t)]. \tag{5.110}$$

We call $\mathscr{H}$ the *Hamiltonian* of our problem. In terms of the Hamiltonian (5.110), necessary for $\mathbf{u}^*(t)$ to be an optimal control are as follows:

$$\dot{\mathbf{x}}^*(t) = \frac{\partial \mathscr{H}}{\partial \mathbf{p}}(\mathbf{x}^*(t), \mathbf{u}^*(t), \mathbf{p}^*(t), t) \tag{5.111a}$$

$$\dot{\mathbf{p}}^*(t) = -\frac{\partial \mathscr{H}}{\partial \mathbf{x}}(\mathbf{x}^*(t), \mathbf{u}^*(t), \mathbf{p}^*(t), t) \tag{5.111b}$$

for all $t \in [t_0, t_f]$, satisfying the boundary conditions $\mathbf{x}^*(t_0) = \mathbf{x}_0$, $\mathbf{x}^*(t_f) = \mathbf{x}(t_f)$. Here we note that $\mathbf{p}^*(t)$ corresponds to $\mathbf{u}^*(t)$ and $\mathbf{x}^*(t)$, so that $\mathbf{p}^*(t)$ and $\mathbf{x}^*(t)$ are a solution of the canonical system (5.111a), (5.111b).

The function $\mathscr{H}(\mathbf{x}^*(t), \mathbf{p}^*(t), \mathbf{u}(t), t)$ has an absolute minimum as a function of $\mathbf{u}(t)$ over Ω at $\mathbf{u}(t) = \mathbf{u}^*(t)$ for t in $[t_0, t_f]$. That is,

$$\min_{\mathbf{u} \in \Omega} \mathscr{H}(\mathbf{x}^*(t), \mathbf{p}^*(t), \mathbf{u}(t), t) = \mathscr{H}(\mathbf{x}^*(t), \mathbf{p}^*(t), \mathbf{u}^*(t), t), \qquad (5.112a)$$

or equivalently

$$\mathscr{H}(\mathbf{x}^*(t), \mathbf{p}^*(t), \mathbf{u}^*(t), t) \leqslant \mathscr{H}(\mathbf{x}^*(t), \mathbf{p}^*(t), \mathbf{u}(t), t) \qquad (5.112b)$$

for all admissible $\mathbf{u}(t)$ in Ω, and

$$\left[\frac{\partial h}{\partial \mathbf{x}}(\mathbf{x}^*(t_f), t_f) - \mathbf{p}^*(t_f) \right]^T \delta \mathbf{x}_f$$

$$+ \left[\mathscr{H}(\mathbf{x}^*(t_f), \mathbf{u}^*(t_f), \mathbf{p}^*(t_f), t_f) + \frac{\partial h}{\partial t}(\mathbf{x}^*(t_f), t_f) \right] \delta t_f = 0. \qquad (5.113)$$

Note that if a maximum is sought, the minimization of $\mathscr{H}$ ($\min \mathscr{H}$) will be replaced by the maximization of $\mathscr{H}$ ($\max \mathscr{H}$) with the inequalities reversed. The minimum principle can now be stated as follows: if $\mathbf{u}^*(t)$ is the optimal control, then there exists a vector $\mathbf{p}^*(t)$, satisfying Eq. (5.111b), such that, at every instant of time t, $t_0 \leqslant t \leqslant t_f$,

$$\mathscr{H}(\mathbf{x}^*(t), \mathbf{u}^*(t), \mathbf{p}^*(t), t) \leqslant \mathscr{H}(\mathbf{x}^*(t), \mathbf{u}(t), \mathbf{p}^*(t), t)$$

for all admissible controls $\mathbf{u}(t)$, that is,

$$\min_{\mathbf{u} \in \Omega} \mathscr{H}(\mathbf{x}^*(t), \mathbf{u}(t), \mathbf{p}^*(t), t) = \mathscr{H}(\mathbf{x}^*(t), \mathbf{u}^*(t), \mathbf{p}^*(t), t).$$

Equations (5.111)–(5.113) constitute a set of *necessary* conditions for optimality. In other words, the minimum principle gives only the necessary conditions for optimality of the control $\mathbf{u}^*(t)$. However, it should be pointed out that these conditions are not, in general, sufficient. The control $\mathbf{u}^*(t)$ causes $\mathscr{H}(\mathbf{x}^*(t), \mathbf{p}^*(t), \mathbf{u}(t), t)$ to assume its *global*, or absolute, minimum.

From the above discussion, the minimum principle, although stated for controls with values in a closed and bounded region, can also be applied to cases where the admissible controls are not bounded. Specifically, in order for $\mathbf{u}^*(t)$ to minimize the Hamiltonian, it is necessary (but not sufficient) that

$$\frac{\partial \mathscr{H}}{\partial \mathbf{u}}(\mathbf{x}^*(t), \mathbf{u}^*(t), \mathbf{p}^*(t), t) = \mathbf{0}. \qquad (5.114)$$

Thus, if Eq. (5.114), is satisfied, and the matrix

$$\frac{\partial^2 \mathcal{H}}{\partial \mathbf{u}^2}(\mathbf{x}^*(t), \mathbf{u}^*(t), \mathbf{p}^*(t), t)$$

is positive definite, this is sufficient to guarantee that $\mathbf{u}^*(t)$ causes $\mathcal{H}$ to be a *local* minimum. Consequently, satisfying Eq. (5.114) and the $m \times m$ matrix $\partial^2 \mathcal{H}/\partial \mathbf{u}^2 > 0$ are necessary and sufficient for $\mathcal{H}(\mathbf{x}^*(t), \mathbf{u}^*(t), \mathbf{p}^*(t), t)$ to be a *global* minimum. We will now illustrate the minimum principle with three examples.

Example 1 Consider a simplified model of the rectilinear motion of a vehicle, such as an automobile, given by

$$\dot{x}(t) = u(t), \tag{1}$$

where $x(t)$ is the vehicle velocity and $u(t)$ is the acceleration (or deceleration). The vehicle is assumed to be moving initially at x_0 ft/sec. Furthermore, it will be assumed that acceleration and braking limitations require that $|u(t)| \leq \xi$ for all t. We wish to determine the optimal control low, $u^*(t)$, which brings the velocity $x(t_f)$ to zero in minimum time, t_f.

The performance index for this minimum-time problem is [2, 62]

$$J = \int_0^{t_f} 1 \, dt. \tag{2}$$

Its Hamiltonian is

$$\mathcal{H} = 1 + p(t)u(t). \tag{3}$$

In order to minimize $\mathcal{H}$, we must have

$$u(t) = \begin{cases} -\xi & \text{for } p(t) > 0, \\ +\xi & \text{for } p(t) < 0. \end{cases} \tag{4}$$

That is,

$$u^*(t) = -\xi \operatorname{sgn}(p(t)). \tag{5}$$

From the minimum principle we have

$$\dot{p}(t) = -\frac{\partial \mathcal{H}}{\partial x} = 0, \tag{6a}$$

$$p(t) = c = \text{constant} \tag{6b}$$

Note that in this simple one-dimensional problem, the determination of the optimal control $u^*(t)$ does not require the computation of the canonical form for $\dot{x}$ (i.e., $\dot{x} = \partial \mathcal{H} / \partial \mathbf{p}$). As is common with differential equations, the value of the constant must be determined from the problem's boundary conditions. In this problem, the boundary conditions are $x(0) = x_0$ and $x(t_f) = 0$. Therefore, the solution takes the form

$$x(t) = x_0 + \int_0^t u(\tau)\, d\tau, \tag{7}$$

However, from Eq. (1) we know that $u(t)$ is a constant, and must be either $+\xi$ or $-\xi$. Therefore, from Eq. (7) we have

$$x(t_f) = x_0 \pm \xi t_f = 0. \tag{8}$$

From the above discussion, we conclude that

$$\text{if} \quad x_0 > 0, \quad u = -\xi \quad \text{(maximum deceleration)}; \tag{9a}$$

$$\text{if} \quad x_0 < 0, \quad u = +\xi \quad \text{(maximum acceleration)}. \tag{9b}$$

Therefore, we can write

$$u^*(t) = -\xi \,\text{sgn}(x_0). \tag{10}$$

Since $u^*(t)$ is expressed in terms of $x(t_0)$, this represents an open-loop control law. As a final remark, we note that the minimum stopping time is given by the expression

$$t_f = \frac{|x_0|}{\xi}. \tag{11}$$

Example 2   For the system

$$\dot{x}_1 = x_2,$$
$$\dot{x}_2 = u$$

with initial conditions $x_1(0) = 10$ and $x_2(0) = 0$, we wish to find the control and trajectory which minimize the performance index

$$J(u) = t_f^2 + \frac{1}{2}\int_0^{t_f} u^2\, dt$$

if the desired final state is:

   (a) $x_1(t_f) = x_2(t_f)$;
   (b) $x_1(t_f) = 0$, $x_2(t_f)$ unspecified.

(a) The minimum principle yields

$$\mathscr{H} = 1/2u^2 + p_1x_2 + p_2u,$$

$$\dot{p}_1 = 0, \qquad \dot{p}_2 = -p_1,$$

$$\dot{x}_1 = x_2, \qquad \dot{x}_2 = u$$

$$x_1(0) = 10, \quad x_2(0) = 0,$$

$$x_1(t_f) = 0, \quad x_2(t_f) = 10,$$

$$\mathscr{H}(t_f) + 2t_f = 0.$$

Solving the $\dot{p}$ and $\dot{x}$ equations subject to the given boundary condition, then the control and trajectory are obtained as follows:

$$x_1(t) = 1/3t^3 - 2t^2 + 10,$$

$$x_2(t) = t^2 - 4t,$$

$$u(t) = 2t - 4.$$

   (b) The $\dot{p}$ and $\dot{x}$ equations remain the same as for part (a). The new boundary conditions give

$$x_1(t) = 5/27\,t^3 - 5/3\,t^2 + 10,$$

$$x_2(t) = 5/9\,t^2 - 10/3\,t,$$

$$u(t) = 10/9t - 10/3t.$$

Example 3   In this example, we will apply the minimum principle to a *fuel-optimal* double-integral plant. Let the system (plant) be given as follows [2, Chapter 8]:

$$\ddot{x}(t) = u(t) \tag{1}$$

$$x(t_f) = \dot{x}(t_f) = 0 \tag{2}$$

with the initial conditions $x(0)$, $\dot{x}(0)$ fixed, while the terminal time t_f is free. It will be assumed that the control $u(t)$ is constrained to be of the form $|u(t)| \leqslant 1$ for all t.

The performance index for this example will be taken to be of the form

$$J(u) = \int_0^{t_f} [k + |u(t)|] \, dt \qquad \text{with} \quad k > 0. \tag{3}$$

We wish to find the optimal control $u^*(t)$, for the given system, that minimizes the given performance index. (In the solution that follows, we will drop the asterisk from the control u.) Now, let

$$x_1(t) = x(t), \tag{4}$$

$$x_2(t) = \dot{x}(t). \tag{5}$$

The Hamiltonian of this problem is (note that as stated earlier, we will use $\mathcal{H}$ with a negative sign)

$$\mathcal{H}(\mathbf{x}(t), \mathbf{u}(t), \mathbf{p}(t), t) = -L[\mathbf{x}(t), \mathbf{u}(t), t] + \mathbf{p}^T(t)[\mathbf{f}(\mathbf{x}(t), \mathbf{u}(t), t],$$
$$\mathcal{H} = -k - |u(t)| + p_1(t) x_2(t) + p_2(t) u(t). \tag{6}$$

The control which maximizes the Hamiltonian is given by

$$u(t) = \begin{cases} -1 & \text{for} \quad p_2(t) < -1, \\ 0 & \text{for} \quad -1 < p_2(t) < 1, \\ +1 & \text{for} \quad p_2(t) > +1. \end{cases} \tag{7}$$

Here we note that singular controls cannot be optimal. Applying the canonical system of equations (5.111a), (5.111b), we obtain the solutions for the state and costate variables $p_1(t)$ and $p_2(t)$ from the differential equations

$$\dot{x}_1(t) = x_2(t), \tag{8}$$

$$\dot{x}_2(t) = u(t), \tag{9}$$

$$\dot{p}_1(t) = -\frac{\partial H}{\partial x_1(t)} = 0, \tag{10}$$

$$\dot{p}_2(t) = -\frac{\partial H}{\partial x_2(t)} = -p_1(t). \tag{11}$$

Letting $p_1(0) = \pi_1$ and $p_2(0) = \pi_2$ we find

$$p_1(t) = \pi_1 = \text{constant}, \tag{12}$$

$$p_2(t) = \pi_2 - \pi_1 t. \tag{13}$$

Equation (13) is the switching function from which the sought optimal control can be obtained. For $\dot{x}_1(t) = x_2(t)$, $\dot{x}_2(t) = u(t)$, the trajectories in the (x_1, x_2) plane are as follows:

$$u(t) = -1: \quad x(t) = -\tfrac{1}{2}x_2^2(t) + \text{constant},$$

$$u(t) = 0: \quad x_2(t) = \text{constant},$$

$$u(t) = +1: \quad x_1(t) = \tfrac{1}{2}x_2^2(t) + \text{constant}.$$

Therefore, for $u(t) = +1$, the trajectories for

$$x_2(t) = x_2(0) + t,$$

$$x_1(t) = x_1(0) + x_2(0) + \tfrac{1}{2}t^2$$

are

$$x_1(t) = x_1(0) - \tfrac{1}{2}x_2^2(0) + \tfrac{1}{2}x_2^2(t)$$

$$= \tfrac{1}{2}x_2^2(t) + \text{constant}. \tag{14}$$

The switching curve S_I is obtained in a closed mathematical form as follows:

$$x_{1I} = -\tfrac{1}{2}x_2|x_2|. \tag{15}$$

Figure 5.15 illustrates the trajectories and switching curves for the given system.

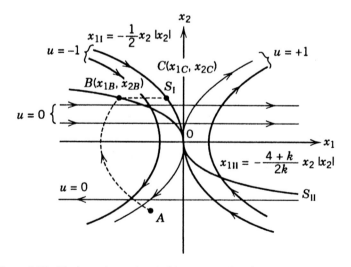

Figure 5.15   Trajectories and switching curves for the system $x(t) = u(t)$.

Next, we develop the mathematical formulation of the switching curve S_{II}. From Figure 5.15, and dropping the argument t, the components x_{1b}, x_{2b} whose dependence we wish to determine for $x_{2b} = x_{2c}$ are given by the expression

$$-\tfrac{1}{2}x_{2b}^2 = x_{1b} + x_{2b}(t_c - t_b). \tag{16}$$

Figure 5.16 shows the function $p_2(t)$ which generates the control sequence $\{-1, 0, +1\}$. From it, the control sequence is given in the form

$$u(t) = \begin{cases} +1, & 0 \leqslant t < t_b, \\ 0, & t_b \leqslant t < t_c, \\ -1, & t_c \leqslant t \leqslant t_f. \end{cases} \tag{17}$$

Clearly, since $p_2(t) = \pi_2 - \pi_1 t$, we must have*

$$\pi_2 > 1 \quad \text{and} \quad \pi_1 > 0. \tag{18}$$

Furthermore, at switch times t_b and t_c we have

$$p_2(t_b) = 1 = \pi_2 - \pi_1 t_b, \tag{19}$$

$$p_2(t_c) = -1 = \pi_2 - \pi_1 t_c. \tag{20}$$

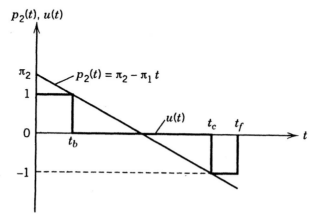

Figure 5.16   The function $p_2(t)$ which generates the control $u(t)$ of Eq. (17).

---

*Note that here π_1 and π_2 are constants, and not the usual meaning of π.

Subtracting Eq. (20) from Eq. (19), we obtain the equation

$$t_c - t_b = 2/\pi_1,$$

or

$$2 = \pi_1(t_c - t_b). \tag{21}$$

Now, the constant π_1 can be determined from the Hamiltonian in the interval $t \in [t_b, t_c]$ and $\dot{p}_2(t) = -p_1(t)$. Thus,

$$\mathscr{H}(t) = -k + p_1 x_{2b} = 0 \qquad \text{for all} \quad t \in [t_b, t_c], \tag{22}$$

$$p_1(t) = k/x_{2b}. \tag{23}$$

Therefore,

$$\pi_1 = -\dot{p}_2(t) = p_1(t) = k/x_{2b}. \tag{24}$$

From Eq. (16), (21), and (24) we have

$$x_{1b} = -\frac{k+4}{2k} x_{2b}^2. \tag{25}$$

Therefore, for the switching curve S_{II} we have the relation

$$x_{1\text{II}} = -\frac{k+4}{2k} x_2 |x_2|. \tag{26}$$

or

$$x_{1\text{II}} = \{(x_1, x_2) : x_1 = -g_k x_2 |x_2|; \ g_k = (k+4)/2k\}.$$

Figure 5.17 illustrates the switching curves for S_{I} and S_{II} in the (x_1, x_2) plane for the control sequence $\{+1, 0, -1\}$.

Curves for different values of k can be generated. We conclude this example with a block diagram for the system (1) and its corresponding performance index (3), (see Figure 5.18).

5.7 DYNAMIC PROGRAMMING AND THE HAMILTON–JACOBI EQUATION

In this section we will briefly treat the *principle of optimality* and the method of *dynamic programming*. As will be shown later, there is a definite connection between the method of dynamic programming and the minimum principle of Pontryagin. The method of dynamic programming, which is important in the

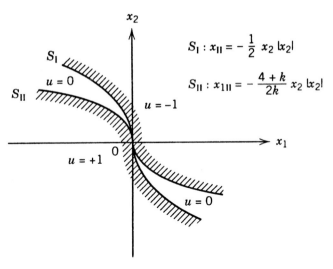

$$S_1 : x_{11} = -\frac{1}{2} x_2 |x_2|$$

$$S_{11} : x_{111} = -\frac{4+k}{2k} x_2 |x_2|$$

Figure 5.17 Switching curves of S_1 and S_{11}.

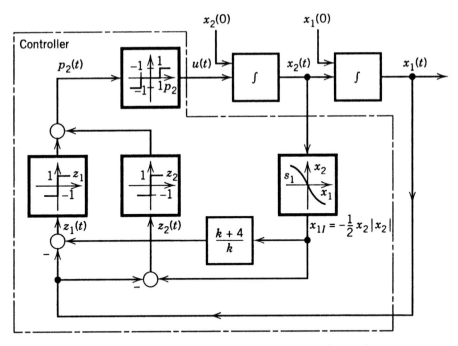

Figure 5.18 Block diagram for the control system $d^2 x(t)/dt^2 = u(t)$.

study of optimal control problems, was developed in the early 1950s by Richard Bellman [6]. In essence, Bellman generalized the Hamilton–Jacobi theory in order to include multistage systems, giving rise to the method of dynamic programming.

Dynamic programming is a computational technique, which extends the optimal decision-making concept, with the aid of the principle of optimality, to *sequences* of decisions which together define an optimal policy (or control law) and trajectory. Dynamic programming is useful in solving multistage optimization problems in which there are only a small number of possible choices for the control at each stage and in which no derivative information is available. The basic result of dynamic programming is a nonlinear partial differential equation known as Bellman's equation. This equation can be derived heuristically, provided certain assumptions are made. On the other hand, the Hamilton– Jacobi equation is, in general, quite difficult to solve. However, when it can be solved, a candidate for an optimal control function is found as a function of the state trajectory. One of the advantages of dynamic programming lies in the insight it provides into the properties of the optimal control function and the optimal performance index. For more details on dynamic programming and the Hamilton–Jacobi equation the reader is referred to references [2, 6, and 12].

In the method of dynamic programming, an optimal policy is found by employing the intuitively appealing concept called *Bellman's principle of optimality*. We will begin this section with a discussion of Bellman's principle of optimality. The principle of optimality can be stated as follows [6]:

> The optimal control sequence for an N-step process $\{u_0^*, u_1^*, \ldots, u_{N-1}^*\}$ is such that, whatever value the first choice u_0^* and hence the value x_1^* is, the choice of the remaining $N - 1$ values in the sequence $\{u_1^*, \ldots, u_{N-1}^*\}$ must constitute the optimal control sequence relative to the state x_1^* (which is now viewed as an initial state).

Stated more simply, the principle of optimality says that the optimal policy has the property that whatever the initial state and initial decision are, the remaining decisions must constitute an optimal policy with regard to the state resulting from the first decision.

The principle of optimality can best be visualized by considering the diagram in Figure 5.19, called a *directed* (or *oriented*) network. The network consists of links of given lengths, joined together at nodes (or points). The length of a link may represent the distance between its terminal nodes, the time taken to perform a task, the cost of transportation between the nodes, etc. The network is said to be oriented because the admissible paths through it are in the same general direction, from left to right.

Given the directed network in Figure 5.19, a *decision* can be defined as a choice among the alternative paths leaving a given node. We would like to find the path from state a to state l with minimum cost. Obviously, a cost is associated with each segment. If we consider the lengths in the diagram to be travel times, then we are looking for the minimum-time path.

Dynamic programming is useful in solving *multistage* optimization problems in which there are only a small number of possible choices of the control at each stage, and in which no derivative information is available. More

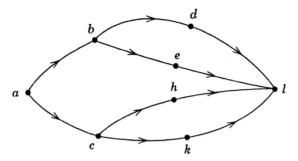

Figure 5.19 Directed network.

specifically, the method may be described in its simplest form for a discrete multistage decision process. For example, at each of a sequence of times $t_1, t_2, \ldots, t_N$ or *stages*, a decision has to be made to follow one of a finite number of paths to the next decision time, each of these paths having a given cost. The problem is to find the overall path from stage t_1 to stage t_N which minimizes the total cost.

Before we proceed in determining the optimal path (or policy), certain definitions are in order. Define J_{ab} as the cost between points a and b, J_{bd} the cost between points b and d, etc. Thus, the total cost from, a to l will be $J = J_{ab} + J_{bd} + J_{dl}$. The optimal path then can be determined from the relation

$$\min J = \min [J_{ab} + J_{bd} + J_{dl}, J_{ab} + J_{be} + J_{el}, J_{ac} + J_{ch} + J_{hl}, J_{ac} + J_{ck} + J_{kl}].$$

From the above discussion, if the initial state is a and if the initial decision is to go to b, then the path from b to l must be selected optimally, if the overall path from a to l is to be optimum. A similar argument can be made if the initial decision is to go to point c, and then from point c to point l. Consequently, the minimum cost in deciding to go from point a to point l via point b and from point a to l via point c can be expressed as follows:

$$g_b = \min [J_{bd} + J_{dl}, J_{be} + J_{el}],$$

$$g_c = \min [J_{ch} + J_{hl}, J_{ck} + J_{kl}].$$

Therefore, from the principle of optimality we have the following algorithm:

$$g_a \triangleq \min J = \min [J_{ab} + g_b, J_{ac} + g_c].$$

The above discussion will now be illustrated with two specific examples.

Example 1   In this example we will apply the above procedure in determining the minimum cost between nodes a and l in Figure 5.20.

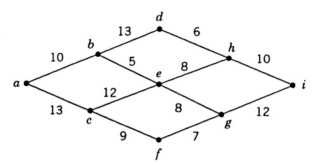

Figure 5.20 Determination of the minimum cost.

We start by noting that at node l, $g_l = 0$. Since there is only one admissible path from node h to l, we will work recursively backwards from node l. The same procedure applies for node j. Thus,

$$g_h = 10 \quad \text{and} \quad g_f = 12,$$
$$g_d = J_{dh} + g_h = 6 + 10 = 16,$$
$$g_f = J_{fj} + g_j = 7 + 12 = 19.$$

Node e is the first point where a decision must be made. Therefore,

$$g_e = \min\,[J_{eh} + g_h, J_{ej} + g_j] = \min\,[8 + 10, 8 + 12] = 18,$$
$$g_b = \min\,[J_{bd} + g_d, J_{be} + g_e] = \min\,[13 + 16, 5 + 18] = 23,$$
$$g_c = \min\,[J_{ce} + g_e, J_{cf} + g_f] = \min\,[12 + 18, 9 + 19] = 28,$$
$$g_a = \min\,[J_{ab} + g_b, J_{ac} + g_c] = \min\,[10 + 23, 13 + 28] = 33.$$

The minimum cost is therefore 33, and it is the best from a to l via a, b, e, h, l.

Example 2 In the five-by-five grid shown in Figure 5.21, it is desired to find the minimum-time path from point A to point B, moving only to the right. (The numbers in the diagram represent times to travel along the legs on the grid.)

The optimal (minimum-time) path is $ACHIJOPUB$, $T_{\min} = 65$ (the path is indicated by $\Rightarrow$).

The maximum-time path is $AGHINOTYB$, and $T_{\max} = 93$ (the path is indicated by $\rightarrow$).

A more formal presentation of the principle of dynamic programming will now be given. Dynamic programming has the property that any optimal control function $\mathbf{u}^*$ found by using the method is usually in a feedback form, that

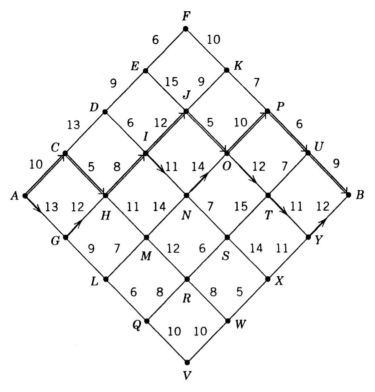

Figure 5.21 Determination of minimum-time path.

is, one that is expressed as a function of the state **x**. As we shall see presently, Bellman's principle of optimality is best utilized with a modern digital computer in a *discretized* state space, where a numerical search for an optimal trajectory is reduced from an N-stage decision process to a sequence of N single-stage decision processes (where N is the number of possible states). We will now develop a recurrence relation of dynamic programming applied to a control process. Let an nth-order time-invariant system be described by the state equation

$$\dot{\mathbf{x}}(t) = \mathbf{f}[\mathbf{x}(t), \mathbf{u}(t), t], \qquad t \in (T_1, T_2), \tag{5.115}$$

where **x** and **f** are n-vectors and **u** is an m-vector with $0 < m < n$. Furthermore, our set of *admissible* controls is the set of all bounded piecewise continuous functions $\mathbf{u}(t)$ on (T_1, T_2) such that

$$\mathbf{u}(t) \in \Omega \qquad \forall\, t \in (T_1, T_2)$$

with Ω a given subset of R^m and

$$\mathbf{u}(t-) = \mathbf{u}(t) \qquad \forall\, t \in (T_1, T_2).$$

We wish now to determine the control law which minimizes the performance index (or cost functional)

$$J = \int_0^T L[\mathbf{x}(t), \mathbf{u}(t), t]\, dt \tag{5.116}$$

with the initial condition $x(0) = x_0$ and the final state $x(T)$ left free. The constraint on the control function is $|u| \leqslant 1$. The above continuous-time problem can be converted into a discrete-time form, which will be more suitable for a numerical solution on a digital computer. There exist a number of special programs to solve differential equations numerically [24, 32]. For simplicity, let us approximate the system by dividing the interval $[0, T]$ into N (large enough to reduce the error) equal steps, $t_0, t_1, \ldots$, with interval $\Delta t = T/N$. Thus the performance index can be approximated as

$$J = \sum_{n=0}^{N-1} L(x_n, u_n)\, \Delta t, \tag{5.117}$$

and the system as

$$\frac{\mathbf{x}(t + \Delta t) - \mathbf{x}(t)}{\Delta t} \approx \mathbf{f}(\mathbf{x}(t), \mathbf{u}(t))$$

or, in the above notation,

$$\frac{x_{n+1} - x_n}{\Delta t} = f(x_n, u_n), \qquad n = 0, 1, 2, \ldots, N. \tag{5.118}$$

Letting

$$f(x_n, u_n)\, \Delta t + x_n = g(x_n, u_n),$$

we have

$$x_{n+1} = g(x_n, u_n). \tag{5.119}$$

An iterative routine for a digital computer can be created to solve this problem. To this end, suppose that the interval $[0, T]$ is small enough so that one step is sufficient. Then $\Delta t = T$ and $N = 1$. Thus,

$$J = L(x_0, u_0)\, \Delta t,$$

$$x_1 = g(x_0, u_0).$$

The optimization procedure for each x_0 is to choose the control variable u_0 so as to minimize cost functional J. Now, let $J_1^*(x_0)$ be the initial value of J. Therefore, for the first step, with x_0 as the starting state, we have the cost of a *one-stage process*

$$J_1^*(x_0) = \min_{u_0 \in \{u_\varepsilon\}} [L(x_0, u_0)\Delta t] \qquad (5.120)$$

where the right-hand side denotes the smallest value of the quantity in the bracket over $2P + 1$ possible values of u_0. Specifically, here we have assumed that the control function u has been quantized in the interval $[-1, +1]$ into $2P + 1$ values of separation ε; that is, $-P\varepsilon, (-P+1)\varepsilon, \ldots, P\varepsilon = 1$. The set of $2P + 1$ values of u has been denoted by $\{u\varepsilon\}$. Therefore, for a two-step control process, that is, $N = 2$, we have

$$J_2^*(x_0) = \min_{u_0 \in \{u_\varepsilon\}} [L(x_0, u_0)\Delta t + J_1^*(x_1)], \qquad (5.121)$$

since $J_1^*(x_1)$ has been found and stored. Performing the above iterative procedure N times, we obtain the set of recurrence relations in the optimum index value as follows:

$$J_1^*(x_0) = \min_{u_0 \in \{u_\varepsilon\}} [L(x_0, u_0)\Delta t], \qquad (5.122a)$$

$$J_2^*(x_0) = \min_{u_0 \in \{u_\varepsilon\}} [L(x_0, u_0)\Delta t + J_{N-1}^*(g(x_0, u_0))], \qquad N = 2, \ldots,. \quad (5.122b)$$

These two equations constitute a set of recurrent formulas in the optimum-performance index value, which will allow us to find the optimal control for an arbitrary integer N. The steps in determining the optimal control sequence is as follows:

1. At time t_{N-1}, treat each state x_{N-1} as an initial state and the optimization problem as a one-step problem from x_{N-1} to x_N.
2. From Eq. (5.122a), calculate $u_{N-1}^* = u^*(x_{N-1})$ and $J_1^*(x_{N-1})$, and store both quantities for each value of x_{N-1}.
3. Continue the above process until time t_0 is reached.
4. In order to find the optimal sequence from an arbitrary starting point x_0, retrieve the stored value $u^*(x_0)$. Then, use this value to generate x_1 by means of the expression $x_{n+1} = g(x_n, u_n)$, that is, Eq. (5.119).
5. Next, retrieve $u^*(x_1)$. Then, using this value, generate x_2 by means of the expression $x_{n+1} = g(x_n, u_n)$.
6. Continue the above process in order to obtain the entire optimal control sequence $\{u_0^*, u_1^*, \ldots, u_{N-1}^*\}$ as well as the optimal trajectory sequence $\{x_0^*, x_1^*, \ldots, x_{N-1}^*\}$.

7. Finally, the optimum performance index is given by Eq. (5.122b), that is, $J_N^*(x_0)$, which can be retrieved from the computer's memory.

The above approach is ideally suited for a digital computer solution. The derivation of the recurrence equation (5.122) reveals another important concept, the *imbedding principle*. For example, if instead $J_N^*(x_0)$ in Eq. (5.122) we let $J_{N-K,N}^*(\mathbf{x}(N-K))$ be the minimum cost possible for, say, the final K stages of an N-stage process, then the optimal policy and minimum costs for a K-stage process are contained (or imbedded) in the results for an N-stage process, provided $N \geqslant K$.

We will now apply the principle of optimality to the optimal control of continuous-time systems. In particular, we shall show that we can derive a nonlinear partial differential equation, called the Hamilton–Jacobi–Bellman equation, that corresponds to the functional equations (5.122).

Again, consider the system

$$\dot{\mathbf{x}} = \mathbf{f}(\mathbf{x}(t), \mathbf{u}(t), t). \tag{5.123}$$

The process described by Eq. (5.123) is to be controlled so as to minimize the performance index

$$J = h(\mathbf{x}(t_f), t_f) + \int_{t_0}^{t_f} L(\mathbf{x}(\tau), \mathbf{u}(\tau), \tau)\, d\tau, \tag{5.124}$$

where h and L are specified functions, t_0 and t_f are fixed, and τ is a dummy variable of integration. Moreover, the function $L(\mathbf{x}(t), \mathbf{u}(t), t)$ is assumed to be continuous with respect to t. In the present analysis, we will let $\mathbf{x}(\mathbf{u}(t), t)$ be the trajectory resulting from the application of $\mathbf{u}(t)$ on the system (5.123), starting from the state x_0 at time $t = t_0$. The imbedding principle can now be used to include the aforementioned problem for a larger class of problems by considering the performance index

$$J\left(\mathbf{x}(t), t, \underset{t \leqslant \tau \leqslant t_f}{\mathbf{u}(\tau)}\right) = h(\mathbf{x}(t_f), t_f) + \int_t^{t_f} L(\mathbf{x}(\tau), \mathbf{u}(\tau), \tau)\, d\tau, \tag{5.125}$$

where t can have any value less than or equal to t_f, and $\mathbf{x}(t)$ can have any *admissible* state value. The reader should note that the performance index will depend on the numerical values for $\mathbf{x}(t)$ and t, and on the optimal control history in the interval $[t, t_f]$.

Next, we need to determine the controls that minimize Eq. (5.125) for all admissible $\mathbf{x}(t)$, and for all $t \leqslant t_f$. Thus, the minimum performance index assumes the form

$$J^*(\mathbf{x}(t), t) = \min_{\substack{\mathbf{u}(\tau) \\ t \leqslant \tau \leqslant t_f}} \left\{ \int_t^{t_f} L(\mathbf{x}(\tau), \mathbf{u}(\tau), \tau)\, d\tau + h(\mathbf{x}(t_f), t_f) \right\}. \tag{5.126}$$

Subdividing the interval $[t, t_f]$ yields

$$J^*(\mathbf{x}(t), t) = \min_{\substack{\mathbf{u}(\tau) \\ t \leqslant \tau \leqslant t_f}} \left\{ \int_t^{t+\Delta t} L d\tau + \int_{t+\Delta t}^{t_f} L d\tau + h(\mathbf{x}(t_f), t_f) \right\}. \tag{5.127}$$

Referring to the principle of optimality stated earlier, we note that it requires that

$$J^*(\mathbf{x}(t), t) = \min_{\substack{\mathbf{u}(\tau) \\ t \leqslant \tau \leqslant t_f}} \left\{ \int_t^{t+\Delta t} L d\tau + J^*(\mathbf{x}(t + \Delta t), t + \Delta t) \right\}, \tag{5.128}$$

where $J^*(\mathbf{x}(t + \Delta t), t + \Delta t)$ is the minimum cost of the process for the time interval $(t + \Delta t) \leqslant \tau \leqslant t_f$ with the initial state $\mathbf{x}(t + \Delta t)$. Next, we assume that the second partial derivatives of J^* exist and are bounded. Expanding $J^*(\mathbf{x}(t + \Delta t), t + \Delta t)$ in a Taylor series about the point $(\mathbf{x}(t), t)$ results in the following expression:

$$J^*(\mathbf{x}(t), t) = \min_{\substack{\mathbf{u}(\tau) \\ t \leqslant \tau \leqslant t_f}} \left\{ \int_t^{t+\Delta t} L d\tau + J^*(\mathbf{x}(t), t) + \left[\frac{\partial J^*}{\partial t}(\mathbf{x}(t), t) \right] \Delta t \right.$$

$$\left. + \left[\frac{\partial J^*}{\partial t}(\mathbf{x}(t), t) \right]^T \left[\mathbf{x}(t + \Delta t) - \mathbf{x}(t) \right] + \text{h.o.t} \right\}, \tag{5.129}$$

where h.o.t. denotes higher-order terms. Assuming Δt to be small, we obtain

$$J^*(\mathbf{x}(t), t) = \min_{\mathbf{u}(t)} \left\{ L(\mathbf{x}(t), \mathbf{u}(t), t) \Delta t + J^*(\mathbf{x}(t), t) + \frac{\partial J^*}{\partial t}(\mathbf{x}(t), t) \Delta t \right.$$

$$\left. + \left[\frac{\partial J^*}{\partial \mathbf{x}}(\mathbf{x}(t), t) \right]^T \mathbf{f}(\mathbf{x}(t), \mathbf{u}(t), t) \Delta t + \text{h.o.t}(\Delta t) \right\}. \tag{5.130}$$

Removing now the terms involving $J^*(\mathbf{x}(t), t)$ and $\partial J^*/\partial t(\mathbf{x}(t), t)$ from the minimization [since they do not depend on $\mathbf{u}(t)$] results in

$$0 = \frac{\partial J^*}{\partial t}(\mathbf{x}(t), t) \Delta t$$

$$+ \min_{\mathbf{u}(t)} \left\{ L(\mathbf{x}(t), \mathbf{u}(t), t) \Delta t \right.$$

$$\left. + \left[\frac{\partial J^*}{\partial \mathbf{x}}(\mathbf{x}(t), t) \right]^T \mathbf{f}(\mathbf{x}(t), \mathbf{u}(t), t) \Delta t + \text{h.o.t}(\Delta t) \right\} \tag{5.131}$$

Dividing Eq. (5.131) by Δt and taking the limit as $\Delta t \to 0$, we obtain the following result:

$$0 = \frac{\partial J^*}{\partial t}(\mathbf{x}(t), t) + \min_{\mathbf{u}(t)} \left\{ L(\mathbf{x}(t), \mathbf{u}(t), t) + \left[\frac{\partial J^*}{\partial \mathbf{x}}(\mathbf{x}(t), t) \right]^T \mathbf{f}(\mathbf{x}(t), \mathbf{u}(t), t) \right\}. \quad (5.132)$$

Equation (5.132) is a first-order, nonlinear partial differential equation and must be solved subject to specified boundary conditions. We elect to set the boundary value for this equation at $t = t_f$. Therefore, from Eq. (5.126) it is obvious that [see also Section 5.3, Eq. (5.36) and accompanying discussion on the terminal state]

$$J^*(\mathbf{x}(t_f), t_f) = h(\mathbf{x}(t_f), t_f). \quad (5.133)$$

Define now the Hamiltonian $\mathcal{H}$ as

$$\mathcal{H}\left(\mathbf{x}(t), \mathbf{u}(t), \frac{\partial J^*}{\partial \mathbf{x}}, t\right) \triangleq L(\mathbf{x}(t), \mathbf{u}(t), t) + \left[\frac{\partial J^*}{\partial \mathbf{x}}(\mathbf{x}(t), t) \right]^T \mathbf{f}(\mathbf{x}(t), \mathbf{u}(t), t); \quad (5.134)$$

then

$$\mathcal{H}\left(\mathbf{x}(t), \mathbf{u}^*\left(\mathbf{x}(t), \frac{\partial J^*}{\partial \mathbf{x}}, t\right), \frac{\partial J^*}{\partial \mathbf{x}}, t\right) = \min_{\mathbf{u}(t)} \mathcal{H}\left(\mathbf{x}(t), \mathbf{u}(t), \frac{\partial J^*}{\partial \mathbf{x}}, t\right), \quad (5.135)$$

since the minimizing control will depend on $\mathbf{x}$, $\partial J^*/\partial \mathbf{x}$, and t. From these definitions, we have thus obtained the Hamilton–Jacobi equation [2,6]

$$0 = \frac{\partial J^*}{\partial t}(\mathbf{x}(t), t) + \mathcal{H}\left(\mathbf{x}(t), \mathbf{u}^*\left(\mathbf{x}(t), \frac{\partial J^*}{\partial \mathbf{x}}, t\right), \frac{\partial J^*}{\partial \mathbf{x}}, t\right), \quad (5.136a)$$

or

$$-\frac{\partial J^*}{\partial t}(\mathbf{x}(t), t) = \mathcal{H}\left(\mathbf{x}(t), \mathbf{u}^*\left(\mathbf{x}(t), \frac{\partial J^*}{\partial \mathbf{x}}, t\right), \frac{\partial J^*}{\partial \mathbf{x}}, t\right). \quad (5.136b)$$

Equation (5.136) is the continuous-time analog of Bellman's recurrence Eq. (5.122) and is known as the *Hamilton–Jacobi–Bellman* equation. If X is a given region in $R^n \times (T_1, T_2)$ containing a target set S, then under the assumption that $J^*(\mathbf{x}(t), t)$ is continuously differentiable in the region X and $(\mathbf{x}(t), t)$ is an element of the region X containing S, Eq. (5.136) is an additional *necessary* condition for optimality. Therefore, dynamic programming constitutes solving the Hamilton–Jacobi–Bellman equation.

Along a specific optimal trajectory, if we define $-\partial J^*/\partial \mathbf{x}$ as the adjoint (or costate; see also Section 5.3) vector $\mathbf{p}(t)$, then $\mathcal{H}$ in Eq. (5.136) is identified as the Hamiltonian for the *minimum principle* (see also Section 5.6). In that case, the quantity $\partial J^*/\partial t$ is clearly the quantity $\mathscr{H}(\mathbf{x}^*, \mathbf{u}^*, t, \mathbf{p})$. Thus,

$$\mathbf{p}(t) = -\frac{\partial J^*}{\partial \mathbf{x}}[\mathbf{x}(t), t],\qquad(5.137)$$

$$\frac{\partial J^*}{\partial t} = \mathscr{H}^* \triangleq \mathscr{H}(\mathbf{x}^*, \mathbf{u}^*, t, \mathbf{p}).\qquad(5.138)$$

Note that the costate corresponds to $\mathbf{u}^*(t), \mathbf{x}^*(t)$. Referring to Eq. (5.136), we observe that for a time-dependent problem with fixed end point, the minimum principle states that

$$\mathscr{H}(\mathbf{x}^*(t), \mathbf{p}^*(t), \mathbf{u}^*(t), t) = \min_{\mathbf{u} \in \Omega} \mathscr{H}(\mathbf{x}^*(t), \mathbf{p}^*(t), \mathbf{u}(t), t),\qquad(5.139)$$

or equivalently

$$\mathscr{H}(\mathbf{x}^*(t), \mathbf{p}^*(t), \mathbf{u}^*(t), t) \leqslant \mathscr{H}(\mathbf{x}^*(t), \mathbf{p}^*(t), \mathbf{u}(t), t) \qquad \forall \mathbf{u} \in \Omega.\qquad(5.140)$$

It should be pointed out that for any function F, we have $-\min F = \max(-F)$. As stated in Section 5.6, in the Russian literature $\mathscr{H}$ is defined with the opposite sign, so that in order to minimize J, one must maximize $\mathscr{H}$.

As stated earlier, the Hamilton–Jacobi–Bellman equation, being a first-order nonlinear partial differential equation, is difficult (if not impossible) to solve. One of the ways to solve it is the *field-of-extremals* approach. In this method, one wishes to know the optimal control function $\mathbf{u}(t)$ from a large number of different initial points to a given terminal *hypersurface*, since one may not know where the system will start from or when it will start. Consequently, one must calculate a family of optimal paths so that all of the possible initial points are on, or at least very close to, one of the calculated optimal paths. In the calculus of variations, such a family of optimal paths is called field of extremals. The hypersurface, mentioned above, is a terminal curve in the (x, t) plane such that $\psi(x, t) = 0$ [12]. Figure 5.22 illustrates optimal paths terminating on a hypersurface.

In Section 5.2 we mentioned the so-called *bang–bang* control problem. Here we will give a more detailed description of the bang–bang principle. Historically, the abrupt switching in the control law was given the name bang–bang because it was easily realized by relays and later by electronic flip-flop devices. The bang–bang principle can be stated using the following theorem [2]:

Theorem. Let $\mathbf{u}^*(t)$ be a time-optimal control for the system

$$\frac{d\mathbf{x}(t)}{dt} = \mathbf{f}[\mathbf{x}(t), t] + B[\mathbf{x}(t), t]\mathbf{u}(t),$$

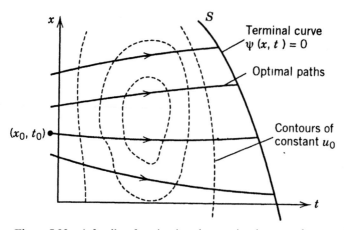

Figure 5.22   A family of optimal paths terminating on a hypersurface.

where $f_i[\mathbf{x}(t), t]$ and $b_{ij}[\mathbf{x}(t), t]$ are continuous in $\mathbf{x}(t)$ and t, and let $\mathbf{x}^*(t)$ and $\mathbf{p}^*(t)$ be the corresponding state trajectory and costate. If the problem is normal, then the components $\mathbf{u}_1^*(t), \mathbf{u}_2^*(t), \ldots, \mathbf{u}_r^*(t)$ of $\mathbf{u}^*(t)$ must be defined by the relation

$$\mathbf{u}_j^*(t) = -\operatorname{sgn}\left\{ \sum_{i=1}^{n} b_{ij}[\mathbf{x}^*(t), t] \mathbf{p}_i^*(t) \right\} \quad \forall t \in [t_0, T^*], \quad j = 1, 2, \ldots, r.$$

Note that the components of $\mathbf{u}^*(t)$ are assumed to be constrained in magnitude by the relation $|u_j(t)| \leqslant 1$, $j = 1, 2, \ldots, r$, $\forall t$. Furthermore, a time-optimal control problem is said to be normal if in the interval $[t_0, T^*]$ there is a countable set of times

$$t_{\gamma j} \in [t_0, T^*], \quad \gamma = 1, 2, 3, \ldots, \quad j = 1, 2, \ldots, r,$$

such that

$$q_j^*(t) = \sum_{i=1}^{n} b_{ij}[\mathbf{x}^*(t), t] \mathbf{p}_i^*(t) = 0 \quad \text{if and only if} \quad t = t_{\gamma j} \quad \text{for some } \gamma, j.$$

The equation for $\mathbf{u}^*(t)$ can be written more compactly as

$$\mathbf{u}^*(t) = -\operatorname{sgn}\{\mathbf{q}^*(t)\}$$

$$= -\operatorname{sgn}\{B^T[\mathbf{x}^*(t), t] \mathbf{p}^*(t)\},$$

where sgn(·) is the *signum* function defined by*

$$\text{sgn}(\varphi) = \begin{cases} +1 & \text{if} \quad \varphi > 0, \\ -1 & \text{if} \quad \varphi < 0. \end{cases}$$

In general, one or more changes in control will occur during the time of operation of the system; thus the name *bang–bang* control, since the controls move suddenly from one point on the boundary of the feasible control region to another point on the boundary. The *bang–bang* control principle will now be demonstrated by means of a specific example.

Example Consider the second-order system governed by

$$\dot{x} = v, \tag{1}$$

$$\dot{v} = u \tag{2}$$

(with given initial conditions $x(0) = x_0, v(0) = v_0$, and t_f), where x, v are scalar variables and u is a bounded scalar control variable, $-1 \leqslant u \leqslant 1$. We wish to find the control $u(t)$ so as to minimize the performance index

$$J(u) = \int_0^{t_f} |u(t)| \, dt \tag{3}$$

with specified terminal conditions $x(t_f) = 0$, $v(t_f) = 0$. Furthermore, we will consider the case $v_0 \geqslant 0$, $x_0 \geqslant -1/2v_0^2$, and assume that $t_f > (t_f)_{\min}$, where $(t_f)_{\min}$ is the minimum time to bring the state from (x_0, v_0) to $(0,0)$ with $-1 \leqslant u \leqslant 1$. Finally, we will show that the control for this case is bang–zero–bang, that is,

$$u = \begin{cases} -1, & 0 \leqslant t < t_1, \\ 0, & t_1 \leqslant t < t_2, \\ +1, & t_2 \leqslant t < t_f. \end{cases} \tag{4}$$

The Hamiltonian for this example is

$$\min \mathcal{H} = \min (|u| + p_x v + p_v u), \tag{5}$$

---

*Note that the function "sgn" is discontinuous, and consequently the system does not possess the qualities needed to insure the existence of smooth solutions from a mathematical standpoint.

which yields the following:

$$\dot{p}_x = 0 \Rightarrow p_x = \text{constant} = v_x,$$

$$\dot{p}_v = -p_x \Rightarrow p_v = -v_x(t - t_f) + v_v,$$

$$\partial \mathcal{H}/\partial u = p_v + \text{sgn } u.$$

The sketch in Figure 5.23 illustrates the slope of $\mathcal{H}$, that is, $\partial \mathcal{H}/\partial u$, in terms of the costates. From this sketch, the optimum u is given by

$$u = \begin{cases} -1 & \text{for } p_v > 1, \\ 0 & \text{for } -1 \leqslant p_v \leqslant 1, \\ +1 & \text{for } p_v > -1. \end{cases} \tag{6}$$

Since p_v is a linearly decreasing function of time, we shall have a bang–zero–bang control if initially $p_v > 1$. Furthermore, since $v_0 > 0$ and $v(t_f) = 0$, we want u to be negative, at least for some time. In view of Eq. (5) and the linearity of p_v, this is possible only if $u = -1$ initially. The switching times t_1, t_2 are indicated in Figure 5.24:

$$\text{for } t < t_1, \dot{v} = -1 \Rightarrow v = -t + v_0 \Rightarrow x = -\tfrac{1}{2}t^2 + tv_0 + x_0; \tag{7}$$

$$\text{for } t < t_2, \dot{v} = +1 \Rightarrow t - t_f \Rightarrow x = -\tfrac{1}{2}(t - t_f)^2; \tag{8}$$

$$\text{for } t_1 < t < t_2, \dot{v} = 0 \Rightarrow v = v(t_1) = v(t_2); \tag{9}$$

and by continuity

$$x = v(t_1)t + C = v(t_2)t + D. \tag{10}$$

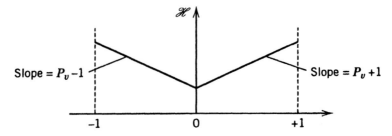

Figure 5.23 Plot of the Hamiltonian in terms of the costates.

Also, by continuity of x at t_1,

$$C = -\tfrac{1}{2}t_1^2 + t_1 v_0 + x_0 - v(t_1)t_1,$$

whence

$$x = v(t_1)t + x(t_1) - v(t_1)t_1, \qquad t_1 < t < t_2. \tag{11}$$

Continuity of x at t_2 yields

$$v(t_1)t_2 + x(t_1) - v(t_1)t_1 = x(t_2) = \tfrac{1}{2}(t_2 - t_f)^2,$$

whence

$$(-t_1 + v_0)t_2 - \tfrac{1}{2}t_1^2 + v_0 t_1 + x_0 - (-t_1 + v_0)t_1 = \tfrac{1}{2}(t_2 - t_f)^2. \tag{12}$$

Also, from (9),

$$-t_1 + v_0 = t_2 - t_f, \qquad \text{or} \quad t_2 = t_f - t_1 + v_0. \tag{13}$$

Equations (12) and (13) are two equations for the switching times t_1 and t_2. This example could be further extended if the times t_1 and t_2 were given as some other function of x_0, v_0 and t_f.

We will now demonstrate the relationship between Bellman's dynamic programming and that of the linear quadratic regulator discussed in Section 5.3. Suppose we are given a plant of the form

$$\dot{x} = A(t)x + B(t)u,$$

whose output is

$$y(t) = C(t)x(t),$$

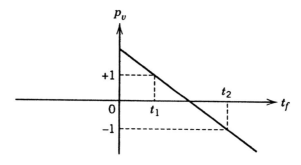

Figure 5.24   Plot of the costate p_v for the switching times t_1, t_2.

where **x** and **y** are of dimension n and m, respectively. Furthermore, suppose we wish to find the optimal linear feedback control for a system with an m-vector input $z(t)$ that is completely specified. In addition, we wish to keep the output close to the input so that not much energy is expended in the process. Define now an error function $e(t) = z(t) - y(t)$. Our performance index will take the form

$$J = e^T(t_2)Se(t_2) + \frac{1}{2}\int_{t_1}^{t_2}[e^T(\tau)Q(\tau)e(\tau) + u^T(\tau)R(\tau)u(\tau)]\,d\tau.$$

As in Section 5.3, the optimal feedback control law $u^*(x, t)$ can by found by minimizing the performance index J. The Hamiltonian of this problem is

$$\mathcal{H} = -\tfrac{1}{2}[z(t) - C(t)x(t)]^TQ(t)[z(t) - C(t)x(t)] - \tfrac{1}{2}[u^T(t)R(t)u(t)]$$
$$+ p(t)\{[A(t)x(t)]^T + [B(t)u(t)]^T\}.$$

Therefore, the optimal control function is obtained from setting

$$\left.\frac{\partial \mathcal{H}}{\partial u}\right|_{u = u^*} = 0.$$

Performing this operation, we obtain

$$u^*(t) = -R^{-1}B^T(t)p(t)$$
$$= -R^{-1}B^T(t)\frac{\partial J^*(x, t)}{\partial x}.$$

The control function $u^*(t)$ is well defined, provided that $R^{-1}(t)$ exists for all t [note the similarity between this equation and Eq. (5.44)].

Finally, the Hamilton–Jacobi–Bellman discussion will be illustrated with the following example. We wish to find the solution to the Hamilton–Jacobi–Bellman equation for the plant (or system) given by

$$\frac{dx}{dt} = Ax + u,$$

$$A^T + A = 0, \qquad \|u\| \leqslant 1,$$

with the cost function given by

$$J = \int_0^{t_f} dt = t_f.$$

Furthermore, we will demonstrate that $J(\mathbf{x}) = \|\mathbf{x}\|$, and determine the optimal control for the given plant. The solution to this problem is as follows: first, we obtain the Hamiltonian

$$\mathcal{H} = 1 + \left(\frac{\partial J}{\partial \mathbf{x}}\right)^T (A\mathbf{x} + \mathbf{u})$$

to minimize $\mathcal{H}$ with respect to $\mathbf{u}$. Thus

$$\mathbf{u} = -\left(\frac{\partial J}{\partial x}\right) \bigg/ \left\|\frac{\partial J}{\partial \mathbf{x}}\right\|.$$

Next, we form the Hamilton–Jacobi–Bellman equation. This is

$$1 + \left(\frac{\partial J}{\partial \mathbf{x}}\right)^T A\mathbf{x} - \left\|\frac{\partial J}{\partial \mathbf{x}}\right\| + \frac{\partial J}{\partial t} = 0.$$

Now, if $J(\mathbf{x}) = \|\mathbf{x}\|$, then

$$\left(\frac{\partial J}{\partial \mathbf{x}}\right) = \frac{\mathbf{x}}{\|\mathbf{x}\|} \quad \text{and} \quad \left(\frac{\partial J}{\partial t}\right) = 0.$$

Therefore, we see that

$$1 + \frac{x^T}{\|\mathbf{x}\|} A\mathbf{x} - \frac{\|\mathbf{x}\|}{\|\mathbf{x}\|} = 0.$$

But since

$$\mathbf{x}^T A\mathbf{x} = \tfrac{1}{2}\mathbf{x}^T(A^T + A)\mathbf{x} = 0,$$

we have $\|\mathbf{x}\|/\|\mathbf{x}\| = 1$, so that $1 - 0 - 1 = 0$, and therefore $J(\mathbf{x}) = \|\mathbf{x}\|$ is a solution to the Hamilton–Jacobi–Bellman equation for the given system. The optimal control $\mathbf{u}^*$ is given by

$$\mathbf{u}^* = -\frac{\partial J}{\partial \mathbf{x}} \bigg/ \left\|\frac{\partial J}{\partial \mathbf{x}}\right\| = -\mathbf{x}/\|\mathbf{x}\|.$$

5.8 CONCLUDING REMARKS

In this chapter we have discussed linear regulators with quadratic costs, the calculus of variations, Pontryagin's minimum principle, and dynamic programming and the Hamilton–Jacobi–Bellman equation. As we saw, the combination of quadratic cost functions and linear state equations provides an important class of problems for which the application of Pontryagin's minimum principle leads to a set of linear equations.

The general form of the linear-state problems with quadratic costs we studied have an n-dimensional state vector $\mathbf{x}$ and an r-dimensional control vector $\mathbf{u}$. The state equation is linear in both the state and control variables and has the general form

$$\frac{d\mathbf{x}}{dt} = A\mathbf{x} + B\mathbf{u},$$

where A is an $n \times n$ matrix and B is an $n \times r$ matrix. The elements in these matrices are, in general, time-dependent. The cost consists of a terminal part and a part that depends quadratically on the state and control variables. Furthermore, we defined the cost as the value of $J(t_0, t_f)$, where the cost function $J(t, t_f)$ is given by

$$J(t, t_f) = \frac{1}{2}\left[\mathbf{x}^T(t_f) S \mathbf{x}(t_f)\right] + \frac{1}{2}\int_{t_0}^{t_f} (\mathbf{x}^T Q \mathbf{x} + \mathbf{u}^T R \mathbf{u})\, d\tau.$$

In this expression, S is a constant $n \times n$ matrix, Q is a time-dependent $n \times n$ matrix, and R is a time-dependent $r \times r$ matrix. The quadratic forms with matrices S and Q are positive semidefinite, and that with the matrix R is positive definite. Since R is nonsingular, R^{-1} exists.

The optimal control system designer should be cautioned that using non-dynamic state feedback requires controllers with *infinite bandwidth*. Furthermore, the appropriate choice of a cost functional (or performance index) is one of the more difficult tasks in formulating the optimal control problem, and certainly one which the control system designer must face. For example, in the classical frequency domain design, the first question that the designer will face is that of stability; performance is usually a later consideration. The control law requires feedback from all state variables. If they are not all directly measurable, a state estimator must be included. In this case, the feedback system and the estimator may be designed separately. The matrix Riccati equation solution can be computed off line, before the control system is placed in operation. With regard to the matrix Riccati equation, the following observations are noted:

1. If $S = 0$ and T (the final time, or t_f) is finite, then the development for the best control is simply found by performing a backward-*swept* (note that the *sweep method* and dynamic programming are one and the same thing for linear–quadratic problems [12]) time integration. The control is then found from Eqs. (5.49), (5.50), and (5.51) with $dP(T)/dt = 0$.
2. If S is finite and $T \to \infty$, then we may apply the stationarity principle, that is, $dP(T)/dt = 0$. In this case, the matrix Riccati equation is a matrix of

constants obtained by solving $dP(t)/dt$ with $dP(t)/dt = 0$. The control can then be obtained by solving the corresponding equations.

3. If $S \rightarrow \infty$, we must use the inverse matrix Riccati equation with $P^{-1}(T) = S^{-1}(T) = 0$.

As mentioned in Section 5.3 in connection with the missile interception, the LQR can command the interceptor missile to the target (evader) along a precalculated or nominal trajectory by making use of a stored nominal state and control history and a set of precalculated varying steering gains. Thus, at each guidance update, the LQR will calculate the difference between vehicle estimated position and velocity (both of these parameters may be obtained from an onboard inertial navigation system) and vehicle precomputed position and velocity. Therefore, when this difference is multiplied by the steering gains, it produces corrections to the nominal control commands. Consequently, the control corrections are subtracted from the nominal controls to produce steering commands, which in turn are sent to the control system. When the LQR algorithm is implemented in an onboard computer, the algorithm uses full missile-state feedback provided by the inertial navigation system. Above all, the algorithm must be able to calculate commands for complex maneuvers, adapting to varying mission requirements, respond to inflight perturbations, and interface with other software subsystems. Real-time implementation of a guidance law demands that these requirements be met. In selecting the optimum policy, the obvious approach would be to compute the total cost index J along all possible path combinations and then choose the best one. The technique presented in this chapter for solving the deterministic linear optimal regulator can be used in conjuction with frequency domain methods in the analysis and design of missile guidance and aircraft flight control systems. Finally, the LQR, used as a guidance law, offers optimal performance and a mature technology.

From a practical point of view, the methods of the calculus of variations described in this chapter are limited basically to control processes that are (1) linear, (2) characterized by quadratic cost criteria, unconstrained in both **x** and **u**, and (3) characterized by low dimensionality. There is clearly a need for methods that will overcome some of these limitations. Bellman's well-known *dynamic programming* has gained considerable popularity as a complement to the classical methods. Dynamic programming attained its greatest practical significance in conjuction with the modern digital computer. As its name implies, it is basically, an ingenious method of computer programming. Since the digital computer accepts only *discrete* data or *data sequences*, it becomes necessary when using Bellman's method to discretize the otherwise continuous control processes. When a continuous control process is viewed in this way, it takes on the characteristics of what is referred to as a *multistage decision process*. Therefore, the essential feature of dynamic programming is that it can reduce the N-stage decision process to sequence of N single-stage decision processes, enabling us to solve the problem in a simple iterative manner on a computer. This reduction is made possible by use of the fundamental principle of optimality.

There exists a large class of variational-type problems which are of great engineering importance but which cannot be handled by the Euler–Lagrange theory. The Pontryagin minimum principle, on the other hand, applies to a much wider class of optimal control problems. There is no restriction to linear state equations, and many different cost functions can be successfully handled. The proof of Pontryagin's minimum principle is quite difficult, and only a brief development was given here. However, it should be pointed out that Pontryagin's minimum principle provides only *necessary* conditions that an optimal control solution must satisfy, if it exists. There is, in general, no guarantee that a given problem will have an optimal solution.

We now summarize some of the important features of optimal control that we have discussed in this chapter:

1. The greatest value of optimal control theory is that it shows us the ultimate capabilities of a specific system subject to specific constraints.
2. It gives us information about how this ultimate achievement can be reached. Suboptimal control may be an acceptable compromise and perhaps represents a more attractive solution from an engineering point of view.
3. Implementation of optimal control theory in hardware places a great burden on the computational equipment, with the overall cost determining its usefulness.

Table 5.2 summarizes the performance indices most frequently used in optimal control theory.

PROBLEMS

5.1 What are the minimum and maximum values with respect to **x** of

$$\langle \mathbf{x}, A\mathbf{x} \rangle$$

subject to the constraint $\langle x, x \rangle = 1$, where A is a given positive definite matrix?

5.2 Find the shortest distance from the origin to the plane $ax + by + cz = d$. [*Hint*: Given the plane $ax + by + cz - d = f(x, y, z) = 0$. The distance of a point from the origin is given by $\theta(x, y, z) \equiv \sqrt{x^2 + y^2 + z^2}$ with the constraint $f = ax + by + cz - d = 0$.]

5.3 An airplane in steady level flight with the thrust tilted at an angle θ from the horizontal has the equilibrium equations

$$L - W - T\sin\theta = 0, \qquad T\cos\theta - D = 0,$$

Table 5.2   Summary of Useful Performance Indices

Physical system (or plant):

$$\dot{x}(t) = A(t)x(t) + B(t)u(t), \qquad x(t_0) = x_0$$
$$y(t) = C(t)x(t)$$

Generalized performance index:

$$J(x_0, t_0, T, u(t)) = \tfrac{1}{2}x^T(T)Sx(T) + \tfrac{1}{2}\int_{t_0}^{T}[x^T(t)Q(t)x(t) + u^T(t)R(t)u(t)]\, dt$$

Hamiltonian:

$$H(x(t), p(t), u(t), (t) = \tfrac{1}{2}x^T(t)Q(t)x(t) + \tfrac{1}{2}u^T(t)R(t)u(t) + p^T(t)[A(t)x(t) + B(t)u(t)]$$

Optimal control:

$$u^*(x(t), t) = -R^{-1}(t)B^T(t)P(t)x(t) \triangleq -L^*(t)x(t)$$

Matrix Riccati equation:

$$\dot{P}(t) = -P(t)A(t) - A^T(t)P(t) + P(t)B(t)R^{-1}(t)B^T(t)P(t) - Q(t)$$

Optimal regulator:

$$J = \tfrac{1}{2}\int_0^{\infty}[x^T(t)Qx(t) + u^T(t)Ru(t)]\, dt$$

Output regulator:

$$J = \tfrac{1}{2}\int_0^{\infty}[y^T(t)Qy(t) + u^T Ru(t)]\, dt$$

Energy-optimal problem:

$$J = \int_0^T u^T(t)Ru(t)\, dt$$

Terminal control problem:

$$J = \tfrac{1}{2}x^T(T)Sx(T) + \tfrac{1}{2}\int_{t_0}^{T}u^T(t)Ru(t)\, dt$$

Problems with Nonquadratic performance indices:

Fuel-optimal problem takes a system from x_1 at t_1 to x_2 at t_2 so as to minimize the fuel loss:

$$J(u) = \int_{t_0}^{T}\sum_{i=1}^{r}|u_i(t)|\, dt \quad |u_i(t)| \leqslant 1, \qquad T \text{ free}, \quad i = 1, 2, \ldots, r.$$

Time-optimal problem:

$$J(u) = \int_{t_0}^{T} dt = T - t_o, \quad T \text{ free}.$$

where the lift $L = (\tfrac{1}{2}\rho S)C_L V^Z$, the drag $D = (\tfrac{1}{2}\rho S)C_D V^2$, and the air density ρ, wing area S, thrust magnitude T, and weight W are given constants. Assume that lift and drag coefficients are related in the usual drag polar form $C_D = C_{D0} + KC_L^2$, where C_{D0} and K are known constants. Show that the tangent of the thrust angle θ for maximum level flight speed is $2KC$. Use Lagrange multipliers in setting up the problem.

5.4   Suppose we want to minimize the integral J of a function of two dependent variables $y(x)$ and $z(x)$ and their derivatives, for fixed boundary

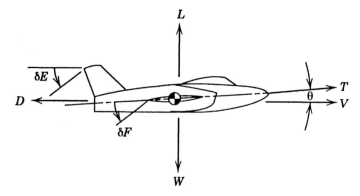

Figure Problem 5.3

conditions,

$$J = \int_{x_1}^{x_2} f(x, y, z, y', z') \, dx.$$

Derive the necessary pair of Euler–Lagrange equations,

$$f_y - \frac{df_y}{dx} = 0, \qquad \frac{df_z}{dx} = 0,$$

by assuming variations of the form

$$Y(x, \alpha) = y(x) + \alpha \eta(x), \qquad Z(x, \alpha) = z(x) + \alpha \zeta(x),$$

and setting $dJ/d\alpha = 0$ for $\alpha = 0$.

5.5 In this problem (see also Example 3, Section 5.3), a simple integral cost will be reformulated as a terminal cost problem. Minimize

$$J(u) = \frac{1}{2} \int_0^T [x(t)^2 + u(t)^2] \, dt$$

where $\dot{x} = -x + u$ with the initial condition $x(0) = x_0$. To solve this problem, define a new state

$$\dot{y}(t) = \frac{1}{2} [x(t)^2 + u(t)^2].$$

Then $y(0) = 0$, and $J(u) = 0 \cdot x(T) + 1 \cdot y(T) = F(x(T), y(T))$. Therefore, the new problem is to minimize $y(T)$, where

$$\dot{x} = -x + u, \qquad x(0) = x_0,$$

$$\dot{y} = \tfrac{1}{2}(x^2 + u^2), \qquad y(0) = 0.$$

5.6 Consider the *time-optimal* case. Minimize

$$J(u) = T = \int_0^T dt$$

subject to the following constraints:

(a) $\dot{x}_1 = x_2,\ \dot{x}_2 = -x + u.$
(b) $x_1(0) = a,\ x_2(0) = b,\ x_1(T) = 0,\ x_2(T) = 0.$
(c) $|u(t)| \leqslant 1.$

5.7 Another fundamental problem in optimal control theory is that the optimal control involves the *optimal state*. Thus, a serious coupling problem occurs. Iterative procedures seem to be the only way to solve such problems. Minimize

$$J(u) = \int_0^T u(t)^2\, dt$$

subject to the following constraints:

(a) $\dot{x}_1 = x_2,\ \dot{x}_2 = -(1 - x_1{}^2)x_2 - x_1 - u.$
(b) $x_1(0) = a,\ x_2(0) = b,\ x_1(T) = 0,\ x_2(T) = 0.$
(c) No constraints on the control.

5.8 Consider the following one-dimensional problem. Minimize

$$J(u) = \int_0^T [x(t)^2 + u(t)^2]\, dt$$

subject to the following constraints:

(a) $\dot{x} = x + u.$
(b) $x(0) = x_0,\ x(T) = 0.$
(c) No constraints on the control.

5.9 Consider the system

$$\dot{x}_1(t) = x_2(t), \qquad x_1(0) = 1,$$
$$\dot{x}_2(t) = x_3(t), \qquad x_2(0) = 2,$$
$$\dot{x}_3(t) = x_4(t), \qquad x_3(0) = 3,$$
$$\dot{x}_4(t) = x_4(t) + u(t), \qquad x_4(0) = 4.$$

Assume that $u(t)$ is unconstrained. Next, consider the cost functional

$$J(u) = \frac{1}{2}x_1{}^4(1) + \frac{1}{2}\int_0^1 \left[x_1{}^2(t) + \frac{1}{2}x_4{}^4(t) + u^2(t) \right] dt.$$

Given the boundary conditions

$$x_2(1) = 0, \qquad x_3{}^2(1) + x_4{}^2(1) = 1 \quad \text{(a circle)}.$$

Suppose that an optimal $u^*(t)$ exists which transfers the initial state to the required boundary conditions and which minimizes the cost functional. State all of the necessary conditions provided by the minimum principle for this problem.

5.10 Consider the first-order system

$$\dot{x}(t) = -x(t), + 2u(t), \qquad |u(t)| \leq 1.$$

It is desired to transfer this system from the initial state $x(0) = 1$ to the terminal state $x(T) = -1$, T given, while minimizing the cost functional

$$J(u) = \int_0^T [\sqrt{|x(t)|} + |u(t)|] dt, \qquad T \text{ fixed}.$$

Can the minimum principle be used to obtain the necessary conditions?

5.11 A first-order system is given in the form $\dot{x}(t) = u(t)$, with $u(t)$ unconstrained. It is desired to force the system from the initial state $x(0) = 1$ to the state 0 at time 1, and minimize the cost functional

$$J(u) = \frac{1}{2}\int_0^1 [x^2(t) + u^2(t)] dt.$$

An engineer claims that the control is $u(t) = -1 \; \forall \, t, \, 0 \leq t \leq 1$. Prove or disprove the engineer's claim.

5.12 This problem considers a simple first-order LQR system. Let the system be described by the scalar, linear, time-invariant model given by

$$\dot{x}(t) = fx(t) + gu(t), \qquad x(0) = x_0,$$

and the quadratic cost function by

$$J = \frac{1}{2} \int_0^\infty (qx^2 + ru^2)dt.$$

Assuming that $F = f$, $G = g$, $R = r$, and $Q = q$, determine the following:

(a) The algebraic Riccati equation.
(b) The optimal control gain L^* and the optimal control law $u^*(t) = -L^*x(t)$, Eq. (5.51).
(c) The solution to the closed-loop system [see Eq. (5.52)]

$$\dot{x}^*(t) = (f - gL^*)x^*(t), \qquad x^*(0) = x_0.$$

5.13 It is common practice in missile guidance to design a wing controlled missile so that the roll attitude is kept at a fixed orientation throughout its flight in order for the guidance system to function properly. For this reason, a roll autopilot is needed to maintain the desired orientation. Assume that the equations of motion for the rolling motion of a wing controlled missile can be expressed by the equations

$$I_x(dp/dt) = (\partial L/\partial p)p + (\partial L/\partial \delta)\delta \qquad (1)$$

$$d\phi/dt = p \qquad (2)$$

where I_x = moment of inertial about the x-axis (the x-axis is commonly taken to be the longitudinal axis of the missile),
ϕ = roll attitude angle (taken to be positive clockwise about the body x-axis),
δ = aileron deflection,
L = rolling moment.

The first term on the right-hand side of Eq. (1) is the roll-damping moment and the second term is the roll moment due to the deflection of the ailerons. Equations (1) and (2) can also be written in the form

$$dp/dt = L_p p + L_\delta \delta \qquad (3)$$

$$d\phi/dt = p \qquad (4)$$

where $L_p = (1/I_x)(\partial L/\partial p)$,
$L_\delta = (1/I_x)(\partial L/\partial \delta)$.

In the state-space notation, Eqs. (3) and (4) can be written as

$$\dot{x} = Ax + Bu$$

$$\begin{bmatrix} \dot{\phi} \\ \dot{p} \end{bmatrix} = \begin{bmatrix} 0 & 1 \\ 1 & L_p \end{bmatrix} \begin{bmatrix} \phi \\ p \end{bmatrix} + \begin{bmatrix} 0 \\ L_\delta \end{bmatrix} \delta \qquad (5)$$

where u is the control. The quadratic performance index to be minimized will have the form

$$J = \frac{1}{2} \int_0^\infty \left[\left(\frac{\phi}{\phi_{max}} \right)^2 + \left(\frac{p}{p_{max}} \right)^2 + \left(\frac{\delta}{\delta_{max}} \right)^2 \right] dt \qquad (6)$$

where ϕ_{max} = maximum desired roll attitude angle.
  p_{max} = maximum desired roll rate.
  δ_{max} = maximum aileron deflection.

Using the quadratic performance index (Eq. (5.36)) with zero terminal constraint, that is,

$$J = \int_0^{t_f} (\mathbf{x}^T \mathbf{Q} \mathbf{x} + \mathbf{u}^T \mathbf{R} \mathbf{u}) dt,$$

it will be noted that in Eq. (6) $(\phi^2/\phi_{max}{}^2) + (p^2/p_{max}{}^2) = \mathbf{x}^T \mathbf{A} \mathbf{x}$ and $(\delta^2/\delta_{max}{}^2) = \mathbf{u}^T \mathbf{B} \mathbf{u}$. Assuming that the $\mathbf{Q}$ and $\mathbf{R}$ matrices are

$$\mathbf{Q} = \begin{bmatrix} (1/\phi_{max}{}^2) & 0 \\ 0 & (1/p_{max}{}^2) \end{bmatrix}, \text{ and } \mathbf{R} = 1/\delta_{max}{}^2$$

and that

$$L_p = -2 \text{ rad/sec},$$
$$L_\delta = 9000 \text{ sec}^{-2},$$
$$\phi_{max} = 10° = 0.174 \text{ rad},$$
$$p_{max} = 300 \text{ deg/sec} = 5.23 \text{ rad/sec},$$

find the optimal control δ. (*Hint:* Solve the steady-state matrix Riccati equation (Eq. (5.38))

$$\mathbf{SA} + \mathbf{A}^T\mathbf{S} - \mathbf{SBR}^{-1}\mathbf{B}^T\mathbf{S} + \mathbf{Q} = 0$$

with the optimal control given by (use Eq. (5.39) for the optimal control gain matrix)

$$\mathbf{u} = -\mathbf{R}^{-1}\mathbf{B}^{T}\mathbf{S}\mathbf{x} = -\mathbf{C}\mathbf{x}.$$

Also, note that since the matrix Riccati equation is symmetric, it is only necessary to solve this equation for the elements generated along and above the diagonal of the matrix).

CHAPTER 6

COVARIANCE ANALYSIS
AND SUBOPTIMAL FILTERING

6.1 COVARIANCE ANALYSIS

Covariance analysis is an analytic technique in which the true error covariance is determined from specification of the filter model and a system (truth) model. Therefore, in order to develop the covariance analysis equations, two models are required, namely, the *truth model* and *filter model*. The truth model (also known as the *system real-world*, or *plant* model) is a description of the dynamics and the statistics of the errors in a given system. It may be meant to represent the best available model of the true system, or it may be hypothesized to test the sensitivity of a particular filter design to modeling errors. The filter model is a suboptimal model (see Section 6.2), usually of reduced order, from which the Kalman gain is computed. The truth and filter models are depicted in block diagram form in Figure 6.1.

The error state, $e(t_i) = x(t_i) - \hat{x}(t_i)$, which is the difference of the true state $x(t_i)$ and the estimated state $\hat{x}(t_i)$, describes the actual performance capabilities of the filter. In view of the above discussion, a statistical characterization of this error can be accomplished by a Monte Carlo analysis (to be discussed later in this chapter), in which many samples of the stochastic process $e(t_i)$ are generated, and then sample statistics are generated as ensemble averages [53]. However, the covariance of $e(t_i)$ can be computed from a simple covariance run. As we shall presently see, this method is used to generate time histories of the filter rms errors.

In recent years, the covariance analysis technique has been commonly applied to inertial navigation systems updated with a Kalman filter that processes external measurements. These systems are ideally suited to covariance

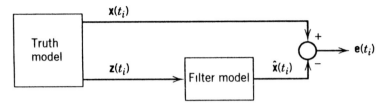

Figure 6.1 Representation of the truth and filter models.

analysis, since the error equations are linear. The error states of an inertial navigation system are zero-mean random variables, and thus their variance is mainly of interest.

As stated above, the difference between the actual state and its estimated state, $\mathbf{x}(t) - \hat{\mathbf{x}}(t)$, is known as the error vector or *estimate error*. By taking the mathematical expectation of the second moment of the estimate error, the covariance matrix is generated, where the diagonal terms are the variances of the state variables and the off-diagonal terms are the covariances. Therefore [see also Section 2.4 and Eq. (4.10)],

$$P(t) = \mathscr{E}\{[(\mathbf{x}(t) - \hat{\mathbf{x}}(t)][\mathbf{x}(t) - \hat{\mathbf{x}})t)]^T\} \tag{6.1}$$

$$= \begin{bmatrix}
p_{11} & p_{12} & p_{13} & \cdots & p_{1n} \\
p_{21} & p_{22} & p_{23} & \cdots & p_{2n} \\
p_{31} & p_{32} & p_{33} & \cdots & \vdots \\
\vdots & \vdots & & \ddots & \vdots \\
p_{n1} & p_{n2} & \cdots & \cdots & p_{nn}
\end{bmatrix}$$

$$= \begin{bmatrix}
\sigma_1^2 & \mathrm{cov}(x_1, x_2) & \cdots & \mathrm{cov}(x_1, x_n) \\
\mathrm{cov}(x_2, x_1) & \sigma_2^2 & \cdots & \mathrm{cov}(x_2, x_n) \\
\vdots & \vdots & \ddots & \vdots \\
\mathrm{cov}(x_n, x_1) & \mathrm{cov}(x_n, x_2) & \cdots & \sigma_n^2
\end{bmatrix} = \begin{bmatrix}
\sigma_1^2 & \sigma_{12} & \cdots & \sigma_{1n} \\
\sigma_{12} & \sigma_2^2 & \cdots & \sigma_{2n} \\
\vdots & \vdots & \ddots & \vdots \\
\sigma_{n1} & \sigma_{n2} & \cdots & \sigma_n^2
\end{bmatrix}.$$

Expressed mathematically, the elements p_{ii} on the diagonal of $P(t)$ are the variances of the state parameters, that is,

$$p_{ii} = \sigma_i^2, \tag{6.2}$$

where σ_i is the standard deviation* of the estimation error of the ith state parameter. Nondiagonal elements give the covariance of the state parameters.

---

*The square roots of the diagonal terms yield the time histories of standard deviations (that is, "1σ values," equal to the rms values provided the processes are zero-mean).

Thus, for $i \neq j$ the correlation coefficient ρ_{ij} between the i and j state elements can be determined from the elements of $P(t)$ as follows:

$$\rho_{ij} = \frac{p_{ij}}{\sqrt{p_{ii}p_{jj}}} = \frac{p_{ij}}{\sigma_i \sigma_j}, \qquad -1 \leqslant \rho_{ij} \leqslant 1. \qquad (6.3)$$

In terms of the correlation coefficient, the covariance matrix $P(t)$ can be expressed as

$$P(t) = \begin{bmatrix} \sigma_1^2 & \rho_{12}\sigma_1\sigma_2 & \rho_{13}\sigma_1\sigma_3 & \cdots & \cdots & \rho_{1n}\sigma_1\sigma_n \\ \rho_{21}\sigma_2\sigma_1 & \sigma_2^2 & \rho_{23}\sigma_2\sigma_3 & \cdots & \cdots & \rho_{2n}\sigma_2\sigma_n \\ \vdots & \vdots & \vdots & & \vdots & \vdots \\ \vdots & \vdots & \vdots & \ddots & \vdots & \vdots \\ \rho_{n1}\sigma_n\sigma_1 & \rho_{n2}\sigma_n\sigma_2 & \rho_{n3}\sigma_n\sigma_3 & \cdots & & \sigma_n^2 \end{bmatrix}.$$

The covariance matrix provides a statistical representation of the uncertainty in the current estimate of the state, and of the correlation between the individual elements of the state. Furthermore, as we have seen earlier, the estimate error covariance is a statistical measure of uncertainty in the estimate of the state vector $x(t)$. The following features of the error covariance matrix relate the matrix to the state vector and its estimate:

1. The error covariance matrix of an n-component state vector is a symmetric $n \times n$ matrix $(p_{ij} = p_{ji})$.
2. The diagonal elements of the error covariance matrix are the mean-square errors of the error vector components.
3. The trace of the error covariance matrix is the mean-square length of the error vector.
4. The off-diagonal terms of the matrix are correlations between the elements of the error vector $x(t) - \hat{x}(t)$.
5. As k goes to infinity, the error covariance matrix $P(k)$ tends to the system dynamics covariance matrix $Q(k)$. That is, as more information is obtained about the state vector through observations, the estimate uncertainty approaches the uncertainty of the environment in which the state vector exists.

Covariance analysis can be used as a tool to evaluate candidate filter designs by considering the system model to be fixed. Sensitivity studies can be performed by holding the filter model constant and varying the system model. For example, if it is found that the performance is a strong function of modeling errors, the filter design should be changed, since system models are in general not known exactly.

In the analysis of inertial navigation systems, it is well known that performance is affected by many instrument error souces. Investigators have identified more than one hundred of these state variables. Most of them, as we shall see in Section 6.2, contribute little to the performance, and thus are not needed for most accuracy analyses. For error analysis, the system analyst has used two approaches: (1) covariance analysis, and (2) Monte Carlo simulation. These methods will now be discussed.

In essence, the objective of a covariance analysis and/or simulation is to develop a mission error analysis tool in order for the analyst to:

1. assess mission performance based on the mission trajectory, the system error budget, and any navigational aids that may be used;
2. serve as a software design tool in determining which error states should be modeled in the Kalman filter;
3. assess the relative critical aspects of individual error components so as to aid in the design and construction of sensor hardware;
4. aid in the design of the navigation system with regard to sensor error calibration, alignment filter, and generation of cost–performance tradeoff studies.

At this point it is appropriate to define what we mean by simulation. In systems analysis, simulation is the verification of the accuracy of the modeling dynamics. Specifically, simulation studies can determine the degree to which the model approximates a known physical process and hence verify its accuracy. Therefore, since many physical processes have certain behavioral responses, one verifies the model by seeing whether its responses coincide with the observed responses of the process.

Simulation studies are commonly used in the design, testing, and evaluation of feedback controllers, whereby the actual development of the control algorithm, the evaluation of the design, and the optimization of parameters, as well as the evaluation of performance in the presence of system disturbances.

The covariance simulation mechanizes a real-world error state model (normally user-specified) which is a subset of the total error sources. Specifically, in simulations of unaided-system performance, the simulation extrapolates the covariance matrix of the real-world error states, utilizing the system state transition matrix. On the other hand, in simulations of aided-system performance, a *design filter* error state vector is defined (again, user-specified), which, in general, is a subset of the true world error sources to be implemented in the simulated onboard-computer Kalman filter mechanization.

One of the requirements of covariance analysis is that all inputs should be of zero mean. For optimal filter performance, all of the real-world error sources must be included in the design filter state vector.

We will now recapitulate the covariance propagation and update equations that are used in covariance analysis. In linear covariance analysis (for example, multisensor Kalman-integrated navigation systems) the error dynamics can be represented to good approximation by the first-order variational equation

$$\delta \dot{\mathbf{x}} = F(t)\, \delta \mathbf{x} + G(t)\, \delta \mathbf{w}, \tag{6.4}$$

where $\delta \mathbf{x}$ = system state vector,
  $\delta \mathbf{w}$ = forcing-function vector,
  $F(t)$ = system error dynamics matrix,
  $G(t)$ = matrix that determines the coupling between the input vector and system states.

Next, let the error vector $\delta \mathbf{z}$ be the difference between the measured value of a navigation variable, such as position or velocity, and the value indicated in the navigating vehicle's computer. Then, the predicted value of this difference, $\delta \hat{\mathbf{z}}$, is related to the system errors modeled in the state vector by the equation

$$\delta \hat{\mathbf{z}} = H(t)\delta \hat{\mathbf{x}} + \delta \mathbf{v}, \qquad (6.5)$$

where $\delta \hat{\mathbf{z}}$ = a vector of measurement variables,
  $\delta \mathbf{v}$ = uncorrelated (white) noise associated with the measurement process
  $H(t)$ = measurement matrix.

The forms of the error dynamics matrix $F(t)$ and the measurement matrix $H(t)$ depend on the navigation system mechanization coordinates and on the navigation sensor error models. For many applications, the covariance analysis approach can be selected as the best method for generating mission navigation data to the specified accuracy and confidence levels and within computer limitations. This approach requires selecting a suitable system matrix $F(t)$ which determines the error propagation of an aircraft or other vehicle navigation system between external updates. The errors modeled in the program are treated initially as uncorrelated normally distributed random variables having zero mean values. An initial covariance matrix of position, velocity, attitude, and modeled sensor errors may be propagated along a reference trajectory and printed out when desired. This covariance matrix is then updated with a Kalman filter whenever position or velocity observations are available. For an aircraft, these updates can be made at regular intervals over any part of the flight. When an external measurement is made, the Kalman filter processes this measurement and updates the navigation system, which then results in a discrete improvement of the estimates of the error states. The covariance-matrix propagation and update will now be discussed (see also Chapter 4, Section 4.3).

Covariance Propagation A matrix Riccati differential equation that describes the time propagation (between updates) of the covariance matrix is

$$\dot{P}(t) = F(t)P(t) + P(t)F^{T}(t) + G(t)Q(t)G^{T}(t) \qquad (6.6)$$

where $P(t)$ = covariance matrix,
  $F(t)$ = system error dynamics matrix,

$G(t) =$ input matrix (which can be an identity matrix),
$Q(t) =$ process noise covariance matrix.

This is Eq. (4.11) where we have assumed that $H(t) = [0]$. Solving this equation constitutes what is commonly referred to a *covariance analysis*. The equation can be numerically integrated to produce the time history of error states, provided that (1) the initial covariance matrix $P(0)$ is given, and (2) a trajectory generator is available to produce position, velocity, and acceleration variation, which are needed in the time-varying elements of the system matrix $F(t)$. We have noted earlier that navigation errors reside in the matrix $P(t)$, whose diagonal terms are the variances of the navigation error states and whose off-diagonal terms are the covariances of these error states. The covariance matrix $P(t)$ can be propagated to $T + DT$ using, for example, a standard fourth-order Runge–Kutta integration.

The Runge–Kutta method is a popular computational scheme for digital computers because of its programming efficiency. The four $P(t)$ derivatives (K_1, K_2, K_3, K_4) are computed and inserted into the Runge–Kutta general equation, which attaches to each derivative a numerical importance with respect to the other derivatives. Taking advantage of symmetry about the matrix diagonal, only the upper diagonal elements of the matrix are integrated. The deviation from the mean is the square root of the variance, so that the square root of the variance element $P(1,1)$ is the standard deviation or navigation error. The element $P(1,2)$ is the covariance of the navigation error between the X and Y axes, etc.

Covariance Update   The general equations for the Kalman filter gain matrix and updated covariance are

$$K(k) = P(k|k-1) H^T(k)[H(k)P(k|k-1)H^T(k) + R(k)]^{-1}, \qquad (6.7)$$

$$P(k|k) = [I - K(k)H(k)]P(k|k-1)$$

$$= [I - K(k)H(k)]P(k|k-1)[I - K(k)H(k)]^T + K(k)R(k)K^T(k) \quad (6.8)$$

or one can use the form

$$P(t^+) = P(t^-) - P(t^-)H^T(t)[H(t)P(t^-)H^T(t) + R]^{-1}H(t)P(t^-), \quad (6.9)$$

where $P(t^-) =$ navigation error covariance matrix prior to update,
$P(t^+) =$ navigation error covariance matrix after update,
$H(t) =$ mapping matrix (also called the measurement sensitivity matrix),
$R =$ covariance matrix of measurement errors.

The time delay between when a measurement is made from, say, a known landmark or a map, and when the navigation system is updated, leads to a delayed observable set of update equations. This algorithm uses the difference between the update time and the time of measurement, inserted into the system state transition matrix, to propagate the measurements to the current time (update time). State corrections are then made at the current time, and thus the covariance $P(t)$ is updated at the measurement time and the effects of this update are propagated to current time. The above calculations are performed for each measurement that is made. Following the update, the updated covariance and trajectory state vector are stored for restarting the updating process. The covariance matrix elements $P(1,1), P(1,2)$, and $P(2,2)$ are stored navigation data. Navigation data can be produced by a time history simulation.

Three of the functions that produce the basic navigation data are as follows: (1) the trajectory generator, (2) the covariance propagation, and (3) the covariance update. Other functions, such as input–output and data handling, will also be required for the navigation process. For most large problems the system dynamics matrix $F(t)$ and the noise sensitivity matrix $G(t)$ are sparse, and this can be used to reduce the computational effort. The system dynamics matrix for inertial system errors is typically a function of navigation parameters such as position, velocity, and acceleration. These parameters can be calculated from an aircraft trajectory model that can simulate maneuvers such as takeoff, climb, cruise, great-circle routes, turns, and landing. Figure 6.2 depicts a functional flow diagram of the above functions.

This loop solves the vehicle flight-path equations in the trajectory generator to provide vehicle position, velocity, and acceleration data required by the covariance propagation and covariance update blocks. The covariance propagation block involves the time solution of a set of navigation error equations which define how the navigation error states propagate during flight between one update point and the next. The covariance update function mechanizes a set of equations which describes the improvement in the navigation error estimates resulting from incorporating an external measurement or fix.

The discussion presented so far in this chapter can be summarized by noting that the covariance analysis provides an effective tool for conducting a tradeoff analysis among the various designs. In this connection, the Runge–Kutta method is presented below for the interested reader [24]. Important to the Runge–Kutta numerical integration algorithm is the step size T. If T is chosen too large, the accuracy will be poor, while if it is chosen small, the computation time is increased and roundoff errors will become pronounced.

Fourth-Order Runge–Kutta Numerical Integration [24]

Given: $dx/dt = f(x, t)$, $x(t_0)$.

Find: $x(t_{n+1})$,   where $t_{n+1} = t_n + \Delta$.

Solution: $x(t_{n+1}) = x(t_n) + \frac{1}{6}(K_1 + 2K_2 + 2K_3 + K_4)$,   where

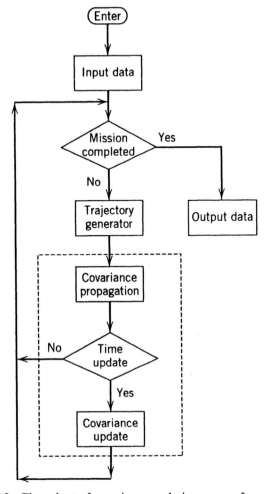

Figure 6.2 Flow chart of covariance analysis program for navigation.

$$K_1 = \Delta \cdot f(x(t_n), t_n),$$

$$K_2 = \Delta \cdot f\left(x(t_n) + \frac{1}{2}K_1, t_n + \frac{\Delta}{2}\right),$$

$$K_3 = \Delta \cdot f\left(x(t_n) + \frac{1}{2}K_2, t_n + \frac{\Delta}{2}\right),$$

$$K_4 = \Delta \cdot f(x(t_n) + K_3, t_n + \Delta).$$

Note:
1. Local truncation error is on the order of Δ^5.
2. The derivative $f(x, t)$ has to be evaluated four times per Δ.

Algorithm for Fourth-Order Runge–Kutta Method
SUBROUTINE RUNGE 4(X, T, DT, N, Y, W1, W2):

1. X is the incoming state vector at time T and the outgoing state vector at time T + DT.
2. Y, W1, W2, are working vectors of length N.
3. The integration step size DT is fixed.
4. This subroutine calls **Subroutine** DERIV(X, T, Y) where the function $f(X, t)$ is evaluated.

SUBROUTINE DERIV (X, T, D)

1. This routine is supplied by the user to evaluate the expression $f(X, t)$.
2. X is the input state vector.
3. D is the output derivative vector $dx/dt = f(X, t)$.
4. T is the time.
5. X, D are $N \times 1$ vectors

The program listing for the Runge–Kutta algorithm is given below:

```
1               SUBROUTINE RUNGE 4 (X, T, DT, N, Y, W1, W2)
2               DIMENSION X(1), Y(1), W1(1), W2(1)
    C
    C           X =  INCOMING STATE AT TIME T AND OUTGOING STATE AT
    C                TIME T + DT
    C           W1, Y, AND W2 ARE WORKING VECTORS OF LENGTH N
    C           STEPSIZE IS FIXED AT DT
    C
3               CALL DERIV (X, T, Y)
4               DO 1 I = 1, N
5               Y(I) = Y(I) DT
6               W1(I) = Y(I)
7     1         W2(I) = X(I) + Y(I)/2.
8               T = T + DT/2.
9               CALL DERIV (W2, T, Y)
10              DO 2 I = 1, N
11              Y(I) = Y(I)*DT
12              W1 (I) = W1(I) + Y(I)*2.
13    2         W2(I) = X(I) + Y(I)/2.
14              CALL DERIV (W2, T, Y)
```

```
15                  DO 3 I = 1, N
16                  Y(I) = Y(I)*DT
17                  W1(I) = W1(I) + Y(I)*2.
18      3           W2(I) = X(I) + Y(I)
19                  T = T + DT/2.
20                  CALL DERIV (W2,T,Y)
21                  DO 4 I = 1, N
22      4           X(I) = X(I) + (W1(I) + Y(I)*DT)/6.
23                  RETURN
24                  END


1                   SUBROUTINE DERIV (X, TIME, D)
2                   DIMENSION X(1), D(1)
3                   D(1) = 2. *(X(2) − X(1)) − X(3) + 1.
(3)     -           WARNING------------SUBSCRIPT OF X IS < 1 OR > THE
                    DIMENSION
4                   D(2) = X(1) − .5*X(2) + X(3)
(4)     -           WARNING----------SUBSCRIPT OF X IS < 1 OR > THE
                    DIMENSION
5       20          RETURN
6                   END
```

In certain dynamical systems where second-order ordinary differential equations govern the vehicle motion, these equations can be integrated numerically using the *Adams*, or *Adams–Moulton*, method. The Adams method is a fourth-order predictor–corrector technique, which can be generally stated as follows [32]:

$$\textit{Predictor:} \quad y_{n+1} = y_n + \frac{h}{24}(55f_n + 59f_{n-1} + 37f_{n-2} + 9f_{n-3}),$$

$$\textit{Corrector:} \quad y_{n+1} = y_n + \frac{h}{24}(9f_{n+1} + 19f_n + 5f_{n-1} + f_{n-2}).$$

where f_i may or may not be a linear function of its arguments.

The predictor–corrector scheme for calculating y_{n+1} is given below [32].

1. Use the predictor to calculate $y_{n+1}^{(0)}$, an initial approximation for y_{n+1}. Set $i = 0$.
2. Evaluate the derivative function and set $f_{n+1}^{(i)} = f(y_{n+1}^{(i)}, t_{n+1})$.
3. Calculate a better approximation $y_{n+1}^{(i)}$ using the corrector formula with $f_{n+1} = f_{n+1}^{(i)}$.

4. If $|y_{n+1}^{(i+1)} - y_{n+1}^{(i)}| >$ a convergence tolerance, then increase i by 1 and go to step 2; otherwise, set $y_{n+1} = y_{n+1}^{(i+1)}$.

The above method can now be applied to the equations of motion of a second-order differential equation having the form

$$\ddot{x} = f(x, \dot{x}, t).$$

The predicted velocity at time t_{n+1} is

$$\dot{x}_{n+1}^{(0)} = \dot{x}_n + \frac{h}{24}(55\ddot{x}_n + 59\ddot{x}_{n-1} + 37\ddot{x}_{n-2} + 9\ddot{x}_{n-3}),$$

whereas the corrected velocity after i iterations is

$$\dot{x}_{n+1}^{(i)} = \dot{x}_n + \frac{h}{24}(9\,\ddot{x}_{n+1}^{(i-1)} + 19\ddot{x}_n + 5\ddot{x}_{n-1} + \ddot{x}_{n-2}),$$

where $h = t_{n+1} - t_n$ is the integration step size. The starting step size can be found by experiment or from physical intuition. Similarly, the position after i iterations is

$$x_{n+1}^{(i)} = x_n + \frac{h}{24}(9\dot{x}_{n+1}^{(i-1)} + 19\dot{x}_n + 5\dot{x}_{n-1} + \dot{x}_{n-2}).$$

The program permits no more than four iterations (i.e., $i \leq 4$) for convergence of the corrector, which is defined to occur when

$$|\dot{x}_{n+1}^{(i)} - \dot{x}_{n+1}^{(i-1)}| \leq \varepsilon_{\max}|\dot{x}_{n+1}^{(i)}|$$

and

$$|x_{n+1}^{(i)} - x_{n+1}^{(i-1)}| \leq \varepsilon_{\max}|x_{n+1}^{(i)}|$$

for each computed vector component of velocity and position. In the above inequalities, $\varepsilon_{\max}$ is the maximum allowed relative error, a user-specified input. Consequently, after convergence is obtained, the truncation error is estimated as

$$E_T = \frac{19}{270}(\dot{x}_{n+1}^{(0)} - \dot{x}_{n+1}^{(i)})$$

and then added to the corrected velocity as a final adjustment:

$$\dot{x}_{n+1} = \dot{x}_{n+1}^{(i)} + E_T.$$

The truncation error is judged to be acceptably small if the following condition is satisfied:

$$|E_T| < \varepsilon_{max} \max(|\dot{x}_{n+1}|, A_4)$$

for each velocity component. A_4 is a reference number, assigned the nominal value of 10^{-4} in the input file. The starting values of the Adams–Moulton method are found by the fourth-order Runge–Kutta procedure, in which the velocity and position at time t_{n+1} are found as

$$\dot{x}_{n+1} = \dot{x}_n + \frac{h}{6}(k_1 + 2k_2 + 2k_3 + k_4),$$

$$x_{n+1} = x_n + h\left[\dot{x}_n + \frac{h}{6}(k_1 + k_2 + k_3)\right],$$

where

$$k_1 = f(t_n, x_n, \dot{x}_n),$$

$$k_2 = f\left(t_n + \frac{h}{2}, x_n + \frac{h}{2}\dot{x}_n + \frac{h}{8}k_1, \dot{x}_n + \frac{k_1}{2}\right),$$

$$k_3 = f\left(t_n + \frac{h}{2}, x_n + \frac{h}{2}\dot{x}_n + \frac{h}{8}k_1, \dot{x}_n + \frac{k_2}{2}\right),$$

$$k_4 = f\left(t_n + h, x_n + h\dot{x}_n + \frac{h}{2}k_3, \dot{x}_n + k_3\right).$$

One advantage of the predictor–corrector techniques is that they readily yield a truncation-error estimate, which may be used as a guide for adjusting the step size throughout the trajectory calculation.

Depending on the particular application, the covariance propagation program will have to use supporting subroutines. Therefore, in a typical computer program implementation of covariance analysis, the program will normally consist of three types of routines: (1) those that implement the general equations, which are not expected to change frequently, (2) those that provide the detailed structure of the system model and the filter model, and (3) service routines which provide matrix arithmetic, plot output, etc. The main routine provides control of the program, which includes data input, initialization of covariance matrices and other data, navigation mode control, mission scenario, and data output. This routine need not be changed in the utilization of the covariance analysis program. The mission scenario simulation can start at the beginning of a mission, or it can be a continuation of a mission which began on a previous execution of the covariance analysis program.

Next, we shall briefly discuss the Monte Carlo simulation method. As stated earlier in this section, a Kalman filter's performance can also be evaluated by the Monte Carlo simulation, which provides a convenient means for the qualitative investigation of the behavior of a stochastic system. The scope of this method is determined by the extent to which the statistical characteristics of the random number generator resemble those of the system variables. There are many well-known techniques by which it is possible to generate random numbers having the required distribution characteristics.

The Monte Carlo simulation incorporates both the truth-model errors and the filter's estimates of these errors. The truth model generates the actual measurements and compares them with the filter's estimates of the measurements. In addition, the Monte Carlo scheme simulates feedback corrections to the truth model and the filter model, and can be used to determine how the system errors propagate under dynamics. Above all, the Monte Carlo simulation is applicable when the disturbance has a nonzero mean (i.e., bias).

Since both the truth model and the filter model are driven by randomly generated noise, each individual Monte Carlo run is expected to be different. Therefore, in order to generate error statistics with a Monte Carlo program, a given case is iterated many times, the iterations differing only in the random number input sequences. The results of the iterations are then averaged to obtain the desired statistics. Consequently, observing the ensemble statistics of several runs gives an indication of the filter's expected performance. Naturally, the more runs are made, the more reliable become the statistics. Between 25 and 50 runs typically is used to determine filter performance. Each run produces a different sequence of random numbers to generate the samples of input white-noise processes. Sample statistics are computed for each time point using the equations

$$\hat{x} = \frac{1}{N} \sum_{i=1}^{N} x_i, \tag{6.10}$$

$$\hat{\sigma}^2 = \frac{1}{N-1} \sum_{i=1}^{N} x_i^2 - \frac{N}{N-1} \hat{x}^2, \tag{6.11}$$

where $\hat{x}$ = estimate of the mean of x,

  $\hat{\sigma}^2$ = estimate of the variance of x,

  N = number of computer runs.

In a Monte Carlo study, on the other hand, the mean and the covariance of the estimation error $\mathbf{e} = \mathbf{x} - \hat{\mathbf{x}}$, compared with the mean of the square root of the diagonal elements of the matrix $P(t)$, are useful in determining if the filter is properly tuned. Therefore, the mean and variance of element j of the vector $\mathbf{e}$ is determined by the following equations:

$$m_j(t_i) = \mathscr{E}\{e_j(t_i)\} \tag{6.12}$$

$$P_{jj}(t_i) = \mathscr{E}\{e_j^2(t_i)\} - m_j^2, \tag{6.13}$$

where the subscript j corresponds to the jth state of the state error vector. As mentioned earlier, in a Monte Carlo simulation the individual samples of a stochastic process can be generated by using a random number generator. A stimple, efficient random number generator subroutine, URAND (uniform random number generator) is presented below [32]:

Random Number Generator (Urand) Program Listing

```
      FUNCTION URAND (IY)
C
      INTEGER IY
C     URAND IS A UNIFORM RANDOM NUMBER GENERATOR.
C
C     THE INTEGER IY SHOULD BE INITIALIZED TO AN
C     ARBITRARY INTEGER PRIOR TO THE FIRST CALL TO URAND. THE
 C    CALLING PROGRAM SHOULD NOT ALTER THE VALUE OF IY
C     BETWEEN SUBSEQUENT CALLS TO URAND. VALUES OF URAND WILL
C     BE IN THE INTERVAL (0, 1).
C
      INTEGER IA, IC, ITWO, M2, M, MIC
      DOUBLE PRECISION HALFM
      REAL S
      DOUBLE PRECISION DATAN, DSQRT
      DATA M2/0/, ITWO/2/
C
      IF(M2.NE.O) GO TO 20
C
C     IF FIRST ENTRY, COMPUTE MACHINE INTEGER WORDLENGTH
C
      M = 1
   10 M2 = M
      M = ITWO*M2
      IF(M.GT.M2) GO TO 10

      HALFM = M2
C
C     COMPUTE MULTIPLIER AND INCREMENT FOR LINEAR
C     CONGRUENTIAL METHOD
C
      IA = 8*IDINT(HALFM*DATAN(1.D0)/8.D0) + 5
      IC = 2*IDINT(HALFM*(0.5DO - DSQRT(3.DO)/6.DO)) + 1
      MIC = (M2 - IC) + M2
C
C     S IS THE SCALE FACTOR FOR CONVERTING TO FLOATING
C     POINT
      S = 0.5/HALFM
C
```

```
C      COMPUTE THE NEXT RANDOM NUMBER
C
  20   IY = IY*IA
C
C      THE FOLLOWING STATEMENT IS FOR COMPUTERS WHICH DO
C      NOT ALLOW INTEGER OVERFLOW ON ADDITION
C
       IF(IY.GT.MIC) IY = (IY – M2) – M2
C
       IY = IY + IC
C
C      THE FOLLOWING STATEMENT IS FOR COMPUTERS WHERE
C      THE WORD LENGTH FOR ADDITION IS GREATER THAN
C      FOR MULTIPLICATION
C
       IF(IY/2.GT.M2) IY = (IY – M2) – M2
C
C      THE FOLLOWING STATEMENT IS FOR COMPUTERS WHERE
C      INTEGER OVERFLOW AFFECTS THE SIGN BIT
C
       IF(IY.LT.O) IY = (IY + M2) + M2
       URAND = FLOAT(IY)*S
       RETURN
       END
```

URAND produces a sequence of integers by setting [32]

$$Y_{n+1} = aY_n + c(\text{modulo } m), \qquad n \geqslant 1, \tag{6.14}$$

on the nth call of URAND, where the values of m (a power of 2), a, and c are computed automatically upon the initial entry. These integers are converted into floating-point numbers in the interval $(0, 1)$ and returned as the value of URAND. The resulting value of Y_{n+1} is returned through the parameter IY. On the first call of URAND, IY should be initialized to an arbitrary integer value. The values of a and c are called in the source code IA and IC, respectively.

Finally, it should be pointed out that Monte Carlo simulations are inefficient and costly. If the models are the same as for covariance analysis, Monte Carlo results should, in the limit, match covariance analysis results.

As an example, we will show how the convariance equation (6.6) is used in practical applications. Consider the theoretical error models used in the statistical analysis of inertial navigation systems. Specifically, we will consider the following models: (1) angular random walk, which is often used to model ring laser gyros, (2) Markov (or exponentially correlated) drift rate, and (3) drift-rate random ramp. Each model will be analyzed to the point of predicting the time behavior of the angle variances. (For more details on these statistical processes, see Section 2.9.)

Angular Random Walk. The error differential equation for the angular random walk is commonly modeled as follows:

$$\frac{d\theta(t)}{dt} = w_\theta(t), \tag{1}$$

where $\theta(t)$ is the angle error and $w_\theta(t)$ is white noise. The corresponding differential equation for the angle error variance P_θ is given by

$$\frac{dP_\theta(t)}{dt} = Q_\theta^2 \tag{2}$$

where Q_θ^2 is the noise power of w_θ. (Note that the power is given by Q_θ^2 rather than Q_θ in difference to the custom of expressing Q_θ in units of degrees/$\sqrt{\text{hour}}$.) Furthermore, Q_θ^2 can be modeled as a constant. Integrating Eq. (2) yields

$$P_\theta = Q_\theta^2 t. \tag{3}$$

Markov (Exponentially Correlated) Drift Rate. The error differential equations for this model are the two-state system given as

$$\frac{d\theta}{dt} = \omega, \tag{4}$$

$$\frac{d\omega}{dt} = -\frac{1}{\tau}\omega + n_\omega, \tag{5}$$

where ω is the drift rate, n_ω is a white noise driving ω, and τ is the correlation time constant. From Eq. (6.6), the covariance differential equation corresponding to this system is given by the following expression:

$$F = \begin{bmatrix} 0 & 1 \\ 0 & -1/\tau \end{bmatrix}, \qquad G = \begin{bmatrix} 0 & 0 \\ 0 & 1 \end{bmatrix},$$

$$\begin{bmatrix} \dot{P}_\theta & \dot{P}_{\theta\omega} \\ \dot{P}_{\theta\omega} & \dot{P}_\theta \end{bmatrix} = \begin{bmatrix} 0 & 1 \\ 0 & -1/\tau \end{bmatrix}\begin{bmatrix} P_\theta & P_{\theta\omega} \\ P_{\theta\omega} & P_\omega \end{bmatrix}$$

$$+ \begin{bmatrix} P_\theta & P_{\theta\omega} \\ P_{\theta\omega} & P_\omega \end{bmatrix}\begin{bmatrix} 0 & 1 \\ 0 & -1/\tau \end{bmatrix} + \begin{bmatrix} 0 & 0 \\ 0 & Q_\omega^2 \end{bmatrix} \tag{6}$$

where P_ω is the drift-rate variance, $P_{\theta\omega}$ is the covariance between drift rate and error angle, and *Q_ω^2 is the noise power of n_ω. Equation (6) can be equivalently

---

*Q_ω is the strength matrix for the white noise required to drive Eq. (6).

written as three scalar equations as follows:

$$\dot{P}_\theta = 2P_{\theta\omega}, \tag{7}$$

$$\dot{P}_{\theta\omega} = -\frac{2}{\tau}P_{\theta\omega} + P_\omega, \tag{8}$$

$$\dot{P}_\omega = -\frac{2}{\tau}P_\omega + Q_\omega^2. \tag{9}$$

In certain applications, Q_ω and τ are modeled as constants. Then the solution to Eq. (9) is given by

$$P_\omega = \frac{\tau}{2}Q_\omega^2 + C_1\exp(-2t/\tau), \tag{10}$$

where C_1 is a constant of integration. Substituting Eq. (10) into Eq. (8) yields

$$P_{\theta\omega} = -C_1\tau\exp(-2t/\tau) + \frac{\tau^2}{2}Q_\omega^2 + C_2\exp(-t/\tau), \tag{11}$$

where C_2 is the new constant of integration. Substituting Eq. (11) into Eq. (7) and using the method of quadratures, yields

$$P_\theta = C_1\tau^2\exp(-2t/\tau) - 2\tau C_2\exp(-t/\tau) + (\tau^2 Q_\omega^2)t, \tag{12}$$

where a zero initial angle error is assumed.

Equation (10) states that, regardless of the initial conditions, the drift-rate variance approaches a constant value of $(\tau/2)Q_\omega^2$ for values of $t \gg \tau$. Assuming the same conditions, Eq. (12) shows that for large t, the exponential terms die out, and the angle error variance grows in proportion to t. That is, its long-term growth is an angular random walk.

Drift-Rate Random Ramp. The error model differential equations for the drift-rate random ramp are as follows:

$$\frac{d\theta}{dt} = \omega, \tag{13}$$

$$\frac{d\omega}{dt} = K, \tag{14}$$

where K is a random constant. Assuming zero initial conditions, and since K is a constant, Eqs. (13) and (14) can be integrated deterministically, yielding

$$\omega = Kt, \tag{15}$$

$$\theta = \tfrac{1}{2}Kt^2. \tag{16}$$

Since these equations are deterministic, the expected value of ω^2, that is, P_ω, is equal to ω^2, and likewise for the error angle θ. Thus,

$$P_\omega = t^2 \mathscr{E}\{K^2\} = t^2 \sigma_K^2 \tag{17}$$

$$P_\theta = t^4 \sigma_K^2 / 4 \tag{18}$$

where σ_K^2 is the variance of the random constant K.

6.1.1 Concluding Remarks

As discussed in Chapter 4, the Kalman filter definition consists of selecting the state variables and specifying the dynamics matrix F, the observation matrix H, the process matrix Q, the measurement noise matrix R, and the initial value of the covariance matrix P. In filter design and performance analysis of multi-sensor systems, the first step is to define error models for each of the sensor systems (e.g., inertial, Global Positioning System, Doppler radar, Loran, Omega, etc.). The system error propagation equations describe mathematically how each of the error sources (in the sensor error models) generates a system error. The second step is to define the system performance, that is, the diagonal elements of the filter's covariance matrix, as predicted by the filter. This is not necessarily the actual system performance.

In order to obtain the actual system performance, as well as its sensitivity to the various error sources, one *propagates* each of the error sources individually and *updates* the resulting system errors with the Kalman gains as computed and stored from the preceding filter run. Since this procedure utilizes the expected rms (root-mean-square) value of each error source, the result is the rms system response to that error source. The rss (root-sum-square) of system responses over all error sources then determines the system performance. This is the *truth model* of system performance and represents actual system performance to the extent that the error model represents the magnitude and behavior of the actual errors. Note that the differential equations describing system error propagation are excited, or *driven*, directly by the individual sensor errors.

One test of a good Kalman filter design is how well its predicted system performance (covariance diagonals) matches the actual system performance. If the two are generally in agreement, the filter design is a good one. If not, one must redefine the Kalman filter by adding states or by changing the matrices Q or R. One then repeats the filter run to generate a new set of Kalman gains and reruns the battery of system error propagations.

Because the filter is necessarily a limited, approximate description of system error behavior, the filter, in general, does not agree exactly with the actual system performance. If the filter-calculated error is much less than the real error, the filter gain will be too small and the filter will not update the real error adequately. This situation can cause filter divergence wherein the real error continues to grow, in contrast to the filter-calculated error, which remains small. To forestall this behavior, it is good design practice to make the filter somewhat pessimistic. That is, the filter computed errors should be somewhat larger than the real errors. This is, however, difficult to do without a reasonably accurate knowledge of the magnitude of the real error. Thus, covariance analysis plays a great role in filter design and performance.

6.2 SUBOPTIMAL FILTERING

As mentioned previously, Kalman filters provide optimum estimates of a linear system by processing noisy measurements of quantities which are functions of the system states. The design of Kalman filters depends upon models of the system (or plant) dynamics and upon the statistical characteristics of the error sources which drive the system and of the errors in the measurements. That is, the optimal Kalman filter equations require an exact mathematical model of the system. However, in most applications these models are not known exactly. Hence, we wish to develop a tool to analyze cases where the Kalman filter model and the real-world model differ, either unintentionally or by design. Because of finite computer memory and throughput (or speed), it is often impractical to model all known states. In addition, the best available models are often so complex that the corresponding filter is not easily implemented. Thus, filters are often implemented using simplified models which approximate the behavior of the system. Stated in other words, the model used for the Kalman filter is an approximation of the real-world system behavior. Fortunately, these suboptimum filters (also known as "reduced order" filters) often provide performance almost as good as the optimum filter based on the exact model.

In navigation systems, the optimal filter model yields the maximum navigation accuracy obtainable with linear filtering of the navigation sensor data, since it models all error sources in the system. The increase in accuracy over that available from conventional* filtering is significant. The increased accuracy is obtained by doubling computer memory and processing-rate requirements for the navigation function over those of, say, a conventional position-referenced Doppler-inertial design. Normally, this does not directly double the computer cost, because the navigation function reflects only a fraction of the total computer memory and burden rate requirements. In present-day weapon system designs, the onboard digital computer performs such tasks as weapon delivery, flight control, guidance, reconnaissance, terrain following and terrain avoidance, system checkout, and navigation functions. These tasks require a computer with a processing time and memory size well within the capability of handling the additional computer requirements im-

---

*By a conventional Kalman filter is meant one with a fixed gain.

posed by the optimal filtering computations. For example, the basic memory size for a weapon system digital computer is 15,000 bits, with a basic computational loop time of once per second. The additional memory and execution time requirements needed to implement optimal filtering will not effect a significant increase in cost for a system of this size. Consequently, the accuracy gains achieved by adding the optimal filter computations result in a direct increase in the cost effectiveness, by factors of about 2 to 3.

From the above discussion, we know that, since the Kalman filter is only optimal when the model corresponds exactly to the real world, the system analyst must examine the effect of mismatch between actual system behavior and the filter model used to represent it. In order to investigate the effect of these modeling errors on the filter estimates, a new set of equations is needed. Specifically, one distinguishes between the true system and the model of the system actually used in the filter.

Examination of the way in which filter results change as the filter model varies is commonly known as *sensitivity analysis*. In sensitivity analyses, a high-order mathematical model is typically used to represent the physical, or *real-world* system as precisely as possible. This model is often referred to as the *truth* or *reference* model. On the other hand, a lower-order simplified model, implemented in the filter, is referred to as the *filter* model (see Section 6.1). A suboptimal filter design which would significantly increase accuracy without increase computer requirements is desirable for application to systems with large computer loading or small computers. The computer requirement for optimal filtering can be reduced significantly by eliminating the airborne computation of the gain matrix K. This will eliminate the instructions and data words for generating the state transition matrix Φ, the covariance matrix P, the measurement matrix H, the measurement noise covariance matrix R, and the optimal gain matrix K. The matrix K will be generated by a smaller set of instructions and data words. The requirements for the suboptimal filter design can probably be relaxed. The degree of flexibility which can be incorporated and the exact computer requirements must be evaluated against specific weapon system and mission requirements.

Let us now define the system state vector $\mathbf{x}_s$, which is assumed to describe the real world as best known by the system analyst. Typically, $\mathbf{x}_s$ is of larger dimension than the filter state vector $\mathbf{x}_f$, in order to provide a more complete description of the physical process. Now, if the system and filter models are of the same dimension, the system model may differ by having various parameters perturbed from the filter values. The system model and observation equations are given by [17, 21]

$$\mathbf{x}_s(k) = \Phi_s(k)\mathbf{x}_s(k-1) + \mathbf{w}_s(k), \tag{6.15a}$$

$$\mathbf{z}_s(k) = H_s(k)\mathbf{x}_s^-(k) + \mathbf{v}_s(k), \tag{6.15b}$$

where $\mathbf{x}_s(k) = $ system (truth) state vector at $t = t_k$,

$\quad \mathbf{x}_s^-(k) = $ system state vector at $t = t_k$, before application of control,

$\Phi_s(k)$ = state-transition matrix for the interval (t_k, t_{k-1}),
$\mathbf{w}_s(k)$ = system noise, assumed to be zero-mean white-noise,
$\mathbf{v}_s(k)$ = system observation noise, assumed to be zero-mean white noise
      and uncorrelated with $\mathbf{x}^-(k)$,
$\mathbf{z}_s(k)$ = actual observations made at $t = t_k$,
$H_s(k)$ = system output matrix at t_k.

The system state vector and the dynamics of Eq. (6.15) (the truth model) are assumed to adequately model all known effects.

Each time a measurement is made, data are processed and used to modify or control the state vector $\mathbf{x}_s(k)$. This can be expressed by the following relations [48]:

$$\mathbf{x}_s^-(k) = \Phi_s(k)\,\mathbf{x}_s^+(k-1) + \mathbf{w}_s(k), \tag{6.16}$$

$$\mathbf{x}_s^+(k) = \mathbf{x}_s^-(k) - D_s(k)\hat{\mathbf{x}}_f(k), \tag{6.17}$$

where $\mathbf{x}_s^+(k)$ = system state vector at $t = t_k$, after application of control,
      $\hat{\mathbf{x}}_f(k)$ = filter estimate at $t = t_k$, based on all data up to and including t_k,
      $D_s(k)$ = control distribution or feedback matrix,
      $\mathbf{x}_f$ = filter model state vector.

The filter estimate $\hat{\mathbf{x}}_f(k)$ is given by the expression

$$\hat{\mathbf{x}}_f(k) = \Phi_f(k)\hat{\mathbf{x}}_f^+(k-1) + K_f(k)\left[\mathbf{z}_s(k) - H_f(k)\Phi_f(k)\hat{\mathbf{x}}_f^+(k-1)\right], \tag{6.18}$$

where $\hat{\mathbf{x}}_f^+(k-1)$ = filter estimate after processing the t_{k-1} data and resetting
      the filter estimate,
      $K_f(k)$ = filter gain matrix,
      $\Phi_f(k)$ = filter state transition matrix.

Upon calculation of $\hat{\mathbf{x}}_f(k)$, we assume that the filter makes a reset (that is, it applies control) of the form [48]

$$\hat{\mathbf{x}}_f^+(k) = \hat{\mathbf{x}}_f(k) - D_f(k)\,\hat{\mathbf{x}}_f(k). \tag{6.19}$$

We now define a composite or augmented state vector as follows:

$$\mathbf{x}_a(k) = \left[\begin{array}{c} \mathbf{x}_s^+(k) \\ \hline \hat{\mathbf{x}}_f^+(k) \end{array}\right]. \tag{6.20}$$

Using the preceding equations, the augmented matrix can be written in the form

$$\mathbf{x}_a(k) = \Phi^*(k)\mathbf{x}_a(k-1) + G^*(k)\mathbf{w}_s(k) + H^*(k)\mathbf{v}_s(k), \tag{6.21}$$

where

$$\Phi^*(k) = \left[\begin{array}{c|c} [I - D_s(k)K_f(k)H_s(k)]\Phi_s(k) & -D_s[I - K_f(k)H_f(k)]\Phi_f(k) \\ \hline [I - D_f(k)]K_f(k)H_s(k)\Phi_s(k) & [I - D_f(k)][I - K_f(k)H_f(k)]\Phi_f(k) \end{array} \right],$$

$$G^*(k) = \left[\begin{array}{c} I - D_s(k)K_f(k)H_s(k) \\ [I - D_f(k)]K_f(k)H_s(k) \end{array} \right], \quad \text{and} \quad H^*(k) = \left[\begin{array}{c} -D_s(k)K_f(k) \\ [I - D_f(k)]K_f(k) \end{array} \right].$$

The covariance matrix for $\mathbf{x}_a(k)$ propagates in accordance with the relation

$$
\begin{aligned}
P_a(t) &= \mathscr{E}\{\mathbf{x}_a(t)\,\mathbf{x}_a^T(t)\} \\
&= \operatorname{cov}\mathbf{x}_a(k) = \Phi^*(k)\operatorname{cov}[\mathbf{x}_a(k-1)]\,\Phi^{*T}(k) \\
&\quad + G^*(k)\operatorname{cov}[\mathbf{w}_s(k)]\,G^{*T}(k) \\
&\quad + H^*(k)\operatorname{cov}[v_s(k)]\,H^{*T}(k)
\end{aligned}
\tag{6.22}
$$

and with the appropriate initial conditions

$$\mathbf{x}_a(t_0) = \left[\begin{array}{c} \mathbf{x}_s(t_o) \\ \hline \hat{\mathbf{x}}(t_0) \end{array} \right] = \left[\begin{array}{c} \mathbf{x}_s(t_0) \\ \hline \mathbf{0} \end{array} \right] \tag{6.23}$$

and

$$P_a(t_0) = \mathscr{E}\{x_a(t_0)\,x_a^T(t_0)\} = \left[\begin{array}{c|c} P_{a0} & 0 \\ \hline 0 & P_0 \end{array} \right]. \tag{6.24}$$

Equation (6.17) and (6.18) are illustrated in Figure 6.3.

In designing a suboptimal Kalman filter, a possible procedure would be as follows:

- Specify the required performance desired.
- Evaluate the performance with the optimal filter.
- Evaluate performance.
- Generate filter-parameter sensitivity curves.
- Evaluate filter cost.

The following points are characteristic of all suboptimal filters:

1. Suboptimal filters are designed so as to minimize the computer burden (e.g., storage capacity) associated with optimal Kalman filters.

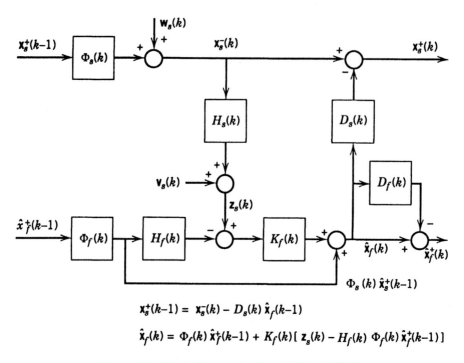

$$\mathbf{x}_g^+(k-1) = \mathbf{x}_g^-(k) - D_g(k)\,\hat{\mathbf{x}}_f(k-1)$$

$$\hat{\mathbf{x}}_f(k) = \Phi_f(k)\,\hat{\mathbf{x}}_f^+(k-1) + K_f(k)[\,\mathbf{z}_g(k) - H_f(k)\,\Phi_f(k)\,\hat{\mathbf{x}}_f^+(k-1)\,]$$

Figure 6.3   Block diagram for Eqs. (6.17) and (6.18).

2. They are designed to display less sensitivity than optimal filters to modeling errors.
3. They are designed to enable near-optimum system performance, subject to the above constraints.

6.2.1   Concluding Remarks

The problem addressed in Section 6.2 is a model reduction, the design of a suboptimal filter. The suboptimal filter design is based on using the most significant subset of states from the real world. That is, the filter design is based on an incomplete and/or incorrect model of the actual system dynamics. Therefore, the real world is commonly replaced by the most complete mathematical model that can be developed, called a *truth model* (also known as *reference model*), as discussed in Section 6.1.1. For instance, if we designate P as the $n \times n$ truth error covariance matrix and P_s as the $m \times m$ filter error covariance matrix, then the dimension of P will be $n > m$.

The basic analysis technique used is called *sensitivity analysis*. A sensitivity analysis evaluates the performance of a suboptimal filter by determining the sensitivity to an incorrect or an incomplete dynamic and statistical modeling.

For this purpose, this analysis uses the reference system of equations (the optimal Kalman filter formulation) and the suboptimal system to derive a set of sensitivity equations relating the two systems. Sensitivity equations generate realistic performance projections for a practical filter mechanization by properly allowing for the unavoidable mismatch between filter design assumptions or simplifications and the real-world environment in which the filter must perform.

The filter design model generates a set of time-varying gains which specify the filter. These time-varying gains are then used in the larger set of reference covariance equations. Appropriate elements of the covariance matrices in these sensitivity equations indicate the performance of the suboptimal filter operating in the real-world reference system. The indications of filter performance in the suboptimal filter covariance equation will be optimistic, because certain error sources which exist in the reference system have been left out of the filter. If the reference system covariance equations are propagated alone, the Kalman gains produced are optimal and so are the performance figures indicated by the covariance matrices. Assuming that an $(n + m)$-state real world exists and that an m-state filter is mechanized, the following assessments can be made:

- *Baseline.* Hypothetical n-state filter in an n-state real world (the best that can be achieved).
- *Best Possible Suboptimal.* Filter performance without selecting suboptimal filter parameters Q and R.
- *Suboptimal.* m-state filter in an $(n + m)$-state real world (what will be achieved).
- *Optimal Filter.* m-state filter in an optimistic m-state real world (what the filter thinks is being achieved).

In the filter design, the following steps are recommended:

- Develop a complete real-world covariance truth model containing all known error sources.
- Identify the dominant error states for the filter modeling by an error budget (see below as well as Section 4.2) experiments.
- Redesign the filter so that it has minimum sensitivity to changes in the real-world statistics.
- Compare the covariance analysis with a system simulation to verify the mechanization equations.

An important feature of the suboptimal filter analysis is the capability to generate an *error budget*. An error budget summarizes the contribution of each error source or group of error sources to the total system error. It is useful in determining the dominant errors contributing to the total error.

Finally, the filter designer should develop a suboptimal filter analysis program for the purpose of

1. generating an error budget to determine dominant error states for use in the filter,
2. performing sensitivity analysis of the filter design to determine the uncertainty in the real-world noise statistics, and
3. studying implementation considerations such as modeling problems, divergence, and round off errors.

PROBLEMS

6.1 Define the error $\tilde{\mathbf{x}}$ in the estimate of a state vector to be the difference between the estimated value $\hat{\mathbf{x}}$ and the actual value $\mathbf{x}$:

$$\tilde{\mathbf{x}} \triangleq \hat{\mathbf{x}} - \mathbf{x}. \tag{1}$$

The error covariance of $\tilde{\mathbf{x}}$, designated by P, is given by

$$P = \mathscr{E}\{\tilde{\mathbf{x}}\,\tilde{\mathbf{x}}^T\}. \tag{2}$$

This equation provides a statistical measure of the uncertainty in the state. Furthermore, this choice of variables allows us to discuss the properties of the covariance matrix independently of the mean of the state. We would like to obtain an estimate at a later time, t_{k+1}, which will have an unbiased error $\tilde{\mathbf{x}}_{k+1}$. To do so, the known state transition matrix $\Phi_k = \Phi(t_{k+1}, t_k)$ is used, giving

$$\hat{\mathbf{x}}_{k+1} = \Phi_k \hat{\mathbf{x}}_k. \tag{3}$$

From the discrete-time difference equation

$$\mathbf{x}_{k+1} = \Phi_k \mathbf{x}_k + \Gamma_k \mathbf{w}_k \tag{4}$$

we can show that the error in the estimate at time t_{k+1}, is unbiased by subtracting Eq. (4) from Eq. (3) to obtain

$$\tilde{\mathbf{x}}_{k+1} = \Phi_k \tilde{\mathbf{x}}_k - \Gamma_k \mathbf{w}_k. \tag{5}$$

Taking the expected value of both sides Eq. (5), show that

$$P_{k+1} = \mathscr{E}\{\tilde{\mathbf{x}}_{k+1}\,\tilde{\mathbf{x}}_{k+1}^T\} = \Phi_k P_k \Phi_k^T + \Gamma_k Q_k \Gamma_k^T. \tag{6}$$

[This is the proof of Eq. (4.17).]

6.2 As an extension of Problem 6.1, the continuous formulation for projecting the error covariance can be obtained through a limiting argument applied to Eq. (2) of Problem 6.1:

$$P_{k+1} = \Phi_k P_k \Phi_k^T + \Gamma_k Q_k \Gamma_k^T. \tag{1}$$

The noise covariance is given by

$$\Gamma_k Q_k \Gamma_k^T = \int_{t_k}^{t_{k+1}} \Phi(t_{k+1}, \tau) G(\tau) Q(\tau) G^T(\tau) \Phi^T(t_{k+1}, \tau) \, d\tau. \tag{2}$$

For very small time intervals, $t_{k+1} - t_k = \Delta t \to 0$ and

$$\Gamma_k Q_k \Gamma_k^T \to G Q G^T \, \Delta t, \tag{3}$$

where terms of the order Δt^2 have been neglected. Show that the continuous-time counterpart of Eq. (1) is given by the matrix Riccati equation

$$\dot{P}(t) = F(t) P(t) + P(t) F^T(t) + G(t) Q(t) G^T(t). \tag{4}$$

[Hint: Start with in the equation $d\Phi(t)/dt = F(t)\Phi(t)$ and take the limit as $t \to 0$; also, use Eq. (4.24).]

6.3 In Problem 4.10, the error equations for a seven-state, single-axis inertial navigation system were given. For a suboptimal (or reduced) filter, the reduced filter state vector can be reduced to a 5×1 state vector as follows:

$$\mathbf{x}_s^T = [\delta v \quad \delta p \quad \delta \theta \quad e_a \quad e_g].$$

The suboptimal matrices F_s, Q_s, R_s, and H_s are

$$F_s = \begin{bmatrix} 0 & 0 & -g & -g & 0 \\ 1 & 0 & 0 & 0 & 0 \\ 1/R_s & 0 & 0 & 0 & 1 \\ 0 & 0 & 0 & -\beta_a & 0 \\ 0 & 0 & 0 & 0 & -\beta_g \end{bmatrix},$$

$$Q_s = \begin{bmatrix} 2\beta_a \sigma_a^2 & 0 \\ 0 & 2\beta_g \sigma_g^2 \end{bmatrix},$$

$$R_s = \begin{bmatrix} 2\beta_v \sigma_v^2 & 0 \\ 0 & 2\beta_p \sigma_p^2 \end{bmatrix}, \quad H_s = \begin{bmatrix} \beta_v & 0 & -g & -g & 0 \\ 1 & \beta_p & 0 & 0 & 0 \end{bmatrix}.$$

Assuming that the real-world model is of dimension $n + m$, and that the suboptimal filter is mechanized with dimension m:

(a) Is there a degradation in performance as a result of the simplification?

(b) Explain the effect (if any) of the mismatch between real world and filter design.

CHAPTER 7

THE $\alpha-\beta-\gamma$ TRACKING FILTERS

7.1 INTRODUCTION

Recently, the problem of target tracking has been receiving considerable attention, as evidenced by numerous publications in technical journals and books. In its most general form, tracking involves estimating the position coordinates of a target and their derivatives of various orders, based on past measurements of some or all of these quantities (commonly at discrete intervals of time). The use of a Kalman filter for tracking necessitates the use of a model in order to describe the target's motion. However, since most available sensors observe targets only as points in space, and usually do not have the capability to observe their dynamics, the pertinent parameters to be modeled are those that relate to the kinematics of the target motion.

In this chapter, alpha (α), beta (β), and gamma (γ) filters for tracking moving (or maneuvering) targets will be discussed. These tracking filters are steady-state versions of the Kalman filter. As discussed earlier, the Kalman filter produces an optimal estimate of the state when a sequence of measurements are corrupted with Gaussian white noise. However, the computational burden in using the Kalman filter can be excessive. In order to reduce it, an approximate filter gain can be used instead of an optimal Kalman gain. This approximate gain is commonly based on the steady-state gains of the Kalman filter and can be obtained by using a fixed gain. That is, in real-time tracking, it is usually necessary to sacrifice optimality for physical feasibility and to use the third-order linear predictor–corrector filter, commonly known as the $\alpha-\beta-\gamma$ filter, instead [18]. More specifically, these filters can be categorized as follows:

1. α-filter: used for tracking position.
2. α–β filter: used for tracking position and velocity.
3. α–β–γ filter: used for tracking position, velocity, and acceleration.

7.2 THE α-FILTER

The α-filter is the simplest of the three filters. In essence, the α-filter is a single coordinate filter, based on the assumption that the target motion is represented by constant position plus noise. Consequently, the state equation of this filter is simply given as

$$\mathbf{x}(k) = [x_1(k) \quad x_2(k) \quad \cdots \quad x_n(k)]^T, \tag{7.1}$$

while the observation and state transition matrices are

$$H(k) = [1], \qquad \Phi(k) = [1], \qquad G(k) = \left[\frac{T^2}{2}\right]^T. \tag{7.2}$$

Furthermore, the gain is given by

$$K(k) = \alpha^T. \tag{7.3}$$

Kalata [39] defined a variable Γ, known as the *target maneuvering index* or simply the *tracking index*, given by the relation

$$\Gamma^2 = T^4 \frac{\sigma_w^2}{\sigma_v^2} = \frac{4\alpha^2}{1-\alpha}$$

or

$$\Gamma = T^2 \frac{\sigma_w}{\sigma_v}, \tag{7.4}$$

where T is the time between state updates, or the radar update interval. Γ is a function of the assumed target maneuverability variance σ_w^2 (i.e., the deviation from the modeled behavior) and of the radar measurement noise variance σ_v^2. The value of α can be computed explicitly form the equation [39].

$$\alpha = \frac{-\Gamma^2 + \sqrt{\Gamma^4 + \Gamma^2}}{8}. \tag{7.5}$$

Equation (7.4) states that the tracking index is proportional to the ratio of the position process noise deviation $\sigma_w T^2/2$ and the measurement noise standard deviation σ_v.

The optimal steady-state filter, known as the α-filter, is given by the expression

$$\hat{\mathbf{x}}(k+1|k+1) = \hat{\mathbf{x}}(k|k) + \alpha[\mathbf{z}(k+1) - \hat{\mathbf{x}}(k|k)]. \qquad (7.6)$$

The steady-state error covariance matrix for the α-filter is

$$P(k|k) = \alpha \sigma_v^2. \qquad (7.7)$$

7.3   THE α–β TRACKING FILTER

The α–β tracking filter algorithm has been used extensively in radar target tracking. In particular, tracking filter algorithms of the α–β type have been successfully applied to two-dimensional *track-while-scan* radars, which measure range and azimuth. (Note that these radars also measure the range rate, whereas others measure only the range, azimuth, and elevation angles.) Therefore, state prediction and state correction operations are done in range–azimuth coordinates. Commonly, the azimuth filter is an α–β filter with range-dependent gains. The range filter is a two-state Kalman filter estimating range and velocity (range rate). This filter provides better response to the following: (1) initialization transients, (2) measurement accuracy fluctuations, and (3) target loss of detection (i.e., signal fading) [3].

It should be noted that the α–β filter mechanization assumes constant sampling intervals and steady-state conditions. Thus, on increasing the sampling time T for missed detections, the computed gains reflect steady-state conditions for larger sampling times, thus weighting observations more heavily in comparison with the optimum gain. As a result, when T exceeds a threshold, with no assigned detections, the track is terminated.

Consider now a second-order plant modeled by the discrete-time state equation with sampling interval T, given in the form

$$\mathbf{x}(k+1) = \Phi(k+1, k)\,\mathbf{x}(k) + \mathbf{w}(k), \qquad k = 1, 2, \ldots,$$

$$\mathbf{x}(0) = \mathbf{x}_0, \qquad (7.8)$$

$$\mathscr{E}\{\mathbf{x}_0\} = \mathbf{\mu}_x,$$

where $\mathbf{x}(k) = $ state vector at time k,

$\mathbf{w}(k) = $ zero-mean, white, Gaussian process noise with covariance matrix $Q(k)$, that is, $\mathbf{w}(k) \sim N(0, Q(k))$,

$\Phi(k+1, k) = $ state transition matrix.

The discrete-time measurement (or observation) equation is given as usual in the form

$$\mathbf{z}(k) = H(k)\mathbf{x}(k) + \mathbf{v}(k), \qquad (7.9)$$

where $\mathbf{z}(k) =$ measurement vector,
$H(k) =$ measurement or observation matrix,
$\mathbf{v}(k) =$ zero-mean, white, Gaussian process noise with covariance matrix
$R(k)$, that is, $\mathbf{v}(k) \sim N(0, R(k))$.

The two-dimensional (i.e., position–velocity) α–β tracking filter for estimating
the state $\mathbf{x}(k)$ of a time-invariant system is commonly expressed in the from
[4]

$$\hat{\mathbf{x}}(k + 1 | k + 1) = \hat{\mathbf{x}}(k | k) + K(k) [\mathbf{z}(k + 1) - \hat{\mathbf{z}}(k + 1 | k)]$$

$$= \hat{\mathbf{x}}(k | k) + \begin{bmatrix} \alpha \\ \beta/T \end{bmatrix} [\mathbf{z}(k + 1) - \hat{\mathbf{z}}(k + 1 | k)], \qquad (7.10)$$

where $K(k)$ is the filter gain and $\hat{\mathbf{z}}(k + 1) - \hat{\mathbf{z}}(k + 1 | k)$ is the *measurement
prediction*. The coefficients α and β are dimensionless, constant filter gains (or
tracking parameters) for position and velocity components of the state, respectively.

In tracking a nonmaneuvering, constant-velocity target with a two-dimensional state vector $\mathbf{x} = [x_1 \ x_2]^T$, where x_1 is the range (e.g., radar range
measurement) and x_2 is the range rate, the state transition matrix takes the
form

$$\Phi(k + 1, k) = \begin{bmatrix} 1 & T \\ 0 & 1 \end{bmatrix},$$

while the observation matrix $H(k)$ takes the form $H(k) = [1 \ 0]^T$. From
Eq. (7.8), the system model dynamics can be written as follows:

$$\begin{bmatrix} x_1(k + 1) \\ x_2(k + 1) \end{bmatrix} = \begin{bmatrix} 1 & T \\ 0 & 1 \end{bmatrix} \begin{bmatrix} x_1(k) \\ x_2(k) \end{bmatrix} + \begin{bmatrix} w_1(k) \\ w_2(k) \end{bmatrix}$$

where T is the tracking time (or sampling period), with initial conditions

$$\begin{bmatrix} x_1(0) \\ x_2(0) \end{bmatrix} = \begin{bmatrix} 0 \\ 0 \end{bmatrix}.$$

Here we have assumed that the system dynamics matrix $F(t)$ in Eq. (4.1) is
given by

$$F(t) = \begin{bmatrix} 0 & 1 \\ 0 & 0 \end{bmatrix},$$

and the state transition matrix has been obtained using Eq. (4.36), that is

$$\Phi(k + 1, k) = \exp\{FT\}.$$

The error covariance matrix equation for this problem becomes

$$P(k+1|k) = \Phi(k+1, k) P(k|k) \Phi^T(k+1, k) + Q(k).$$

A simulation can be carried out by assuming the initial conditions for $P(0)$ and $Q(0)$.

For the $\alpha-\beta$ filter, the coefficients* α and β are determined from the relations

$$\Gamma^2 = T^4 \frac{\sigma_w^2}{\sigma_v^2} = \frac{\beta^2}{1-\alpha}, \tag{7.11}$$

$$\alpha = \sqrt{2\beta} - \tfrac{1}{2}\beta, \tag{7.12}$$

$$\beta = 2(2-\alpha) - 4\sqrt{1-\alpha}. \tag{7.13}$$

From the dynamics model (7.8) and measurement model (7.9), the time update and measurement update equations can be expressed as follows [see also Section 4.3, (4.26)–(4.30)]:

Time update:

$$\hat{\mathbf{x}}(k|k-1) = \Phi(k, k-1)\,\hat{\mathbf{x}}(k-1|k-1), \tag{7.14}$$

$$P(k|k-1) = \Phi(k, k-1) P(k-1|k-1) \Phi^T(k, k-1) + Q(k-1), \tag{7.15}$$

$k = 1, 2, \dots$.

Measurement update:

$$K(k) = P(k|k-1) H^T(k) [H(k) P(k|k-1) H^T(k) + R(k)]^{-1}, \tag{7.16}$$

$$\hat{\mathbf{x}}(k|k) = \hat{\mathbf{x}}(k|k-1) + K(k)\,[\mathbf{z}(k) - H(k)\hat{\mathbf{x}}(k|k-1)], \tag{7.17}$$

$$P(k|k) = [I - K(k) H(k)]P(k|k-1). \tag{7.18}$$

In the above equations, we use the notation $(k|j)$ to denote the state estimate for time k, given the measurements through time j. A recursive, inverse form of the covariance matrix (7.15), also known as the information matrix, is given the equation

$$P^{-1}(k+1|k+1) = P^{-1}(k+1|k) + H^T(k+1) R^{-1}(k+1) H(k+1). \tag{7.19}$$

---

* The constant coefficients α and β are usually predetermined, although multiple sets and some criteria, such as range, operator selection, etc., could be used to select the proper set of values.

The Kalman gain (7.16) can be expressed in an alternative form by making use of the well-known matrix inversion lemma* [58]; thus,

$$
\begin{aligned}
K(k) &= P(k|k-1)H^T(k)[H(k)P(k|k-1)H^T(k)+R(k)]^{-1} \\
&= [I - P(k|k-1)H^T(k)R^{-1}(k)H(k)]^{-1}P(k|k-1)H^T(k)R^{-1}(k) \\
&= \{I - P(k|k-1)H^T(k)[H(k)P(k|k-1)H^T + R^{-1}(k)]^{-1}H(k)\} \\
&\quad P(k|k-1)H^T(k)R^{-1}(k) \\
&= [I - K(k)H(k))]P(k|k-1)H^T(k)R^{-1}(k) \\
&= P(k|k)H^T(k)R^{-1}(k).
\end{aligned} \tag{7.20}
$$

As mentioned earlier, in order to reduce the computational requirements to maintain each track, the steady-state Kalman filter is used. Thus, in the steady state the components of the state estimation covariance matrix are expressed as follows [4]:

$$
\lim_{k \to \infty} P(k|k) = \lim_{k \to \infty} P(k+1|k+1) = [p_{ij}], \tag{7.21}
$$

while the components of the "one-step" prediction covariance are given by

$$
\lim_{k \to \infty} P(k+1|k) = [m_{ij}]. \tag{7.22}
$$

The Kalman filter gain assumes the form

$$
\lim_{k \to \infty} K(k) \triangleq \begin{bmatrix} \alpha \\ \beta/T \end{bmatrix}. \tag{7.23}
$$

The input–output relationship in terms of the coefficients α and β can be expressed as [10]

$$
\hat{x}(k|k) = \begin{bmatrix} 1-\alpha & (1-\alpha)T \\ -\beta/T & 1-\beta \end{bmatrix} \hat{x}(k-1|k-1) + \begin{bmatrix} \alpha \\ \beta/T \end{bmatrix} z(k). \tag{7.24}
$$

Summarizing the above discussion, it is clear that in order for the Kalman filter to achieve the aforementioned steady-state conditions, the statistics of the noise processes $w(k)$ and $v(k)$ must be stationary. On the other hand, if the

---

* The matrix inversion lemma can be summarized as follows; let A_1 be an $n \times m$ nonsingular matrix, A_2 and $n \times m$ matrix, A_3 an $m \times m$ nonsingular matrix, and A_4 an $m \times n$ matrix. Then

$$
(A_1 + A_2 A_3 A_4)^{-1} = A_1^{-1} - A_1^{-1}A_2(A_4 A_1^{-1}A_2 + A_3^{-1})^{-1}A_4 A_1^{-1}.
$$

noise processes are not stationary or the data rate is not constant, then a filter using the steady-state gains will provide suboptimal estimates.

Example Consider the problem of a shipborne tracking radar (compare this example with the radar tracking example given in Section 4.3), which is tracking a maneuverable target. We wish to estimate the target's position, velocity, and acceleration. The observation vector $\mathbf{z}(t)$ is normally composed of position measurements in the radar coordinate system. That is, the observation vector will consist of measurements of position [range $r \equiv x(k)$, elevation $\varepsilon \equiv y(k)$, and azimuth $\theta \equiv z(k)$]. The state vector in this case can be expressed as

$$\mathbf{x}(k) = [x(k) \quad y(k) \quad z(k) \quad \dot{x}(k) \quad \dot{y}(k) \quad \dot{z}(k)]^T. \tag{1}$$

The process (or model of the dynamics) is assumed to be represented by the linear vector differential equation

$$x(k+1) = \Phi(k+1, k)\,\mathbf{x}(k) + \Gamma(k)\,\mathbf{w}(k) + \psi(k), \tag{2}$$

$$z(k) = H\mathbf{x}(k) + \mathbf{v}(k), \tag{3}$$

with

$$\mathbf{x}(0) \sim N(\hat{\mathbf{x}}(0), P(0)),$$

$$\mathbf{w}(k) \sim N(0, Q(k)),$$

$$\mathbf{v}(k) \sim N(0, R(k)),$$

where $\psi(k)$ is a fixed bias included in the dynamics in order to allow for errors due to nonlinearities, reduction in system dimension, etc. Here $\psi(k)$ can be treated as a random sequence, and can be included in the system noise $\mathbf{w}(k)$ by changing $Q(k)$. Note that the true value of $Q(k)$ is unknown; however, we can assume a certain *a priori* value such as $Q(0) = \sigma_{Q(0)}^2$.

The target (or vehicle) equation of motion can be represented by the following equation:

$$\frac{d}{dt}\ddot{x}(t) = -\frac{1}{\tau}\ddot{x}(t) + w(t), \tag{4}$$

where $w(t)$ is a zero-mean, driving white noise process and τ is the correlation time (in this case, it is the duration of the target maneuver). Moreover, the model represented by Eq. (4) is of single spatial dimension, consisting of an exponentially correlated acceleration expressed in terms of white noise in accordance with the Wiener–Kolmogorov whitening procedure. Now, if $a(t)$ is the acceleration, then we can write

$$\frac{d}{dt} a(t) = -\frac{1}{\tau} a(t) + w(t). \tag{5}$$

(Note that in missile guidance if one models acceleration rate is a target maneuver, this acceleration rate is known as the *jerk* of the target, and is given in units of m/s^3.). In Eq. (5), the acceleration rate $\dot{a}(t)$, being a process correlated in time, has an autocorrelation function

$$\varphi_f(\tau) = \sigma_a^2 \, e^{-\alpha|\tau|} \tag{6}$$

where σ_a^2 is the variance of the target acceleration and α is the reciprocal of the correlation time constant. In this example, the value of $Q(k)$ can be taken as

$$Q(k) = 2\alpha\sigma_m^2 Q,$$

where

$$Q = \begin{bmatrix} q_{11} & q_{12} & q_{13} \\ q_{21} & q_{22} & q_{23} \\ q_{31} & q_{32} & q_{33} \end{bmatrix}.$$

The model represented by Eq. (5) has been used extensively in the literature. The usual Kalman filter steady-state equations are

$$\hat{x}(k|k-1) = \Phi(k, k-1) \, \hat{x}(k-1|k-1) \tag{7}$$

$$\hat{x}(k|k) = \hat{x}(k|k-1) + K[z(k) - H\hat{x}(k|k-1)], \tag{8}$$

where $K = 3 \times 1$ Kalman gain matrix,
$H = [1 \ \ 0 \ \ 0] =$ system observation (or measurement) matrix.

For the present example, the target (or system) state vector will take the form

$$x(k) = [x(k) \quad \dot{x}(k) \quad \ddot{x}(k)]^T.$$

The state transition matrix $\Phi(k, k-1)$ turns out to be [71]

$$\Phi(k, k-1) = \begin{bmatrix} 1 & T & 1/\alpha^2(-1 + \alpha T + e^{-\alpha T}) \\ 0 & 1 & 1/\alpha(-1 + e^{-\alpha T}) \\ 0 & 0 & e^{-\alpha T} \end{bmatrix}, \tag{9}$$

where T is the uniform sampling rate and $\alpha = 1/\tau$. For small αT, Eq. (9) reduces to the following simpler form:

$$\Phi(k, k - 1) = \begin{bmatrix} 1 & T & T^2/2 \\ 0 & 1 & T \\ 0 & 0 & 1 \end{bmatrix}. \tag{10}$$

The filter model represented by Eqs. (7)–(9) can be easily programmed and implemented in real time tracking performed on board a vehicle (in this case, a ship).

The steady-state Kalman gain depends on four parameters as follows:

- τ, the mean duration of the target maneuver (from 0.1 to 100 sec),
- σ_Q^2, the variance of the magnitude of the target maneuver, calculated using a typical probability density,
- T, the sampling rate,
- σ_R^2, the variance of the observation noise.

The discussion of the example presented here is meant only as a basis for further investigation, whose solution will depend on the user's intended purpose.

7.4   THE $\alpha-\beta-\gamma$ TRACKING FILTER

Adding an acceleration state to the $\alpha-\beta$ tracking filter, the estimate of the state $x(t)$ becomes

$$\hat{x}(k + 1|k + 1) = \hat{x}(k|k) + \begin{bmatrix} \alpha \\ \beta/T \\ \gamma/T^2 \end{bmatrix} [z(k + 1) - \hat{z}(k + 1|k)]. \tag{7.25}$$

In this case, the acceleration estimate improves the position and velocity predictions [40]. For a constant-acceleration target, the state vector takes the form $x = [x \ \dot{x} \ \ddot{x}]^T$, so that the state transition matrix is given by (see example in Section 7.3)

$$\Phi(k + 1, k) = \begin{bmatrix} 1 & T & \frac{1}{2}T^2 \\ 0 & 1 & T \\ 0 & 0 & 1 \end{bmatrix},$$

where in this case the dynamics matrix $F(t)$ has been taken to be of the form (Reference [71] discusses the case for a maneuvering target)

$$F(t) = \begin{bmatrix} 0 & 1 & 0 \\ 0 & 0 & 1 \\ 0 & 0 & 0 \end{bmatrix}.$$

The observation matrix in this case becomes $H(k) = [1 \ 0 \ 0]^T$. For the $\alpha-\beta-\gamma$ filter, the coefficients α, β, γ satisfy

$$\Gamma^2 = T^4 \frac{\sigma_w^2}{\sigma_v^2} = \frac{\gamma^2}{1-\alpha}, \tag{7.26}$$

$$\beta = 2(2-\alpha) - 4\sqrt{1-\alpha}, \tag{7.27}$$

$$\gamma = \frac{\beta^2}{2\alpha}. \tag{7.28}$$

If the dynamics model (7.8) includes the input dynamics matrix $G(k)$, then

$$\mathbf{x}(k+1) = \Phi\mathbf{x}(k) + G(k)\,\mathbf{w}(k). \tag{7.29}$$

In this case, the input matrix $G(k)$ and covariance matrix $P(k|k)$ take the following form [10]:

α-filter:

$$G(k) = [T^2/2]^T,$$

$$P(k|k) = \alpha\sigma_v^2. \tag{7.30}$$

$\alpha-\beta$ filter:

$$G(k) = [T^2/2 \quad T]^T,$$

$$P(k|k) = \sigma_v^2 \begin{bmatrix} \alpha & \dfrac{\beta}{T} \\ \dfrac{\beta}{T} & \dfrac{\beta(2\alpha - \beta)}{2(1-\alpha)T^2} \end{bmatrix}. \tag{7.31}$$

$\alpha-\beta-\gamma$ filter:

$$G(k) = [T^2/2 \quad T \quad 1]^T,$$

$$P(k|k) = \sigma_v^2 \begin{bmatrix} \alpha & \dfrac{\beta}{T} & \dfrac{\gamma}{T^2} \\[2.5ex] \dfrac{\beta}{T} & \dfrac{4\alpha\beta + \gamma(\beta - 2\alpha - 4)}{4(1-\alpha)T^2} & \dfrac{\beta(\beta-\gamma)}{2(1-\alpha)T^3} \\[2.5ex] \dfrac{\gamma}{T^2} & \dfrac{\beta(\beta-\gamma)}{2(1-\alpha)T^3} & \dfrac{\gamma(\beta-\gamma)}{(1-\alpha)T^4} \end{bmatrix} \qquad (7.32)$$

Note that for all three filter types, $Q(k) = \sigma_w^2$ and $R(k) = \sigma_v^2$. The general time-update covariance matrix takes the form

$$P(k|k-1) = \Phi(k, k-1)P(k-1|k-1)\,\Phi^T(k, k-1)$$

$$+ \, G(k-1)\,Q(k-1)G^T(k-1).$$

In terms of the coefficients α, β, and γ, the input–output relationship between the measurements and the state estimate $\hat{\mathbf{x}}(k|k)$ is given by [10]

$$\hat{\mathbf{x}}(k|k) = \begin{bmatrix} 1-\alpha & (1-\alpha)T & (1-\alpha)\dfrac{T^2}{2} \\[2.5ex] -\dfrac{\beta}{T} & (1-\beta) & \left(1-\dfrac{\beta}{2}\right)T \\[2.5ex] -\dfrac{\gamma}{T^2} & -\dfrac{\gamma}{T} & \left(1-\dfrac{\gamma}{2}\right) \end{bmatrix}$$

$$\hat{\mathbf{x}}(k-1|k-1) + \begin{bmatrix} \alpha \\[1ex] \dfrac{\beta}{T} \\[1ex] \dfrac{\gamma}{T^2} \end{bmatrix} \mathbf{z}(k). \qquad (7.33)$$

7.5 CONCLUDING REMARKS

Target tracking can be accomplished using Doppler and non-Doppler types of radar. Doppler frequency tracking, for example, provides clutter rejection that enables continuous range and angle tracking of a single moving target. Specifically, since the velocity, or more precisely the Doppler frequency, is measured directly, predictions of future target position can be made more accurately.

As discussed throughout this book, the benefits of Kalman filtering have been known for some time, and it has been applied successfully in aerospace and industrial processes. With regard to target tracking as discussed in this chapter, Kalman filters can greatly improve the accuracy. Furthermore, Kalman filters permit higher-order modeling of the expected target motion and can adaptively weight the current observation and track history. In particular, tracking a moving target from a radar or other measurements of its position is often accomplished by a simple three-state estimator of predictor–corrector type, in which the estimates are alternately extrapolated by the linear equation

$$\hat{\mathbf{x}}(k|k-1) = \Phi\hat{\mathbf{x}}(k-1|k-1)$$

and corrected (or updated) by the incorporation of a position measurements $\mathbf{z}_t$ according to the relation

$$\hat{\mathbf{x}}(k|k) = \hat{\mathbf{x}}(k|k-1) + K\left[\mathbf{z}_t - H\hat{\mathbf{x}}(k|k-1)\right].$$

In the above equation, $\hat{\mathbf{x}}$ is the three-dimensional vector of estimates of position, velocity, and acceleration along a single axis (see example in Section 7.3).

The α–β tracker is widely used because of the simplicity of the calculations as compared to the common Kalman filter. However, the α–β tracker may be viewed as a special case of the Kalman filter, having nonadaptive weights. In practice, Kalman filters and α–β trackers are widely used in both track-while-scan and single-target tracking modes. Both methods are readily implemented using recursive digital computer algorithms. For certain applications, suboptimal filters (see Chapter 6) have been developed in order to reduce the computational load.

Two technological factors which led to growth in Kalman filter applications during the 1980s are: (1) low-cost, high-speed digital computing capability made Kalman filters practical for a variety of applications, and (2) an increase in electronically scanned antennas has resulted in more multiple-target tracking systems.

PROBLEMS*

7.1   In Section 7.2 we have seen that the α-filter is the simplest tracking filter. Its time-update error covariance matrix is given by

$$P(k|k-1) = \Phi P(k-1|k-1)\Phi^T + GQG^T, \qquad (1)$$

and its state estimate by

$$\hat{\mathbf{x}}(k|k) = \hat{\mathbf{x}}(k|k-1) + K(k)[\mathbf{z}(k) - H\hat{\mathbf{x}}(k|k-1)]. \qquad (2)$$

---

* Problems 7.4–7.7 have been contributed by Dr. Guanrong Chen, Associate Professor, Department of Electrical & Computer Engineering, University of Houston, Houston, Texas.

Its measurement-update error covariance matrix is given by

$$P(k|k) = [I - KH]P(k|k-1). \tag{3}$$

Furthermore, we have seen that the steady-state error covariance matrix given by Eq. (7.7) is

$$P(k+1|k+1) = P(k|k) = P = \alpha \sigma_v^2. \tag{4}$$

(a) Using Eq. (1) in Eq. (3) and inserting Eq. (4) for P, determine P in terms of α, σ_w^2, and σ_v^2. Also, determine the tracking index Γ^2 in terms of α. For this problem, assume that

$$\Phi = [1], \qquad H = [1], \qquad G = [T^2/2]^T, \qquad K(k) = [\alpha]^T,$$
$$R = \sigma_v^2, \quad \text{and} \quad Q = \sigma_w^2$$

where T is the sampling time.

(b) Write Eq. (3) in terms of α.

7.2   Given the model for a target tracking system

$$x(k+1) = F(k)x(k) + G(k)w(k)$$

and the observation model as

$$z(k) = H(k)x(k) + v(k).$$

Show that the error covariance matrix $P(k|k)$ for the α–β filter is given by

$$P(k|k) = P = \begin{bmatrix} p_{11} & p_{12} \\ p_{12} & p_{22} \end{bmatrix} = \sigma_v^2 \begin{bmatrix} \alpha & \dfrac{\beta}{T} \\ \dfrac{\beta}{T} & \dfrac{\beta(2\alpha - \beta)}{2(1-\alpha)T^2} \end{bmatrix}$$

[*Hint:* From Eq. (7.20) where

$$K = PH^T R^{-1} = PH^T \sigma_v^{-2} = \begin{bmatrix} p_{11} \\ p_{12} \end{bmatrix}$$

with $p_{11} = \alpha\sigma_v^2$, $p_{12} = (\beta/T)\sigma_v^2$, use the relations

$$P(k|k-1) = F(k-1)P(k-1|k-1)F^T(k-1) + G(k-1)Q(k-1)G^T(k-1),$$

$$P = [I - K(k)H(k)]P(k|k-1)$$

in the form

$$(I - KH)^{-1} = F(k-1)PF^T(k-1) + G(k-1)\sigma_w^2 G^T(k-1),$$

where it can be assumed that $P(k|k) = P(k-1|k-I) = P.]$

7.3 Given Eqs. (7.26)–(7.28) and the equations

$$\mathbf{x}(k) = [x(k) \quad \dot{x}(k) \quad \ddot{x}(k)]^T, \tag{1}$$

$$F(k) = \begin{bmatrix} 1 & T & 0.5T^2 \\ 0 & 1 & T \\ 0 & 0 & 1 \end{bmatrix} \tag{2}$$

$$\mathbf{G}(k) = \begin{bmatrix} \dfrac{T^2}{2} & T & 1 \end{bmatrix}^T, \tag{3}$$

$$\mathbf{H}(k) = [1 \quad 0 \quad 0], \tag{4}$$

$$R(k) = \sigma_v^2, \tag{5}$$

$$Q(k) = \sigma_w^2, \tag{6}$$

$$\mathbf{K}(k) = [\alpha \quad \beta/T \quad \gamma/T^2]^T. \tag{7}$$

Show that for the α–β–γ filter, the steady-state covariance matrix $P(k|k)$ is given by [Eq. (7.32)]

$$P(k|k) = \sigma_v^2 \begin{bmatrix} \alpha & \dfrac{\beta}{T} & \dfrac{\gamma}{T^2} \\ \dfrac{\beta}{T} & \dfrac{4\alpha\beta + \gamma(\beta - 2\alpha - 4)}{4(1-\alpha)T^2} & \dfrac{\beta(\beta-\gamma)}{2(1-\alpha)T^3} \\ \dfrac{\gamma}{T^2} & \dfrac{\beta(\beta-\gamma)}{2(1-\alpha)T^3} & \dfrac{\gamma(\beta-\gamma)}{(1-\alpha)T^4} \end{bmatrix}. \tag{8}$$

Hint: Let

$$\mathbf{K} = PH^T(k)\sigma_v^{-2} = \begin{bmatrix} p_{11} \\ p_{12} \\ p_{13} \end{bmatrix}. \tag{9}$$

Thus, using Eq. (7) gives

$$p_{11} = \alpha\sigma_v^2, \tag{10}$$

$$p_{12} = (\beta/T)\sigma_v^2, \tag{11}$$

$$p_{13} = (\gamma/T^2)^{-1}\sigma_v^2. \tag{12}$$

Also, as in Problem 7.2, equate the elements of P (i.e., $p_{11}, p_{12}, \ldots, p_{33}$) with the elements of the product

$$(I - \mathbf{KH})^{-1}P = FPF^T + G\sigma_w^2 G^T.$$

7.4 A popular tracker used by radar engineers is the α–β–γ tracker. This is a suboptimal filter and can be described by (here we will follow the terminology given in reference [21])

$$\hat{\mathbf{x}}_k = A\hat{\mathbf{x}}_{k-1} + \mathbf{H}(\mathbf{v}_k - CA\hat{\mathbf{x}}_{k-1}), \tag{1}$$

$$\hat{\mathbf{x}}_0 = \mathscr{E}\{\mathbf{x}_0\},$$

where $\hat{\mathbf{x}}_k$ is the suboptimal estimate of $\mathbf{x}_k$, and

$$\mathbf{H} = [\alpha \quad \beta/h \quad \gamma/h^2]^T \tag{2}$$

In practice, the α, β, γ values are chosen according to the physical model and depending on the user's experience. In this problem, we consider the case where

$$A = \begin{bmatrix} 1 & h & h^2/2 \\ 0 & 1 & h \\ 0 & 0 & 1 \end{bmatrix} \text{ and } C = [1 \quad 0 \quad 0], \qquad h > 0. \tag{3}$$

The *limiting* (or *steady-state*) Kalman filter with constant matrices independent of time, is given by

$$\vec{\mathbf{x}} = \Phi\vec{\mathbf{x}}_{k-1} + \mathbf{Gv}_k, \tag{4}$$

$$\vec{\mathbf{x}}_0 = \mathscr{E}\{\mathbf{x}_0\},$$

where $\Phi = (I - \mathbf{GC})A$ with

$$\mathbf{G} = \begin{bmatrix} g_1 \\ g_2 \\ g_3 \end{bmatrix} = \frac{PC}{C^T PC + \sigma_m} = \frac{1}{P[1,1] + \sigma_m} \begin{bmatrix} P[1,1] \\ P[2,1] \\ P[3,1] \end{bmatrix} \tag{5}$$

and with $P = [P[i,j]]_{3\times 3}$ the positive definite solution of the following matrix Riccati equation:

$$P = A[P - PC(C^T PC + \sigma_m)^{-1}C^T P]A^T + \Gamma Q\Gamma^T$$

$$= A[P - PC(C^T PC + \sigma_m)^{-1} C^T P] A^T + \begin{bmatrix} \sigma_p & 0 & 0 \\ 0 & \sigma_v & 0 \\ 0 & 0 & \sigma_a \end{bmatrix}. \quad (6)$$

A necessary condition for the α–β–γ tracker, (1), to be limiting Kalman filter is $\mathbf{H} = \mathbf{G}$, or equivalently,

$$\begin{bmatrix} \alpha \\ \beta/h \\ \gamma/h^2 \end{bmatrix} = \frac{1}{p_{11} + \sigma_m} \begin{bmatrix} p_{11} \\ p_{21} \\ p_{31} \end{bmatrix},$$

so that

$$\begin{bmatrix} p_{11} \\ p_{21} \\ p_{31} \end{bmatrix} = \frac{\sigma_m}{1 - \alpha} \begin{bmatrix} \alpha \\ \beta/h \\ \gamma/h^2 \end{bmatrix}. \quad (7)$$

Using the relations (the matrix P is assumed to be symmetric)

$$P = \begin{bmatrix} p_{11} & p_{21} & p_{31} \\ p_{21} & p_{22} & p_{32} \\ p_{31} & p_{32} & p_{33} \end{bmatrix}$$

$$= A \left(P - \frac{1}{p_{11} + \sigma_m} P \begin{bmatrix} 1 & 0 & 0 \\ 0 & 0 & 0 \\ 0 & 0 & 0 \end{bmatrix} P \right) A^T + \begin{bmatrix} \sigma_p & 0 & 0 \\ 0 & \sigma_v & 0 \\ 0 & 0 & \sigma_a \end{bmatrix} \quad (8)$$

and

$$\mathbf{G} = \frac{PC}{C^T PC + \sigma_m} = \frac{1}{p_{11} + \sigma_m} \begin{bmatrix} p_{11} \\ p_{21} \\ p_{31} \end{bmatrix}, \quad (9)$$

determine the matrix P, that is, Eq. (8). This is strictly an algebraic problem, and the answer can be obtained by substituting Eqs. (7) and (9) into Eq. (8). Note that by setting

$$g_1 = \alpha, \qquad g_2 = \beta/h, \quad \text{and} \quad g_3 = \gamma/h^2,$$

we have a decoupled filter. Under certain conditions on the α, β, γ values, α–β–γ tracker for the time-invariant system (1) is a limiting Kalman filter, so that these conditions will guarantee near-optimal performance of the tracker.

7.5 As an extension of Problem 7.4, determine the matrix P in terms of the parameters α, β, γ. Substitute Eq. (7) of Problem 7.4 into the answer for P, yielding [21]

$$\frac{h^4}{\gamma^2}\sigma_a = P_{11} + \sigma_m,$$

$$P_{11} = \frac{h^4(\beta+\gamma)^2}{\gamma^3(2\alpha+2\beta+\gamma)}\sigma_a - \frac{h}{\gamma(2\alpha+2\beta+\gamma)}\sigma_v,$$

$$2hp_{21} + h^2p_{31} + h^2p_{22} \tag{1}$$

$$= \frac{h^4}{4\gamma^2}(4\alpha^2 + 8\alpha\beta + 2\beta^2 + 4\alpha\gamma + \beta\gamma)\sigma_a + \frac{h^2}{2}\sigma_v - \sigma_p,$$

$$P_{31} + P_{22} = \frac{h^4}{4\gamma^2}(4\alpha + \beta)(\beta+\gamma)\sigma_a + \tfrac{3}{4}\sigma_v$$

and

$$P_{22} = \frac{\sigma_m}{(1-\alpha)h^2}\left[\beta\left(\alpha+\beta+\frac{\gamma}{4}\right) - \frac{\gamma(2+\alpha)}{2}\right],$$

$$P_{32} = \frac{\sigma_m}{(1-\alpha)h^2}\gamma\left(\alpha+\frac{\beta}{2}\right), \tag{2}$$

$$P_{33} = \frac{\sigma_m}{(1-\alpha)h^4}\gamma(\beta+\gamma).$$

7.6 A simple two-dimensional real-time tracking system is given by [21]

$$\mathbf{x}_{k+1} = \begin{bmatrix} 1 & h \\ 0 & 1 \end{bmatrix}\mathbf{x}_k + \xi_k,$$

$$v_k = \begin{bmatrix} 1 & 0 \end{bmatrix}\mathbf{x}_k + \eta_k,$$

where $h > 0$, and $\{\xi_k\}$, $\{\eta_k\}$ are both uncorrelated zero-mean Gaussian white noise sequences.

(a) Write the equation for the α–β tracker system in the form of Eq. (1) of Problem 7.4.

(b) Derive the decoupled Kalman filtering algorithm for this α–β tracker. Use the fact that $\mathbf{x}_k^T = [x_k \ \dot{x}_k]$.

7.7 Given the linear system [21]

$$\mathbf{x}_{k+1} = A\mathbf{x}_k + \Gamma \xi_k,$$
$$\mathbf{v}_k = C\mathbf{x}_k + \eta_k, \tag{1}$$

Here, A, Γ, and C are known $n \times n$, $n \times p$, and $q \times n$ constant matrices, respectively, with $1 \leqslant p, q \leqslant n$; and $\{\xi_k\}$ and $\{\eta_n\}$ are zero-mean Gaussian white noise sequences with

$$\mathcal{E}\{\xi_k \xi_k^T\} = Q\delta_{kl}, \qquad \mathcal{E}\{\eta_k \eta_k^T\} = R_{kl}, \quad \text{and} \quad \mathcal{E}\{\xi_k \eta_l^T\} = 0,$$

where Q and R are known $p \times p$ and $q \times q$ nonnegative and positive definite symmetric matrices, respectively, independent of k. The system (1) is *completely* controllable if the matrix

$$M_{A\Gamma} = [\Gamma \quad A\Gamma \quad \cdots \quad A^{n-1}\Gamma] \tag{2}$$

has full rank, and *completely* observable if the matrix

$$N_{CA} = \begin{bmatrix} C \\ CA \\ \vdots \\ CA^{n-1} \end{bmatrix} \tag{3}$$

has full rank. Determine if the tracking system

$$\mathbf{x}_{k+1} = \begin{bmatrix} 1 & h & h^2/2 \\ 0 & 1 & h \\ 0 & 0 & 1 \end{bmatrix} \mathbf{x}_k + \xi_k, \tag{4}$$

$$\mathbf{v}_k = [1 \quad 0 \quad 0]\mathbf{x}_k + \eta_k,$$

where $h > 0$ is the sampling time, is completely controllable and observable.

CHAPTER 8

DECENTRALIZED KALMAN FILTERS

8.1 INTRODUCTION

It is well known that the conventional Kalman filtering algorithms, although globally optimal, diverge and tend to be numerically unreliable. More numerically stable and better-conditioned implementations of the Kalman filtering algorithms can be obtained using, for example, the Bierman U–D and square-root formulations [9]. In recent years, *decentralized* estimation for linear systems has been an active area of research in which decentralized and parallel versions of the Kalman filter have been reported in the literature [16, 19, 20, 35, 36 and 65]. For example, the approaches by Speyer [65] and Willsky et al. [73] are based on decomposing a central estimation problem in- to smaller, local ones. These works show that the global estimates can be obtained using linear operations on the local filter.

This chapter develops the concepts of decentralized Kalman filtering with parallel processing capabilities, for use in distributed multisensor systems. In essence, the study of decentralized filters began as an attempt to reduce throughput. At this point, we can define decentralized filtering as a two-stage data-processing technique which processes data from multisensor systems. In the first stage, each *local* processor uses its own data to make a best local estimate. These estimates are then obtained in a parallel processing mode. The local estimates are then fused by a master filter to make a best *global* estimate of the state vector of the master system.

Decentralized estimation offers numerous advantages in many applications. Specifically, it provides significant advantages for real-time multisensor applications, such as integrated inertial navigation systems, where the inertial navigation system may be aided by a number of different sensors. In this design, a number of sensor-dedicated *local filters* run in parallel, the outputs being fused into a *master filter*, yielding estimates that are globally optimal. As a result, the filtering duties are divided so that each measurement is processed in a local filter, which contains only those states that are directly relevant to the measurement. The state estimates and covariances of a given local filter are then weighted and combined in the master filter, whose own state is the fusion of all the local filter states. This *hierarchical* filtering structure provides some important improvements: (1) increasing the data throughput (i.e., rate) by local parallel processing, (2) reducing the required bandwidth for information transmission to a central processor, and (3) allowing for a more fault-tolerant system design.

In a multisensor system, each individual sensor has its own *built-in* Kalman filter. One is interested in combining the estimates from these independent data sources (filters) in order to generate a global estimate that will, ideally, be optimal. Furthermore, decentralization makes for easy fault detection and isolation (FDI), since the output of each local sensor filter can be tested, and if a sensor should fail, it can expeditiously be removed from the sensor network before it affects the total filter output. Also, decentralization increases the input data rates significantly and yields moderate improvements in throughput.

8.2 FILTER MECHANIZATIONS AND ARCHITECTURES

Kalman filtering techniques have been successfully utilized in a multitude of aerospace applications, especially in inertial navigation and guidance systems. Furthermore, and as stated earlier (see Chapter 4), it is a well-known fact that the conventional Kalman filter provides the best sequential linear, unbiased estimate, or a globally optimal estimate, when noise processes are jointly Gaussian. However, in practice the use of a large *centralized* Kalman filter may not be feasible due to such factors as (1) computational burden (i.e., excessive processing time), (2) high data rate, and (3) accuracy specifications. As a result, decentralized or parallel versions of the standard Kalman filter began to receive increasing attention in recent years. For example, in inertial navigation applications, decentralized Kalman filtering implementations enable one to allocate a multitude of Kalman filters for fast multisensor and multitarget threat tracking [20].

In reference [9], Bierman proposed a federated square-root information filter (SRIF). This filter architecture was designed to provide good throughput and an optimal estimate. In the SRIF architecture, a bank of local filters feeds a master filter. Furthermore, each local filter is assigned to a sensor and processes a single measurement or group of measurements. Here, the local filters contain all or part of the master filter state. (Note that all filters are implemented in SRIF form.) The Bierman estimate is optimal, and requires constant feedback between the local and master filters. The master filter must run at the highest data rate, that is, whenever data are acquired, the master filter must be propagated and updated. Also, the local filters must be specially constructed so that one filter keeps track of process noise information and the other local filters have no information about process noise. Consequently, the federated SRIF inputs all the process noise information into one of the local filters and no process noise into the other filters. Therefore, the absence of process noise information means that there is no dynamics model information in the other local filters from time step to time step, or infinite process noise injected into the dynamical nonbias state.

From the above discussions, we see that the centralized (or *monolithic*) Kalman filter is undoubtedly the best (optimal) estimator for simple, well-behaved linear systems. As the model increases in size and complexity, however, this strains the processing resources. Due to the large size of the relevant aircraft models and severe constraints on airborne computer throughput capabilities, the mathematical ideal of a complete centralized Kalman filter has never been fully realized in the avionics environment. Therefore, centralized filtering in a multisensor environment can suffer from severe computational loads and must run at high input–output rate. For this reason, and because of the rapidly increasing computational burden associated with large filter state size, real-time inertial navigation filters generally employ reduced-order (i.e., suboptimal) models. Such models are carefully designed to reflect the dominant error sources and to represent the remaining error sources by noise statistics or other simplified means.

A comparison between the centralized and decentralized filters is given in Figure 8.1.

Consider the conventional discrete-time* linear dynamic system (or global model) of the form

$$\mathbf{x}(k+1) = \Phi\mathbf{x}(k) + G\mathbf{w}(k), \tag{8.1}$$

$$\mathbf{z}(k) = H\mathbf{x}(k) + \mathbf{v}(k), \tag{8.2}$$

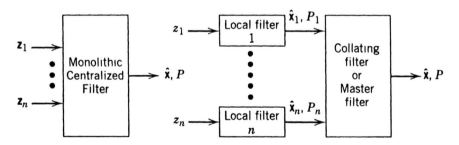

CENTRALIZED ARCHITECTURE   DECENTRALIZED ARCHITECTURE

Figure 8.1 Centralized and decentralized filters.

---

*It is well known that, even though the conventional Kalman filtering algorithms are theoretically globally optimal, they can be numerically unreliable. As discussed earlier in the text, this problem can be remedied by using more-numerically-stable and better-conditioned implementations of the Kalman filtering algorithms, such as the square-root or $U-D$ methods.

where the following usual assumptions apply:

$$\mathbf{x}(k) \in R^n, \quad \mathbf{w}(k) \in R^p, \quad \mathbf{z}(k), \mathbf{v}(k) \in R^m,$$

$$\Phi(k) \in R^{n \times n}, \quad G \in R^{n \times p}, \quad \text{and} \quad H \in R^{m \times n},$$

and with the initial conditions

$$\mathbf{x}_0 \sim N(\bar{\mathbf{x}}_0, P_0), \quad \mathbf{w}(k) \sim N(0, Q), \quad \mathbf{v}(k) \sim N(0, R).$$

Here $\mathbf{w}(k)$ and $\mathbf{v}(k)$ are white-noise processes independent of each other and of the initial state estimate $\mathbf{x}_0$. We assume that P_0, Q, and R are symmetric and positive definite. For the above assumed model, the results of the conventional Kalman filter algorithm are summarized below [see also Eq. (4.26)–(4.30)]:

Propagation (or prediction):

$$\hat{\mathbf{x}}(k|k-1) = \Phi(k)\hat{\mathbf{x}}(k-1|k-1), \quad k = 0, 1, 2, \ldots \quad \text{(state propagation)},$$

$$P(k|k-1) = \Phi(k)P(k-1|k-1)\Phi^T(k) + GQG^T, \quad \text{(covariance propagation)}$$

$$P(0|0) = P(0).$$

Residual and gain computation:

$$\eta(k) = \mathbf{z}(k) - H\hat{\mathbf{x}}(k|k-1) \quad \text{(innovations or measurement residual)},$$

$$S(k) = HP(k|k-1)H^T + R \quad \text{(innovations covariance)},$$

$$K(k) = P(k|k-1)HS^{-1}(k) \quad \text{(Kalman gain matrix)}.$$

Measurement update:

$$\hat{\mathbf{x}}(k|k) = \hat{\mathbf{x}}(k|k-1) + K(k)\eta(k) \quad \text{(state update)},$$

$$P(k|k) = P(k|k-1) - K(k)S(k)K^T(k) \quad \text{(covariance update)}.$$

8.3 PARALLEL KALMAN FILTER EQUATIONS

Decentralized or parallel concepts for Kalman filtering are gaining practical significance for real-time implementation as a result of the recent advances in chip technology. In particular, the development and implementation of the very high speed integrated circuit (VHSIC) program resulted in Kalman filter architectures for maximum efficiency. A decentralized Kalman filtering system consists of a bank of N local subsystems (i.e., local Kalman filters), each local filter generating estimates of the full state vector. All decentralized Kalman filtering mechanizations require some processing at the local level, that is, at the subsystem level. Depending on the type of application, these local processed data may then be sent to a central processing unit (also referred to as a *master filter, fusion center,* or *collating unit*) that combines the local results in order to generate a global or central result. The decentralized mechanizations discussed here are the most attractive ones, since they require no interprocessor communications and no communication from the collating filter back to the local filters (that is, bidirectional communication). In other words, the local Kalman filters are *stand alone*, in that each of them generates estimates based solely on its own available raw data. This allows the local filters to run in parallel at a faster rate.

The central collating filter combines the estimates from the local filters to produce the central estimate. References [19] and [35] provide a definitive formulation of these ideas. Specifically, it is shown in these references that the central estimate is globally optimal, as if all measurements were available in one location to the central collating unit to feed a centralized Kalman filter. Stated another way, theoretically there is no loss in performance of these decentralized structures as compared to the optimal centralized filter.

Decentralized or parallel ideas for the Kalman filter started to emerge more than a decade ago, but it was not until recently that these ideas formed a solid discipline [9, 14, and 15]. References [19] and [35] have demonstrated that each local processor depends on its own information to generate its state estimates and covariances using its own local Kalman filter. Thus, local estimates are then combined at a central (or master) processor to produce the global optimal estimates. In the case where there is *hierarchy* in the filtering structure (as, for example, local processors being at a lower level and a central processor at a higher level), it is natural to employ multirate filtering, whereby lower levels can run at a faster rate than higher ones. Multirate filtering may also be desirable in situations where there are sensors of different nature, such as an inertial navigation system being aided by such navigation aids as Doppler radar, the Global Positioning System, Loran, Omega, and Tercom.

In decentralized or parallel filtering, the local filter architectures are completely autonomous. That is, they do not communicate with each other and are not interrupted by the central processor with any feedback information. As a result, they can be run entirely in parallel, all communication being unidirectional, upwards in the hierarchy to the central processor [57]. This architecture is useful for the purposes of FDI, as for example in integrated inertial navigation systems.

In order to parallelize the conventional Kalman filter, consider again the standard discrete-time linear dynamic system

$$\mathbf{x}(k+1) = \Phi \mathbf{x}(k) + G \mathbf{w}(k), \tag{8.1}$$

$$\mathbf{z}(k) = H \mathbf{x}(k) + \mathbf{v}(k), \tag{8.2}$$

where, as before, $\mathbf{w}(k)$ and $\mathbf{v}(k)$ are the process noise and the measurement noise with covariances Q and R, respectively, and with the usual statistical assumptions on $\mathbf{w}(k)$, $\mathbf{v}(k)$, and the initial condition $\mathbf{x}_0$. If the system is observed by N independent sensors, then

$$\mathbf{z}(k) = [\mathbf{z}_1^T(k), \mathbf{z}_2^T(k), \ldots, \mathbf{z}_N^T(k)]^T,$$

$$\mathbf{w}(k) = [\mathbf{w}_1^T(k), \mathbf{w}_2^T(k), \ldots, \mathbf{w}_N^T(k)]^T,$$

$$\mathbf{v}(k) = [\mathbf{v}_1^T(k), \mathbf{v}_2^T(k), \ldots, \mathbf{v}_N^T(k)]^T,$$

$$H = [H_1^T, H_2^T, \ldots, H_N^T]^T,$$

$$G = [G_1, G_2, \ldots, G_N].$$

Suppose now that there is a bank of N Kalman filters available, each having access to a subblock of the measurement vector $\mathbf{z}$, that is,

$$\mathbf{z}_j = H_j \mathbf{x} + \mathbf{v}_j,$$

where $\mathbf{z}_j$ and $\mathbf{v}_j$ are the jth blocks of $\mathbf{z}$ and $\mathbf{v}$, respectively. H_j is the jth row block of H, and R_j is the jth diagonal block of R. The crucial assumption here is that the sub-noise-processes $\mathbf{v}_j$ are uncorrelated, so that the matrix R is block diagonal of the form

$$R = \text{block diag}\,(R_1, \ldots, R_N).$$

This is always true if the orginal measurement noise covariance matrix is diagonal, that is, the components of $\mathbf{v}$ are uncorrelated. In reference [35] it was shown that instead of assigning a large, centralized Kalman filter processing the entire measurement vector, one can use N smaller Kalman filters running in parallel and a master or central filter that collates the local processed data to generate the final estimates. Thus, in the above scheme, the jth filter processes the measurements $\mathbf{z}_j$, and generates estimates $\mathbf{x}_j$ and the associated covariances P_j at every instant of time. Note that the jth processor is just running a conventional Kalman filter, but is processing a measurement of smaller dimension. Consequently, the global estimates and covariances at each step of the way are generated through the following formulas [35]:

Measurement update:

$$P^{-1}(k|k) = P^{-1}(k|k-1) + \sum_{j=1}^{N} [P_j^{-1}(k|k) - P_j^{-1}(k|k-1)],$$

$$P(k|k) = \left[(1-N)P(k|k-1)^{-1} + \sum_{j=1}^{N} P_j(k|k)^{-1} \right]^{-1} \qquad (8.3a)$$

and

$$\hat{\mathbf{x}}(k|k) = P(k|k) \left(P^{-1}(k|k-1)\,\hat{\mathbf{x}}(k|k-1) \right.$$

$$\left. + \sum_{j=1}^{N} [P_j^{-1}(k|k)\,\hat{\mathbf{x}}_j(k|k) - P_j^{-1}(k|k-1)\,\hat{\mathbf{x}}_j(k|k-1)] \right), \qquad (8.3b)$$

$$\hat{\mathbf{x}}(k|k) = P(k|k) \left[(1-N)P(k|k-1)^{-1}\,\hat{\mathbf{x}}(k|k-1) + \sum_{j=1}^{N} P_j(k|k)^{-1}\hat{\mathbf{x}}(k|k) \right].$$

Time update:

$$\hat{\mathbf{x}}(k) = \Phi(k)\,\hat{\mathbf{x}}(k-1), \qquad (8.4a)$$

$$P(k) = \Phi(k)\,P(k-1)\Phi^T(k) + G(k)\,Q(k)\,G^T(k). \qquad (8.4b)$$

From the above result, we note that parallelism manifests itself only at the measurement update level, that is, the time update equations are the same as for the centralized filter. Reference [35] discusses how parallelism can also be utilized at the time update level when the process noise and the measurement noise are correlated. The above scheme uses the covariance filter for time update and the information filter for measurement update.

Figure 8.2 illustrates the decentralized (or parallel) filter architecture concept [35].

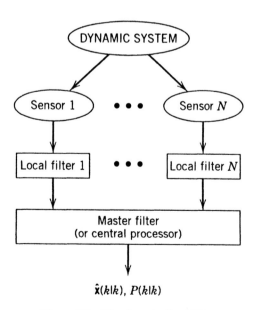

Figure 8.2 The decentralized filter.

8.4   DECENTRALIZED KALMAN FILTERING ARCHITECTURES USING REDUCED-ORDER MODELS

In the discussion of decentralized mechanization in Section 8.3, we noted that each local filter generates estimates of the full state vector. However, this may cause problems if the state dimension is large. For this reason, a reduced-order filter may be more advantageous to use than a full-order filter. We begin the discussion by assuming the jth local filter of a bank of N local filters can be described by the model [57]

$$\mathbf{x}_j(k+1) = \Phi_j(k)\,\mathbf{x}_j(k) + \mathbf{w}_j(k), \qquad (8.5a)$$

$$\mathbf{z}_j(k) = C_j\mathbf{x}_j(k) + \mathbf{v}_j(k), \qquad (8.5b)$$

where $\mathbf{x}_j$ is an n_j-vector $(m_j \leqslant n_j \leqslant n)$ and $\mathbf{z}_j$ is an m_j-subvector of $\mathbf{z}(k)$. Furthermore, assume as before that

$$\mathbf{w}_j(k) \sim N(0, Q_j),$$

$$\mathbf{v}_j(k) \sim N(0, R_j)$$

and that $\mathbf{w}_j(k), \mathbf{w}_i(k)$ and $\mathbf{v}_j(k), \mathbf{v}_i(k)$ are uncorrelated for $i \neq j$. From the above, we note that if Q and R are positive definite,

$$Q = \text{diag}\,(Q_1, \ldots, Q_N),$$

$$R = \text{diag}\,(R_1, \ldots, R_N).$$

In the present case, since the local filters are of reduced order, Eqs. (8.3) and (8.4) are no longer valid. However, this fact can be remedied by assuming that if there exists a matrix $D_j \in R_j^{n \times n}$ such that

$$H_j = C_j D_j, \qquad (8.6)$$

then Eqs. (8.3) and (8.4) take the form

$$P^{-1}(k|k) = P^{-1}(k|k-1) + \sum_{j=1}^{N} D_j^T [P_j^{-1}(k|k) - P_j^{-1}(k|k-1)] D_j,$$

$$P(k|k) = \left\{ P^{-1}(k|k-1) + \sum_{j=1}^{N} D_j^T [P_j^{-1}(k|k) - P_j^{-1}(k|k-1)] D_j \right\}^{-1} \qquad (8.7)$$

and

$$\hat{\mathbf{x}}(k|k) = P(k|k)\left\{P^{-1}(k|k-1)\,\hat{\mathbf{x}}(k|k-1) + \sum_{j=1}^{N} D_j^T\left[P_j^{-1}(k|k)\,\hat{\mathbf{x}}(k|k)\right.\right.$$

$$\left.\left. - P_j^{-1}(k|k-1)\,\hat{\mathbf{x}}(k|k-1)\right]D_j\right\}. \tag{8.8}$$

These equations are in *information-filter* form. They give the fusing algorithm of the master filter, based on the local-filter estimates. The global estimate of the state vector $\mathbf{x}(k|k)$ given in Eq. (8.8) is the same as the estimate given in Eq. (8.3b). The above derivation can be found in [65]. Here we note that each local filter must communicate only an n_j-vector of the state estimate and its corresponding $n_j \times n_j$ covariance matrix.

One important aspect of this algorithm is that the designer can choose C_j to be the $m_j \times m_j$ identity matrix (i.e., $n_j = m_j$) and set $D_j = H_j$ in Eq. (8.7). The collating-filter time update remains the same as in the centralized Kalman filter equations, that is,

$$\hat{\mathbf{x}}(k|k) = \Phi(k)\,\hat{\mathbf{x}}(k|k-1) \tag{8.9a}$$

$$P(k|k) = \Phi(k)\,P(k|k-1)\,\Phi^T(k) + G(k)\,Q(k)\,G^T(k). \tag{8.9b}$$

In implementing decentralized Kalman filters for multisensor systems, the desired attributes are following: (1) speed, (2) robustness, (3) global optimality, (4) that the master filter carry less computational load than the centralized filter, (5) that the master filter be able to run at a selectively lower rate than the local filters, (6) that the local filters be completely autonomous (i.e., run in parallel) and (7) the ability to reject faulty data [i.e., fault detection and isolation (FDI)]. The design of a robust integrated navigation system must address more than just the integration of the sensors and their associated local filters in an unfailed situation: it also must address the inclusion of FDI and reconfiguration capabilities into the filter structure.

8.5 PROBABILITY FUSION EQUATIONS

The use of data from multiple sensors occurs in many military and civilian applications. Specifically, the basic distributed estimation technique has been applied together with the *probabilistic data association* (PDA) scheme in order to handle the data association problem when measurement origins are uncertain [16]. In this way, fusion algorithms have been derived for: (1) the *joint PDA*, (2) *interactive multiple mode* (IMM), and (3) the PDA with *multiple-model* algorithms.

Several investigators have examined the *multiple-model adaptive estimation* (MMAE) method with regard to distributed filtering as a means of addressing the tremendous computational complexity of multiple, centralized Kalman filters, and as a means of improving the fault tolerance of the estimator. MMAE techniques appear well suited for use with a distributed architecture, in particular the decentralized or distributed Kalman filter. Other researchers —in particular, Chang [16], Bar-Shalom [3], and Chong, Mori, and Chang [20]—have investigated the decentralized MMAE problem, focusing primarily on the probability fusion equations, that is, the computation of the global probability density of the *global* state estimate from the local probability density of the local filter's state estimate. These probability densities are computed from the likelihood functions associated with local-filter measurement residuals, and are themselves used in turn to compute the global parameter and state estimates. The computational burden associated with this collection of local filters will generally be much less than that associated with a set of centralized filters, one for each combination of parameter values. However, in order to duplicate the performance of the centralized MMAE, one must have a fusion of the local-filter results for each possible configuration of centralized filters.

Each of the parameter estimation schemes discussed in Chapter 3 [maximum likelihood (ML), maximum *a posteriori* (MAP) and mean-square error (MSE)] requires the computation of a conditional density function, or a conditional expectation in the case of the MSE estimate. The conditional expectation of the MSE estimate can be computed given the appropriate conditional density, so that if one has the ability to compute the appropriate conditional density, one can then obtain any of the above estimates.

In the case of a multiple-model estimator, rather than compute a continuous density function, what is usually computed is a discrete probability function. For example, if we have a bank of filters, indexed on $k = 1,\ldots, N$, and each filter is constructed with the assumption that the parameter vector has the value θ_k, we would like to compute the probability that $\theta = \theta_k$ for each k, given that we have observed the i measurements $Z_i = [z_0,\ldots, z_i]$. Using the residuals and other quantities computed by the kth matched, centralized Kalman filter, we can compute

$$p(z_1 | \theta_k, z_{i-1}, z_{i-2}, \ldots) = p(z_i | \theta_k, z_{i-1}). \tag{8.10}$$

Equation (8.10) is the probability density of the ith measurement, given that we are assuming θ to have the value θ_k, and given that we have observed the preceding measurements $z_{i-1}, z_{i-2}, \ldots$. From this conditional probability density, produced by each filter at each measurement update, we can recursively compute the conditional probability $p(\theta = \theta_k | Z_i)$, where Z_i represents the entire measurement set up to time i. The aforementioned calculation is carried out by the equation

$$p(\theta = \theta_k | z_i, z_{i-1}, \ldots) = \frac{p(z_i | \theta_k, z_{i-1}, \ldots) \; p(\theta = \theta_k | z_{i-1}, z_{i-1}, \ldots)}{\sum_j p(z_j | \theta_j, z_{i-1}, \ldots) \; p(\theta = \theta_j | z_{i-1}, z_{i-2}, \ldots)}. \tag{8.11}$$

From Eq. (8.11), we note that in order to compute $p(\theta = \theta_k | Z_i)$ for a given k, we need not only $p(\theta = \theta_k | Z_{i-1})$ and $p(z_i | \theta_k, Z_{i-1})$, but also information from the other filters as well in order to compute the denominator of the above equation. Consequently, given this probability, we can compute the conditional expectation needed for the MSE estimate of θ, by probability weighting according to the equation [see also Eq. (3.57)]

$$\theta_{\text{MSE}} = \mathcal{E}\{\theta | \mathbf{z}\} = \sum_j \theta_j \, p(\theta = \theta_j | \mathbf{z}). \tag{8.12}$$

On the other hand, we can also compute the MAP estimate by choosing the estimate of θ to be the value of θ_k that is associated with the filter producing the largest value for $p(\theta = \theta_k | Z)$. Finally, the ML estimate can be determined directly from $p(z_i | \theta_k, z_{i-1}, z_{i-2}, \ldots)$, again by choosing the filter producing the largest value for this conditional probability density.

8.6 CONCLUDING REMARKS

In this chapter we have considered the centralized and decentralized estimation of linear systems with distributed sensors. After a discussion of the general nature of the problem, we considered systems where the measurements are given by the linear combinations of the signal and noises. The centralized Kalman filter has been the most common and successful filter design implementation for more than three decades now. The centralized Kalman filter, which has been used extensively in the integrated multisensor systems, receives all available measurements and combines the information contained in those measurements to obtain an optimal solution. In particular, for simple, well-modeled linear systems, the centralized Kalman filter is unquestionably the optimal estimator. However, when one considers tradeoffs of data flow, algotithmic requirements, processing speed versus optimality, and fault tolerance in a multisensor environment the problem is often best treated as one of decentralized (or distributed) estimation.

For this reason, in many large-scale decentralized systems, measurements on the state of the system are often obtained from geographically distributed sensors, such as surveillance systems, power systems, and transportation networks. Moreover, on many occasions, one may wish to perform some pre-processing at each sensor and communicate the local estimates to the central computer where the master filter resides and from which a global estimate is obtained.

The decentralized architecture has the following advantages [19]:

1. Local estimates are available at the sensor level, where they may be useful for local decision making and monitoring.
2. The cost of communication may be reduced, since only processed data are transmitted.
3. The decentralized architecture insures survivability in the event that the central processor fails; that is, the decentralized filter offers better fault tolerance.
4. The local processor may be realized by means of inexpensive microprocessors.

It should be noted that this class of problems can be found under various names in the literature, such as decentralized estimation, distributed estimation, hierarchical estimation, and data fusion. We have used the term decentralized estimation to bring out the independent, multilevel nature of the problem.

PROBLEMS*

8.1   For discrete measurements, a dynamic system can be modeled in the form

$$\mathbf{x}_k = \Phi \mathbf{x}_{k-1} + \mathbf{w}_k, \tag{1}$$

where $\mathbf{w}_k$ is the system noise with covariance Q_k. At an instant of time t_k, the measurements are given in the form

$$\mathbf{z}_k = H_\lambda \mathbf{x}_k + \mathbf{v}_k, \tag{2}$$

where $\mathbf{v}_k$ is the measurement noise with covariance R_k. For multisensor systems, the measurements $\mathbf{z}_k$ can be written as

$$\mathbf{z}_k^T = [z_{1k}, z_{2k}, ..., z_{ik}], \tag{3}$$

where the subscript i denotes the *local* system and $\mathbf{z}_{ik}$ is the vector of measurements from local system i at time t_k. We can now write the measurement equation for the measurements $\mathbf{z}_{ik}$ in the form

$$\mathbf{z}_{ik} = H_{ik} \mathbf{x}_{ik} + \mathbf{v}_{ik}. \tag{4}$$

---

* Problems 8.6 and 8.7 have been contributed by Dr. Kuo-Chu Chang, Associate Professor, Department of Systems Engineering, George Mason University, Fairfax, Virginia.

Each local system i can be described by

$$\mathbf{x}_{ik} = \Phi^i_{k,k-1} \mathbf{x}_{i,k-1} + \mathbf{w}_{ik} \tag{5}$$

with the corresponding measurement equation

$$\mathbf{z}_{ik} = A_{ik} \mathbf{x}_{ik} + \mathbf{v}_{ik}. \tag{6}$$

For a centralized system described by Eqs. (1) and (2), the Kalman filter prediction and update equations are presented below [see Eqs. (4.26)–(4.30) in Chapter 4, Section 4.3]:

Time update (prediction or propagation):

$$\hat{\mathbf{x}}_k(-) = \Phi_{k,k-1} \hat{\mathbf{x}}_{k-1}(+), \tag{7}$$

$$P_k(-) = \Phi_{k,k-1} P_{k-1}(+) \Phi^T_{k,k-1} + Q_k; \tag{8}$$

Measurement update:

$$\hat{\mathbf{x}}_k(+) = \hat{\mathbf{x}}_k(-) + K_k [\mathbf{z}_k - H_k \hat{\mathbf{x}}_k(-)], \tag{9}$$

$$K_k = P_k(-) H_k^T [H_k P_k(-) H_k^T + R_k]^{-1}, \tag{10}$$

$$P_k(+) = (I - K_k H_k) P_k(-), \tag{11}$$

where $(-)$ and $(+)$ denote the estimates before and after the update. (Note that the standard Kalman filter equations are written without the subscript i.) Assuming now that the measurements between local systems are independent, the Kalman filter makes a best estimate of the state vector $\mathbf{x}_k$ in terms of the measurements $\mathbf{z}_{ik}$ as follows:

Update:

$$\hat{\mathbf{x}}_k(+) = \hat{\mathbf{x}}_k(-) + \sum_i K_{ik} [\mathbf{z}_{ik} - H_{ik} \hat{\mathbf{x}}_k(-)], \tag{12}$$

$$K_{ik} = P_k(+) H_{ik}^T R_{ik}^{-1}, \tag{13}$$

$$P_k(+) = \left(I - \sum_i K_{ik} H_{ik} \right) P_k(-). \tag{14}$$

Reformulate the update Eqs. (9) and (11) to reflect the local sensors ik.

8.2 In the decentralized Kalman filter, we have seen that the local estimates, which are obtained in a parallel processing model, are fused (i.e., combined) by a master filter in order to make a best global estimate of the state vector $\mathbf{x}_k$ of the master system. For the ith local system, which is described by Eqs. (5) and (6) of Problem 8.1, the best local estimate of the state vector $\mathbf{x}_{ik}$ can be obtained using Kalman filtering based on the local measurements $\mathbf{z}_{ik}$ and is

$$\hat{\mathbf{x}}_{ik}(+) = P_{ik}(+) P_{ik}^{-1}(-) \hat{\mathbf{x}}_{ik}(-) + P_{ik}(+) A_{ik}^T R_{ik}^{-1} \mathbf{z}_{ik}, \tag{1}$$

$$P_{ik}^{-1}(+) = P_{ik}^{-1}(-) + A_{ik}^T R_{ik}^{-1} A_{ik}, \tag{2}$$

where $\hat{\mathbf{x}}_{ik}(+)$ is the best estimate after update with the error covariance matrix $P_{ik}(+)$, and $\hat{\mathbf{x}}_{ik}(-)$ is the best prediction of the local state vector with the corresponding error covariance matrix $P_{ik}(-)$. Equations (1) and (2) describe a parallel processing structure. A best global estimate for the state vector $\mathbf{x}_k$ can be obtained in terms of the local estimates $\hat{\mathbf{x}}_{ik}$. The transformation from the local states $\mathbf{x}_{ik}$ to the global states $\mathbf{x}_k$ is usually described by a linear transformation of the form

$$\mathbf{x}_{ik} = M_i \mathbf{x}_k. \tag{3}$$

Using the transformation matrix M_i in Eq. (3), the design matrices H_{ik} for the master system and A_{ik} for local system i assume the relation

$$H_{ik} = A_{ik} M_i. \tag{4}$$

Using the transformation (3), determine the best estimate $\hat{\mathbf{x}}_k(+)$ in terms of the local estimates $\hat{\mathbf{x}}_{ik}$ and the corresponding inverse error covariance matrix $P_k^{-1}(+)$ for the fusion algorithm.

8.3 By substituting the Kalman gain matrices

$$K_{ik}^d = P_k(+) M_i^T P_{ik}^{-1}(+), \tag{1}$$

$$K_{ik}^d(-) = P_k(+) M_i^T P_{ik}^{-1}(-) \tag{2}$$

into the results of Problem 8.2 [that is, the global or best estimate $\hat{\mathbf{x}}_k(+)$], reformulate the best estimate $\hat{\mathbf{x}}_k(+)$ in terms of the above gains.

8.4 In this problem we consider parallel filtering architectures using reduced-order models. Consider a global model of the form

$$\mathbf{x}(k+1) = \Phi\mathbf{x}(k) + G\mathbf{w}(k), \tag{1}$$

$$\mathbf{z}(k) = H\mathbf{x}(k) + \mathbf{v}(k), \tag{2}$$

where the usual noise statistics apply [i.e., $\mathbf{x}_0 \sim N(\bar{\mathbf{x}}_0, P_0)$, $\mathbf{x}(k) \in R^n$, $\mathbf{w} \sim N(0, Q)$, etc.]. Here $\mathbf{w}(k)$ and $\mathbf{v}(k)$ are white-noise processes independent of each other and of the initial state estimate $\mathbf{x}_0$. The covariances P_0, Q, and R are symmetric and positive definite. In order to simplify the notation, the time index k will be dropped. Let $\tilde{\mathbf{x}}$ and $\tilde{P}$ denote the time-update estimate and error covariance matrix, respectively. Let $\hat{\mathbf{x}}$ and $\hat{P}$ be the corresponding measurement update quantities. Using the above notation, the conventional centralized Kalman filter equations are:

$$\tilde{\mathbf{x}} = \Phi\hat{\mathbf{x}}, \tag{3}$$

$$\tilde{P} = \Phi\hat{P}\Phi^T + GQG^T, \tag{4}$$

$$\hat{\mathbf{x}} = \tilde{\mathbf{x}} + K(z - H\tilde{\mathbf{x}}) = (I - KH)\tilde{\mathbf{x}} + Kz, \tag{5}$$

$$\hat{P} = (I - KH)\tilde{P}(I - KH)^T + KRK^T \tag{6a}$$

$$= (I - KH)\tilde{P}, \tag{6b}$$

$$K = \tilde{P}H^T(H\tilde{P}H^T + R)^{-1} \tag{7a}$$

$$= \hat{P}H^T R^{-1}. \tag{7b}$$

(a) Write Eqs.(5)–(6) in the information-filter from (see Sections 4.3 and 8.4).
(b) Assume that there is a bank of N local subsystems, each with the following model:

$$\mathbf{x}_j(k+1) = \Phi_j\mathbf{x}_j(k) + B_j\mathbf{w}_j(k), \tag{8}$$

$$\mathbf{z}_j(k) = C_j\mathbf{x}_j(k) + \mathbf{v}_j(k). \tag{9}$$

Here the conventional algorithms for the jth local filter are given by [57]

$$\tilde{\mathbf{x}}_j = \Phi_j\hat{\mathbf{x}}_j, \tag{10}$$

$$\tilde{P}_j = \Phi_j\hat{P}_j\Phi_j^T + B_jQ_jB_j^T, \tag{11}$$

$$\hat{\mathbf{x}}_j = (I - K_j C_j)\tilde{\mathbf{x}}_j + K_j\mathbf{z}_j, \tag{12a}$$

or

$$\hat{P}_j^{-1} \hat{\mathbf{x}}_j = \tilde{P}_j^{-1} \tilde{\mathbf{x}}_j + C_j^T R_j^{-1} \mathbf{z}_j. \tag{12b}$$

Also

$$\hat{P}_j = (I - K_j C_j) \tilde{P}_j, \tag{13a}$$

or

$$\hat{P}_j^{-1} = \tilde{P}_j^{-1} + C_j^T R_j^{-1} C_j, \tag{13b}$$

where

$$K_j = \tilde{P}_j C_j^T (C_j \tilde{P}_j C_j^T + R_j)^{-1} \tag{14a}$$

$$= \hat{P}_j C_j^T R_j^{-1}, \tag{14b}$$

$$I - K_j C_j = \hat{P}_j \tilde{P}_j^{-1}. \tag{15}$$

(Note that the observation matrix for the jth local model is $C_j \in R^{m_j \times n_j}$ and note $H_j \in R^{m_j \times n}$: similarly, the matrix associated with the system noise at the jth subsystem is B_j and not G_j). Write the measurement-update equations in information-filter form. For this part, use the transformation (8.6), $H_j = C_j D_j$.

8.5 In Problem 8.4, the centralized time-update equations in covariance form were given by Eqs. (3)–(4). Assuming that [57]

$$Q = \text{diag}\,(Q_1, \ldots, Q_N)$$

holds, we have

$$GQG^T = \sum_{j=1}^{N} G_j Q_j G_j^T. \tag{1}$$

Now, suppose that there exists a matrix $E_j \in R^{n \times n}$ such that

$$G_j = E_j B_j. \tag{2}$$

From Eqs. (4) and (11) of Problem 8.4 and Eq. (1), deduce the equation for $\tilde{P}$. For this problem, define

$$\bar{P} = \Phi \hat{P} \Phi^T \quad \text{and} \quad \bar{P}_j = \Phi_j \hat{P}_j \Phi_j^T.$$

8.6 For further appreciation of the basic fusion methodology, consider a single target being tracked by two sensors (e.g., radar), where at time k, each local node has local prior and updated estimates $\hat{x}_{k|k-1}^i$, $P_{k|k-1}^i$ and $\hat{x}_{k|k}^i$, $P_{k|k}^i$, $i = 1, 2$. Assume that the fusion processor has its own prior estimate $\hat{x}_{k|k-1}$, $P_{k|k-1}$. In this problem it is desired to compute the global estimates $\hat{x}_{k|k}$, $P_{k|k}$, using only those local estimates and the global prior estimates. Note that the global estimates can be obtained using linear operations on the local estimates. The result is

$$\hat{x}_{k|k} = P_{k|k} \big[P_{k|k}^{1}{}^{-1} \hat{x}_{k|k}^1 + P_{k|k}^{2}{}^{-1} \hat{x}_{k|k}^2$$
$$- P_{k|k-1}^{1}{}^{-1} \hat{x}_{k|k-1}^1 + P_{k|k-1}^{2}{}^{-1} \hat{x}_{k|k-1}^2 + P_{k|k-1}^{-1} \hat{x}_{k|k-1} \big], \qquad (1)$$

where

$$P_{k|k} = \big[P_{k|k}^{1}{}^{-1} + P_{k|k}^{2}{}^{-1} - P_{k|k-1}^{1}{}^{-1} - P_{k|k-1}^{2}{}^{-1} + P_{k|k-1}^{-1} \big]^{-1}. \qquad (2)$$

When the local processors and fusion processor have the same prior estimates, the fusion equations can be further simplified to

$$\hat{x}_{k|k} = P_{k|k} \big[P_{k|k}^{1}{}^{-1} \hat{x}_{k|k}^1 + P_{k|k}^{2}{}^{-1} \hat{x}_{k|k}^2 - P_{k|k-1}^{-1} \hat{x}_{k|k-1} \big], \qquad (3)$$

where

$$P_{k|k} = \big[P_{k|k}^{1}{}^{-1} + P_{k|k}^{2}{}^{-1} - P_{k|k-1}^{-1} \big]^{-1}. \qquad (4)$$

From the above equations, we can see that the common prior estimates (i.e., the redundant information) are subtracted in the linear fusion operation. For the general nonlinear case, the fusion equations can be expressed in linear form. Derive Eqs. (1) and (2). [*Hint*: Assuming that $jl \neq 0$, use the concept of the information filter

$$P_{k|k,j}^{i}{}^{-1} = P_{k|k-1}^{i}{}^{-1} + H_k^{i\prime} (R_k^i) \, H_k^i, \qquad (5)$$

$$P_{k|k,j,l}^{-1} = P_{k|k,j}^{-1} + H_k^{2\prime} (R_k^2)^{-1} \, H_k^2,$$

$$= P_{k|k-1}^{-1} + H_k^{1\prime} (R_k^1)^{-1} \, H_k^1 + H_k^{2\prime} (R_k^2)^{-1} \, H_k', \qquad (6)$$

where ($'$) denotes matrix transposition, and where $P_{k|k,j}$ is the associated error covariance matrix of the global estimate updated by the jth measurement from sensor (1).]

8.7 As a continuation of Problem 8.6, determine the fusion equations, similar to Eqs. (3) and (4) of Problem 8.6, given the lth measurement from sensor

2. The global estimate given the jth measurement from sensor 2 is

$$\hat{\mathbf{x}}_{k|k,j,l} = \hat{\mathbf{x}}_{k|k,j} + P_{k|k,j,l} + H_k^{2\prime}(R_k^2)^{-1}(z_{k,l}^2 - H_k^2 \hat{x}_{k|k,j}), \tag{1}$$

and the error covariance matrix is

$$P_{k|k,j,l} = P_{k|k,j} - P_{k|k,j} H_k^{2\prime}(H_k^2 P_{k|k,j} + H_k^{2\prime} + R_k^2)^{-1} H_k^2 P_{k|k,l}. \tag{2}$$

Here it is desired to determine $\hat{\mathbf{x}}_{k|k,j,l}$ and $P_{k|k,j,l}^i$ when the two nodes interchange information every scan, in which case they will have the same prior estimates as the global node (a node is defined as a distributed network having its own sensors).

REFERENCES

[1] Åström, K. J.: *Introduction to Stochastic Control Theory*, Academic Press, Inc., New York, 1970.

[2] Athans, M. and Falb, P. L.: *Optimal Control: An Introduction to the Theory and Its Applications*, McGraw-Hill, New York, 1966.

[3] Bar-Shalom, Y. (ed.): *Multitarget–Multisensor Tracking: Advanced Applications*, Artech House, Norwood, Massachusetts, 1990.

[4] Bar-Shalom, Y. and Formann, T. E.: *Tracking and Data Association*, Mathematics in Science and Engineering, Vol. 179, Academic Press, Inc., Scan Diego, California, 1988.

[5] Battin, R. H.: *Astronautical Guidance*, McGraw-Hill, New York, 1964.

[6] Bellman, R.: *Dynamic Programming*, Princeton University Press, Princeton, New Jersey, 1957.

[7] Bellman, R. (ed.): *Mathematical Optimization Techniques*, University of California Press, Berkeley, California, 1963.

[8] Bellman, R., Glicksberg, I., and Gross, O.: *On the Bang–Bang Control Problem*, Quart. Appl. Math., Vol. 14, pp. 11–18, 1956.

[9] Bierman, G. J.: *Factorization Methods for Discrete Sequential Estimation*, Academic Press, Inc., New York, 1977.

[10] Blair, W. D.: *Two-Stage $\alpha-\beta-\gamma-\lambda$ Estimator for Continuous $\alpha-\beta$ Target Trackers*, Proceedings of the American Control Conference (ACC), June 1992.

[11] Bliss, G. A.: *Lectures on the Calculus of Variations*, University of Chicago Press, Chicago, Illinois, 1963.

[12] Bryson, A. E., Jr., and Ho, Y. C.: *Applied Optimal Control. Optimization, Estimation, and Control*, 2nd Edition, Blaisdell Publishing Co., Waltham, Massachusetts, 1975.

[13] Carlson, N. A.: *Fast Triangular Formulation of the Square Root Filter*, AIAA Journal, Vol. 11, No. 9, pp. 1259–1265, September 1973.

[14] Carlson, N. A.: *Federated Square Root Filter for Decentralized Parallel Processing*, Proceedings of the National Aerospace and Electronics Conference, NAECON, Dayton, Ohio, pp. 1448–1456, 18–22 May 1987.

[15] Carlson, N. A. and Berarducci, M.: *Federated Kalman Filter Simulation Results*, Proceedings of the 49th Annual Meeting of the Institute of Navigation (ION), Boston, Massachusetts, 21–23 June 1993.

[16] Chang, Kuo-Chu: *Distributed Estimation in Distributed Sensor Networks*, Doctor of Philosophy Dissertation, The University of Connecticut, Storrs, Connecticut, 1986. (Also as Chapter 2 in *Large-Scale Systems, Detection, Estimation, Stability and Control*, Marcel Dekker, Inc., New York, 1992.)

[17] Chen, Guanrong (ed.).: *Approximate Kalman Filtering*, World Scientific Publishing Company, Teaneck, New Jersey, 1993.

[18] Chen, G. and Chui, C. K.: *Design of Near-Optimal Linear Digital Tracking Filters with Colored Input*, J. Comput. Appl. Math., Vol. 15, pp. 353–370, 1986.

[19] Chong, C. Y.: *Hierarchical Estimation*, Proceedings of the 2nd MIT/ONR C^3 Workshop, Monterey, California, 1979.

[20] Chong, C. Y., Mori, S., and Chang, K. C.: *Distributed Multitarget–Multisensor Tracking*, Chapter 8 in *Distributed Multisensor–Multitarget Tracking: Advanced Applications*, Artech House, Norwood, Massachusetts, 1990.

[21] Chui, C. K. and Chen, G.: *Kalman Filtering with Real-Time Applications*, 2nd Edition, Springer-Verlag, Berlin, Heidelberg, New York, 1991.

[22] Chui, C. K. and Chen, G.: *Linear Systems and Optimal Control*, Springer-Verlag, Berlin, Heidelberg, New York, 1989.

[23] Chui, C. K., Chen, G. and Chui, H. C.: *Modified Extended Kalman Filtering and a Real-Time Paralled Algorithm for System Parameter Identification*, IEEE Trans. Automatic. Control, Vol. 35, No. 1, pp. 100–104, January 1990.

[24] Conte, S. D.: *Elementary Numerical Analysis, an Algorithmic Approach*, McGraw-Hill, New York, 1965.

[25] Davenport, W. B. and Root, W. L.: *An Introduction to the Theory of Random Signals and Noise*, McGraw-Hill, New York, 1958.

[26] D'Azzo, J. J. and Houpis, C. H.: *Linear Control System Analysis and Design*, 4th Edition, McGraw-Hill, New York, 1995.

[27] Doob, J. C.: *Stochastic Processes*, John Wiley and Sons, Inc., New York, 1953.

[28] Doyle, J. C.: *Guaranteed Margins for LQG Regulators*, IEEE Trans. Automat. Control, Vol. AC-23, No. 4, pp. 756–757, August 1978.

[29] Fisher, R. A.: *On the Mathematical Foundations of Theoretical Statistics*, Phil. Trans. Roy. Soc. London, Vol. 222, p. 309, 1922.

[30] Fisher, R. A.: *Theory of Statistical Estimation*, Proc. Cambridge Phil. Soc., Vol. 22, p. 700, 1925.

[31] Fitzgerald, R. J.: *Divergence of the Kalman Filter*, IEEE Trans. Automat. Control, Vol. AC-16, No. 6, pp. 736–747, December 1971.

[32] Forsythe, G. E., Malcolm, M. A. and Moler, C. B.: *Computer Methods for Mathematical Computations*, Prentice-Hall, Inc., Englewood Cliffs, New Jersey, 1977.

[33] Friedland, B.: *Optimum Steady-State Position and Velocity Estimation Using Noisy Sampled Position Data*, IEEE Trans. Aerospace and Electronic Systems, Vol. AES-9, No. 6, pp. 906–911, November 1973.

[34] Gelfand, I. M. and Fomin, S. V.: *Calculus of Variations*, Prentice-Hall, Inc., Englewood Cliffs, New Jersey, 1963.

[35] Hashemipour, H. R., Roy, S., and Laub, A. J.: *Decentralized Structures for Parallel Kalman Filtering*, IEEE Trans. Automat. Control, Vol. 33, No. 1, pp. 88–94, January 1988.

[36] Hong, L.: *Distributed Filtering Using Set Models*, IEEE Trans. Aerospace and Electronic Systems, Vol. 28, No. 4, pp. 1144–1153, October 1992.

[37] Jazwinski, A. H.: *Stochastic Processes and Filtering Theory*, Academic Press, Inc., New York, 1970.

[38] Johnson, C. D.: *On the Optimal Control of Stochastic Linear Systems*, IEEE Trans. Automat. Control, Vol. AC-16, No. 6, pp. 776–785, December 1971.

[39] Kalata, P. R.: *The Tracking Index: A Generalized Parameter for $\alpha-\beta$ and $\alpha-\beta-\gamma$ Target Trackers*, IEEE Trans. Aerospace and Electronic System, Vol. AES-20, No. 2, pp. 174–182, March 1984.

[40] Kalata, P. R.: *$\alpha-\beta$ Target Tracking Systems: A Survey*, Proceedings of the American Control Conference (ACC), pp. 832–835, June 1982.

[41] Kalman, R. E.: *A New Approach to the Linear Filtering and Prediction Problems*, Trans. ASME J. Basic Eng. Ser. D, Vol. 82, pp. 35–45, March 1960.

[42] Kalman, R. E. and Bucy, R.: *New Results in Linear Filtering and Prediction*, Trans. ASME J. Basis Eng. Ser. D, Vol. 83, pp. 95–108, March 1961.

[43] Kaminski, P. G., Bryson, A. E., Jr., and Schmidt, S. F.: *Discrete Square Root Filtering: A Survey of Current Techniques*, IEEE Trans. Automat. Control, Vol. AC-16, No. 6, pp. 727–735, December 1971.

[44] Kerr, T. H.: *Computational Techniques for the Matrix Pseudoinverse in Minimum Variance Reduced-Order Filtering and Control*, in *Control and Dynamic Systems, Advances in Algorithms and Computational Techniques for Dynamic Control Systems*, Vol. XXVII, C. T. Leondes (ed.), Academic Press, Inc., New York, 1989.

[45] Kushner, H. J.: *Stochastic Stability and Control*, Academic Press, Inc., New York, 1967.

[46] Leitmann, G. (ed.): *Optimization Techniques*, Academic Press, Inc., New York, 1962.

[47] Luenberger, D. G.: *An Introduction to Observers*, IEEE Trans. Automat. Control, Vol. AC-16, No. 6, pp. 596–602, December 1971.

[48] Maybeck, P. S.: *Stochastic Models, Estimation, and Control*, Vol. 1, Academic Press, Inc., New York, 1979.

[49] Maybeck, P. S.: *Stochastic Models, Estimation, and Control*, Vol. 2, Academic Press, Inc., New York, 1982.

[50] McGee, L. A.: *Effect of Reduced Computer Precision on a Midcourse Navigation and Guidance System Using Optimal Filtering and Linear Prediction*, NASA TN D-3382, 1966.

[51] Oshman, Y.: *Gain-Free Square Root Information Filtering Using the Spectral Decomposition*, AIAA J. Guidance and Control, Vol. 12, No. 5, pp. 681–690, September–October 1989.

[52] Padulo, L. and Arbib, M. A.: *System Theory: A Unified State-Space Approach to Continuous and Discrete Systems*, W. B. Saunders Company, Philadelphia, Pennsylvania, 1974.

[53] Papoulis, A.: *Probability, Random Variables, and Stochastic Processes*, 2nd Edition, McGraw-Hill, New York, 1984.

[54] Plant, J. B.: *On the Computation of Transtition Matrices for Time-Invariant Systems*, Proc. IEEE, pp. 1377–1398, August 1968.

[55] Pontryagin, L. S., Boltyanskii, V. G., Gamkrelidge, R. V., and Mishchenko, E. F.: *The Mathematical Theory of Optimal Processes*, Interscience Publishers, Inc., New York, 1962.

[56] Potter, J. E.: *W Matrix Augmentation*, M.I.T. Instrumentation Laboratory Memo SGA 5–64, Cambridge, Massachusetts, 1964.

[57] Roy, S., Hashemi, R. H., and Laub, A. J.: *Square Root Parallel Filtering Using Reduced-Order Local Filters*, IEEE Tans. Aerospace and Electronic Systems, Vol. 27, No. 2, pp. 276–289, March 1991.

[58] Sage, A. P. and Melsa, J. L.: *Estimation Theory with Applications to Communication and Control*, McGraw-Hill, New York, 1971.

[59] Schlee, F. H., Standish, C. J., and Toda, N. F.: *Divergence in the Kalman Filter*, AIAA J., Vol. 5, No. 6, pp. 1114–1120, June 1967.

[60] Schmidt, S. F.: *Application of State-Space Methods to Navigation Problems*, in *Advances in Control Systems*, Vol. 3, C. T. Leondes (ed.), Acdemic Press, Inc., 1966. (Originally published as Philco WDL Technical REport No. 4, July 1964.)

[61] Siouris, G. M.: *Aerospace Avionics Systems: A Modern Synthesis*, Academic Press, Inc., San Diego, California, 1993.

[62] Siouris, G. M. and Leros, A. P.: *Minimum-Time Intercept Guidance for Tactical Missiles*, Control-Theory and Adv. Technol. (C-TAT), Vol. 4, No. 2, pp. 251–263, June 1988.

[63] Smith, G. L., Schmidt, S. F., and McGee, L. A.: *Application of Statistical Filter Theory to the Optimal Estimation of Position and Velocity On-Board a Circumlunar Mission*, NASA TN D-1208, 1962.

[64] Soong, T. T.: *Probabilistic Modeling and Analysis in Science and Engineering*, John Wiley and Sons, Inc., New York, 1981.

[65] Speyer, J. L.: *Computation and Transmission Requirements for a Decentralized Linear-Quadratic-Gaussian Control Problem*, IEEE Trans. Automat. Control, Vol. AC-24, No. 2, pp. 266–269, April 1979.

[66] Stein, G. and Athans, M.: *The LQG/LTR Procedure for Multivariable Feedback Control Design*, IEEE Trans. Automat. Control, Vol. AC-32, No. 2, pp. 105–114, February 1987.

[67] Thornton, C. L.: *Triangular Covariance Factorization for Kalman Filtering*, Jet Propulsion Laboratory, Technical Memorandum 33–798, October 1976.

[68] Thornton, C. L. and Bierman, G. J.: *Gram-Schmidt Algorithms for Covariance Propagation*, Proceedings of the IEEE Conference on Decision and Control, Institute of Electrical and Electronics Engineers, New York, 1975, pp. 489–498.

[69] Thornton, C. L. and Bierman, G. J.: UDU^T *Covariance Factorization for Kalman Filtering*, in *Control and Dynamic Systems*, Academic Press, Inc., New York, pp. 177–248, 1980.

[70] Tse, E. and Athans, M.: *Observer Theory for Continuous-Time Linear Systems*, Inform. and Control, Vol. 22, pp. 405–434, 1973.

[71] Whang, I. H., Sung, T. K., and Lee, J. G.: *A Modified Target Maneuver Estimation Technique Using Pseudo-Acceleration Information*, Proceedings of the 1992 AIAA Guidance Navigation and Control Conference, Hilton Head, South Carolina, pp. 1249–1254, August 1992.

[72] Wiener, N.: *The Extrapolation, Interpolation and Smoothing of Stationary Time Series*, John Wiley and Sons, Inc., New York, 1949.

[73] Willsky, A., Bello, M., Castanon, B., Levy, G., and Verghese, G.: *Combining and Updating of Local Estimates and Regional Maps Along Sets of One-Dimensional Tracks*, IEEE Trans. Automat. Control, Vol. AC-27, No. 4, pp. 799–813, August 1982.

APPENDIX A

MATRIX OPERATIONS AND ANALYSIS

A.1 INTRODUCTION

The purpose of this appendix is to provide the reader with the basic concepts of matrix theory. The theory of matrices plays an important role in the formulation and solution of problems in mathematics, engineering, and optimal control and estimation theory. Matrix theory not only provides an extremely helpful tool for designing a mathematical model of a system with many variables, but also affords a practical and convenient method of adapting the data for processing by a digital computer. Moreover, since many of the models of systems considered in this book are rather complicated, consisting for the most part of coupled systems of difference or differential equations, it becomes apparent that matrix theory must be used. Matrix theory provides a convenient shorthand notation for treating sets of simultaneous linear algebraic equations.

A.2 BASIC CONCEPTS

Consider a system of m linear equations in the n unknowns $x_1, x_2, \ldots, x_n$, of the form

$$a_{11}x_1 + a_{12}x_2 + \cdots + a_{1n}x_n = c_1,$$

$$a_{21}x_1 + a_{22}x_2 + \cdots + a_{2n}x_n = c_2,$$
$$\vdots$$
$$a_{i1}x_1 + a_{i2}x_2 + \cdots + a_{in}x_n = c_i, \tag{A.1}$$
$$\vdots$$
$$a_{m1}x_1 + a_{m2}x_2 + \cdots + a_{mn}x_n = c_m,$$

where the real numbers a_{ik}, c_i occurring as coefficients are given. Here $a_{ik}(1 \leqslant i \leqslant m, 1 \leqslant k \leqslant n)$ is the coefficient of x_k in the ith equation; the first subscript of a_{ik} thus gives the equation, and the second, the unknown to which a_{ik} belongs. c_i $(1 \leqslant i \leqslant m)$ is the so-called *constant term* of the ith equation.

This system of linear equations can be visualized as representing a linear transformation in which the set of n numbers $\{x_1, x_2, \ldots, x_n\}$ is transformed into the set of m numbers $\{c_1, c_2, \ldots, c_m\}$. Equation (A.1) can be represented in matrix notation by the equation [8]

$$\begin{bmatrix} a_{11} & a_{12} & \cdots & a_{1n} \\ a_{21} & a_{22} & \cdots & a_{2n} \\ \vdots & \vdots & & \vdots \\ a_{m1} & a_{m2} & \cdots & a_{mn} \end{bmatrix} \begin{bmatrix} x_1 \\ x_2 \\ \vdots \\ x_n \end{bmatrix} = \begin{bmatrix} c_1 \\ c_2 \\ \vdots \\ c_m \end{bmatrix}. \tag{A.2}$$

The sets of quantities $x_i (i = 1, 2, \ldots, n)$ and $c_i (i = 1, 2, \ldots, m)$ are commonly represented as matrices of one column each (column vectors). In order to emphasize the fact that such a matrix consists of only one column, it is denoted by a lower case boldface letter, or by a typical element enclosed in braces. That is,

$$\mathbf{x} \equiv \{x_i\} \equiv \begin{bmatrix} x_1 \\ x_2 \\ \vdots \\ x_n \end{bmatrix}, \qquad \mathbf{c} \equiv \{c_i\} \equiv \begin{bmatrix} c_1 \\ c_2 \\ \vdots \\ c_m \end{bmatrix}.$$

An abbreviated symbolism for Eq. (A.2) is the simple matrix equation

$$A\mathbf{x} = \mathbf{c} \tag{A.3}$$

where

$$A \equiv [a_{ij}] \equiv \begin{bmatrix} a_{11} & a_{12} & \cdots & a_{1n} \\ a_{21} & a_{22} & \cdots & a_{2n} \\ \vdots & \vdots & & \vdots \\ a_{m1} & a_{m2} & \cdots & a_{mn} \end{bmatrix},$$

is an $m \times n$ matrix. The symbol a_{ij} represents a typical element, the first subscript i denoting the row and the second subscript j denoting the column occupied by the element. Equation (A.3) may also be written in the form

$$\sum_{k=1}^{n} a_{ik} x_k = c_i \qquad (i = 1, 2, \ldots, m), \tag{A.4}$$

where a_{ij} is the general element in the ith row and jth column of an ordered array of mn scalars arranged in m rows and n columns. Moreover, the scalar a_{ij} in row i and column j is called the ij entry of the matrix A. If the matrix has m rows and n columns, then the matrix is said to be (of order) $m \times n$. An $m \times 1$ matrix is called a column matrix or column vector, and a $1 \times n$ matrix is called a row matrix or row vector.

A matrix that has the same number of rows and columns is called a *square matrix*. A square matrix with n rows and n columns is called a matrix of *order n*. The *principal* or *main diagonal* of a square matrix consists of those entries which appear in the same row and column. The matrix $A = (a_{ij})$ is said to be *diagonal* if $a_{ij} = 0$ for all $i \neq j$. Finally, the *trace* of a square matrix is defined as the sum of the main diagonal elements,

$$\text{tr}(A) = a_{11} + a_{22} + a_{33} + \cdots + a_{nn} = \sum_{i=1}^{n} a_{ii}.$$

At this point, a brief discussion of determinants is in order. Associated with any square matrix $[a_{ij}]$ of order n we define the determinant $= |a_{ij}|$ by

$$|A| = \begin{bmatrix} a_{11} & a_{12} & \cdots & a_{1n} \\ a_{21} & a_{22} & \cdots & a_{2n} \\ \vdots & \vdots & & \vdots \\ a_{n1} & a_{n2} & \cdots & a_{nn} \end{bmatrix}$$

as a number obtained as a certain sum of all possible products in each of which there appears one and only one element from each row and each column. More specifically, the determinant is a polynomial in the elements of A defined by

$$|A| = \sum (\pm a_{1i_1} a_{2i_2} \cdots a_{ni_n}),$$

where the summation extends over all $n!$ permutations $i_1, i_2, \ldots, i_n$ of the subscripts $1, 2, \ldots, n$, and the sign before a term is $+$ or $-$ according as the permutation $i_1, i_2, \ldots, i_n$ is even or odd.

If the row and column containing an element a_{ij} in a square matrix A are deleted, the determinant of the remaining square array is called the *minor* of a_{ij}, and is denoted by M_{ij}. The *cofactor* of a_{ij}, denoted by A_{ij}, is then defined by the relation

$$A_{ij} = (-1)^{i+j} M_{ij}.$$

Thus, if the sum of the row and column indices of an element is even, then cofactor and the minor of that element are identical; otherwise they differ in sign. If in any matrix A several rows and/or columns are deleted, the remaining elements form a rectangular array which is called a *submatrix* of A.

From the theorem of expansion of determinants by minors we have [6]:

If $A = (a_{ij})$ is any square matrix, then for $i, j = 1, 2, \ldots, n$

$$|A| = (-1)^{i+1} a_{i1} |A_{i1}| + (-1)^{i+2} a_{i2} |A_{i2}| + \cdots + (-1)^{i+n} a_{in} |A_{in}|$$

$$= (-1)^{1+j} a_{1j} |A_{1j}| + (-1)^{2+j} a_{2j} |A_{2j}| + \cdots + (-1)^{n+j} a_{nj} |A_{nj}|.$$

On the other hand, if $A = (a_{ij})$ is a square matrix and if $k \neq i$, then

$$(-1)^{k+1} a_{i1} |A_{k1}| + (-1)^{k+2} a_{i2} |A_{k2}| + \cdots + (-1)^{k+n} a_{in} |A_{kn}| = 0.$$

Similarly, if $k \neq j$,

$$(-1)^{1+k} a_{ij} |A_{1k}| + (-1)^{2+k} a_{2j} |A_{2k}| + \cdots + (-1)^{n+k} a_{nj} |A_{nk}| = 0.$$

In general, determinant expansion can be performed about any row or column. The equation for the expansion of an arbitrary determinant about row i is

$$|A| = \sum_{j=1}^{n} (-1)^{i+j} a_{ij} |A_{ij}|$$

and about column j is

$$|A| = \sum_{i=1}^{n} (-1)^{i+j} a_{ij} |A_{ij}|$$

where $|A_{ij}|$ is the corresponding minor.

The following properties of determinants are easily established:

1. If all elements of any row or column of a square matrix are zero, its determinant is zero.
2. The value of the determinant is unchanged if the rows and columns of the matrix are interchanged; that is, the determinant of a matrix A is equal to the determinant of the transposed matrix A^T (see discussion in connection with Eq. (A.8)).
3. If two rows (or two columns) of a square matrix are interchanged, the sign of its determinant is changed.
4. If two rows (or two columns) of A are identical, then $|A| = 0$.

5. If all elements of one row (or column) of a square matrix are multiplied by a number k, the determinant is multiplied by k.
6. The determinant of the negative of a square matrix is not necessarily the negative of the determinant, but one has

$$|-A| = (-1)^n |A|,$$

where n is the order of the matrix A.

From the above discussion, we can now state the well-known *Cramer's rule* for solving a set of n linear equations in n unknown quantities, of the form

$$\sum_{k=1}^{n} a_{ik} x_k = c_i \qquad (i = 1, 2, \ldots, n). \tag{A.5}$$

In the case when the determinant of the matrix of coefficients is not zero ($|A| \neq 0$), the only possible solution of Eq. (A.5) is given by

$$x_r = \frac{1}{|A|} \sum_{i=1}^{n} A_{ir} c_i \qquad (r = 1, 2, \ldots, n). \tag{A.6}$$

Cramer's rule can be stated in words as follows:

When the determinant $|A|$ of the matrix of coefficients in a set of n linear algebraic equations in n unknowns $x_1, \ldots, x_n$ is not zero, that set of equations has a unique solution. The value of any x, can be expressed as the ratio of two determinants, the denominator being the determinant of the matrix of coefficients, and the numerator being the determinant of the matrix obtained by replacing the column of the coefficients of x, in the coefficient matrix by the column of the right-hand members.

When A is a square matrix, the matrix obtained from A by replacing each element by its cofactor and then interchanging rows and columns is called the *adjoint* of A [3, 5]:

$$\text{Adj } A = \begin{bmatrix} A_{11} & A_{21} & \cdots & A_{n1} \\ A_{12} & A_{22} & \cdots & A_{n2} \\ \vdots & \vdots & & \vdots \\ A_{1n} & A_{2n} & \cdots & A_{nn} \end{bmatrix} = [A_{ji}]. \tag{A.7}$$

The adjoint of a product is found to be equal to the product of the adjoints in the reverse order.

The *identity matrix* (or *unit matrix*) of order n, denoted by I, is the square $n \times n$ matrix having unit entries in its principal diagonal and zeros elsewhere.

Its ij entry is called the *Kronecker delta* and is designated by δ_{ij}. Thus,

$$I = \begin{bmatrix} 1 & 0 & \cdots & 0 \\ 0 & 1 & \cdots & 0 \\ \vdots & \vdots & & \vdots \\ 0 & 0 & \cdots & 1 \end{bmatrix} = (\delta_{ij}), \qquad \delta_{ij} = \begin{cases} 1 & \text{if } i = j, \\ 0 & \text{if } i \neq j. \end{cases}$$

For instance, the 3×3 identity matrix is given as

$$I = \begin{bmatrix} 1 & 0 & 0 \\ 0 & 1 & 0 \\ 0 & 0 & 1 \end{bmatrix}.$$

A *zero matrix* 0 (also known as the null matrix) has zeros for all its elements. It is easily verified that

$$A - A = 0.$$

Similarly, it is readily verified that for any matrix A there follow

$$AI = A, \qquad IA = A$$

and

$$A0 = 0, \qquad 0A = 0.$$

(matrix multiplication to be discussed in Section A.2.1).

A *triangular matrix* is one in which the entries either above or below the principal diagonal are all equal to zero. More specifically, a 3×3 *upper triangular* matrix assumes the form

$$A = \begin{bmatrix} a_{11} & a_{12} & a_{13} \\ 0 & a_{22} & a_{23} \\ 0 & 0 & a_{33} \end{bmatrix}, \qquad a_{ij} = 0 \quad \text{for } i > j,$$

while a lower triangular matrix has zero entries above the principal diagonal as follows:

$$A = \begin{bmatrix} a_{11} & 0 & 0 \\ a_{21} & a_{22} & 0 \\ a_{31} & a_{32} & a_{33} \end{bmatrix}, \qquad a_{ij} = 0 \quad \text{for } i < j.$$

The matrix which is obtained from $A = [a_{ij}]$ by interchanging rows and columns is called the *transpose* of A, and is denoted by A^T:

$$A^T = \begin{bmatrix} a_{11} & a_{21} & \cdots & a_{m1} \\ a_{12} & a_{22} & \cdots & a_{m2} \\ \vdots & \vdots & & \vdots \\ a_{1n} & a_{2n} & \cdots & a_{mn} \end{bmatrix}. \tag{A.8}$$

Thus, the transpose of an $m \times n$ matrix is an $n \times m$ matrix. In general, if A is an $n \times n$ matrix which is the same as its transpose, that is,

$$A = A^T,$$

then we call A a *symmetric matrix*. [Note that $(A^T)^T = A$].

The above discussion can be expanded to products of two matrices. For example, if A is an $m \times q$ matrix and B is an $q \times n$ matrix, then both the products AB and $B^T A^T$ exist, the former being an $m \times n$, and the latter an $n \times m$ matrix. Furthermore, the transpose of the product of AB is the product of the transposes in reverse order:

$$(AB)^T = B^T A^T.$$

Assume now that $A = (a_{ij})$ satisfies the relation

$$A = -A^T \tag{A.9}$$

or, equivalently,

$$a_{ij} = -a_{ji} \qquad \text{for } i = 1, 2, \ldots, n, \ j = 1, 2, \ldots, n.$$

Then we say that A is a *skew-symmetric* matrix.

An *inverse* of a matrix A is a matrix denoted by A^{-1}, having the property that both $A^{-1}A$ and AA^{-1} are unit (identity) matrices; that is, $A^{-1}A = AA^{-1} = I$. A matrix A has an inverse if and only if it is *nonsingular*, that is, its determinant is not zero. The inverse of the matrix A, if it exists, can be expressed as

$$A^{-1} = \frac{1}{|A|} \text{Adj } A, \qquad |A| \neq 0. \tag{A.10}$$

To determine the inverse of a *product* of nonsingular square matrices, consider the product

$$AB = C.$$

Premultiplying both sides of this equation successively by A^{-1} and B^{-1}, there follows

$$I = B^{-1}A^{-1}C,$$

and hence, by postmultiplying both sides of this equation by C^{-1} and replacing C by AB, we obtain*

$$(AB)^{-1} = B^{-1}A^{-1}.$$

This inversion formula can be extended to several matrices. If $A_1, A_2, ..., A_r$ are nonsingular matrices of the same order, then their product is nonsingular and

$$(A_1 A_2, ... A_r)^{-1} = A_r^{-1} ... A_2^{-1}A_1^{-1}.$$

Positive integral powers of an $n \times n$ matrix A, and the zero power, are defined as follows:

$$A^2 = A\,A, \qquad A^n = A^{n-1}A, \qquad A^o = I.$$

The matrix A is called *idempotent* if $A^2 = A$. From the preceding result, if the matrix A is nonsingular, we have

$$(A^n)^{-1} = (A^{-1})^n.$$

In order to illustrate the use of the inverse matrix, consider Eq. (A.3). Premultiplying both sides by A^{-1}, we obtain

$$\mathbf{x} = A^{-1}\mathbf{c},$$

$$\begin{bmatrix} x_1 \\ x_2 \\ \vdots \\ x_n \end{bmatrix} = \frac{1}{|A|} \begin{bmatrix} A_{11} & A_{21} & \cdots & A_{n1} \\ A_{12} & A_{22} & \cdots & A_{n2} \\ \vdots & \vdots & & \vdots \\ A_{1n} & A_{2n} & \cdots & A_{nn} \end{bmatrix} \begin{bmatrix} c_1 \\ c_2 \\ \vdots \\ c_n \end{bmatrix}, \qquad (A.11)$$

or

$$x_i = \frac{1}{|A|}(A_{1i}c_1 + A_{2i}c_2 + \cdots + A_{ni}c_n) \qquad (i = 1, 2, ..., n). \qquad (A.12)$$

---

*In the general case, $[AB \cdots MN]^{-1} = N^{-1}M^{-1} \cdot B^{-1}A^{-1}$

More specifically, let us determine the inverse for a general 2×2 matrix A:

$$A = \begin{bmatrix} a_{11} & a_{12} \\ a_{21} & a_{22} \end{bmatrix},$$

$$|A| = \begin{vmatrix} a_{11} & a_{12} \\ a_{21} & a_{22} \end{vmatrix} = a_{11}a_{22} - a_{21}a_{12},$$

$$A^{-1} = \frac{1}{|A|} \begin{bmatrix} a_{22} & -a_{12} \\ -a_{21} & a_{11} \end{bmatrix}.$$

(Note that an $n \times n$ matrix A is an array of n^2 entries, whereas its determinant is a single scalar function of these entries.)

Another important concept in matrix theory is that of the *rank* of a matrix. We define here the rank of any matrix A as the order of the largest square submatrix of A (formed by deleting certain rows and/or columns of A) whose determinant does not vanish. Assume now that a certain matrix A is of rank r. Now, if a set of r rows of A containing a nonsingular $r \times r$ submatrix R is selected, then any other row of A is a linear combination of those r rows. Furthermore, assume that the square array R of order r in the upper left corner of the matrix A has a nonvanishing determinant, and consider the following submatrix of A:

$$M = \begin{bmatrix} a_{11} & \cdots & a_{1r} & a_{1s} \\ \vdots & & \vdots & \vdots \\ a_{r1} & \cdots & a_{rr} & a_{rs} \\ a_{q1} & \cdots & a_{qr} & a_{qs} \end{bmatrix} = \left[\begin{array}{ccc|c} & & & a_{1s} \\ & R & & \vdots \\ & & & a_{rs} \\ \hline a_{q1} & \cdots & a_{qr} & a_{qs} \end{array} \right],$$

where $s > r$ and $q > r$. Therefore, since the matrix A is of rank r, the determinant of this square submatrix must vanish for all such s and q. It is obvious that the process of interchanging rows and columns does not affect the rank of a matrix, so that the two matrices A and A^T have the same rank.

In many practical applications, the matrix A is real and *symmetric*, so that two elements which are symmetrically placed with respect to the principal diagonal are equal. That is,

$$a_{ji} = a_{ij}.$$

However, if the coefficients are complex, the most important cases are those in which symmetrically situated elements are complex conjugates. Mathematically, this concept is expressed as

$$a_{ji} = a_{ij}^{*}. \tag{A.13}$$

Matrices with this symmetry property are called *Hermitian* matrices [1].

In general, if A and B are any two $n \times n$ matrices for which there is a nonsingular matrix P such that

$$B = P^{-1}AP = QAQ^{-1},$$

then A and B are called *similar* matrices. Similar matrices correspond to the same linear transformation, but with a different choice of basis.

The *pseudoinverse* of a matrix, often called the Moore–Penrose generalized inverse, is defined as

$$A^+ = C^T(CC^T)^{-1}(B^TB)^{-1}B^T \qquad \text{(A.14)}$$

with

$$0^+ = 0^T.$$

Therefore, if A is nonsingular, $A^+ = A^{-1}$. There are several advantages to employing the pseudoinverse rather than the more inclusive generalized inverse. These properties are as follows:

1. The pseudoinverse of a pseudoinverse yields the original matrix. That is, $(A^+)^+ = A$.
2. AA^+ and A^+A are symmetric matrices.
3. The pseudoinverse of a matrix is unique.

Finally, if A^TA is of full rank*, then

$$A^+ = (A^TA)^{-1}A^T. \qquad \text{(A.15a)}$$

Similarly, if AA^T is of full rank, then

$$A^+ = A^T(AA^T)^{-1}. \qquad \text{(A.15b)}$$

A matrix of interest in estimation theory is the exponential matrix function e^A. As we have seen in Section 4.4 [Eq. (4.49)], the exponential matrix is defined by the infinite series [6]

$$e^A = I + A + \frac{1}{2!}A^2 + \frac{1}{3!}A^3 + \cdots = \sum_{i=0}^{\infty} \frac{A^i}{i!}.$$

---

*If $\mathbf{A}$ is $n \times n$, then it is of rank n (or "full rank") if and only if it is nonsingular.

A.2.1   Matrix Algebra

In this subsection we will consider the rules of matrix algebra. The addition of two matrices A and B is defined if and only if they are of the same order. Let

$$A = \begin{bmatrix} a_{11} & a_{12} & \cdots & a_{1n} \\ a_{21} & a_{22} & \cdots & a_{2n} \\ \vdots & \vdots & & \vdots \\ a_{m1} & a_{m2} & \cdots & a_{mn} \end{bmatrix} \quad \text{and} \quad B = \begin{bmatrix} b_{11} & b_{12} & \cdots & b_{1n} \\ b_{21} & b_{22} & \cdots & b_{2n} \\ \vdots & \vdots & & \vdots \\ b_{m1} & b_{m2} & \cdots & b_{mn} \end{bmatrix}.$$

Then the sum $A + B$ is given by

$$A + B = \begin{bmatrix} a_{11}+b_{11} & a_{12}+b_{12} & \cdots & a_{1n}+b_{1n} \\ a_{21}+b_{21} & a_{22}+b_{22} & \cdots & a_{2n}+b_{2n} \\ \vdots & \vdots & & \vdots \\ a_{m1}+b_{m1} & a_{m2}+b_{m2} & \cdots & a_{mn}+b_{mn} \end{bmatrix}.$$

That is, the entries in the sum

$$C = A + B$$

are equal to the algebraic sums of the corresponding entries in A and B:

$$c_{ij} = a_{ij} + b_{ij}.$$

Matrix addition is *commutative* and *associative* that is,

$$A + B = B + A$$

and

$$(A + B) + C = A + (B + C).$$

The scalar multiplication of a matrix A by a number (i.e., a scalar) r is defined by multiplying each entry in A by r. The result rA is called the product of r and A:

$$rA = \begin{bmatrix} ra_{11} & ra_{12} & \cdots & ra_{1n} \\ ra_{21} & ra_{22} & \cdots & ra_{2n} \\ \vdots & \vdots & & \vdots \\ ra_{m1} & ra_{m2} & \cdots & ra_{mn} \end{bmatrix}$$

More briefly,

$$rA = r[a_{ij}] = [ra_{ij}].$$

The subtraction of two matrices is commonly defined in terms of matrix addition and scalar multiplication. Thus, the subtraction of a matrix B from a matrix A is obtained by adding to get

$$C = A + kB = A - B$$

where $k = -1$.

Matrix multiplication is somewhat more complicated than scalar multiplication. It should be noted at the outset that matrices of arbitrary order cannot be multiplied indiscriminately. Furthermore, matrix multiplication is not generally a commutative operation. The *commutative law* $AB = BA$ holds for some matrix products, but not all.

Consider now the product of two matrices A and B. Let A be of order $m \times n$ and B of order $r \times s$. The product of these two matrices is*

$$C = AB = [a_{ik}][b_{kj}] \equiv \left[\sum_{k=1}^{n} a_{ik} b_{kj} \right],$$

which is defined if and only if $n = r$. That is, the number of columns of A must equal the number of rows of B. The product matrix C is of order $m \times s$. The rule for matrix multiplication prescribes the (i,j) entry of C to be as follows:

$$c_{ij} = a_{i1} b_{1j} + a_{i2} b_{2j} + \cdots + a_{in} b_{nj},$$

where the entries a_{ik} $(k = 1, 2, \ldots, n)$ are of the ith row of A, and the entries b_{kj} are of the jth column of B. That is, the entry c_{ij} of the product matrix C is defined as the row-column product of the ith row of A,

$$A = [a_{i1} a_{i2} \cdots a_{in}]$$

with the jth column of B,

$$B = [b_{1j} b_{2j} \cdots b_{nj}]^{T}.$$

Although AB and BA are not always equal, they do have the same trace whenever AB is square:

$$\text{tr}(AB) = \sum_{i=1}^{m} \sum_{j=1}^{n} a_{ij} b_{ji} = \text{tr}(BA).$$

---

*Note that $|AB \cdots MN| = |A||B| \cdots |M||N|$.

Matrix multiplication is *distributive*

$$A(B + C) = AB + AC,$$

$$(A + B)C = AC + BC$$

whenever both sides of the equations are defined. Furthermore, matrix multiplication is *associative*, that is,

$$(AB)C = A(BC).$$

Hence, the rules for matrix algebra differ from those of scalar algebra only in the commutative property. Matrix algebra does not admit division; however, a substitute operation, matrix inversion, has already been discussed above.

Consider now two vectors $\mathbf{x}$ and $\mathbf{y}$, where $\mathbf{x}$ is an $n \times 1$ column vector and $\mathbf{y}$ is a $1 \times n$ row vector. Then their product

$$\mathbf{x}^T\mathbf{y} = \mathbf{y}^T\mathbf{x} = x_1 y_1 + x_2 y_2 + \cdots x_n y_n$$

is a scalar (1×1) matrix called the *scalar product* (or *inner product*) of $\mathbf{x}$ and $\mathbf{y}$. Note that for $n = 3$ this relationship is also the same as the dot product of vector analysis and field theory.) A generalization of the inner product leads to the weighted inner product defined by

$$\mathbf{x}^T W \mathbf{y} = \mathbf{y}^T W^T \mathbf{x}$$

for any symmetric $n \times n$ matrix $W = W^T$.

A special case of the inner product $\mathbf{x}^T\mathbf{y}$ is when $\mathbf{y} = \mathbf{x}$. Then

$$\mathbf{x}^T\mathbf{x} = x_1^2 + x_2^2 + \cdots + x_n^2 \equiv |\mathbf{x}|^2.$$

In view of the above discussion, we will now briefly state the so-called Schwarz inequality. For the two vectors $\mathbf{x}$ and $\mathbf{y}$, the Schwarz inequality can be stated as follows [4]:

$$|\mathbf{x}^T\mathbf{y}| \leqslant |\mathbf{x}|\,|\mathbf{y}|, \qquad x, y \in R^n.$$

Furthermore, if $\mathbf{y} \neq 0$, then

$$\mathbf{x}^T\mathbf{x} \geqslant (\mathbf{y}^T\mathbf{x})^T(\mathbf{y}^T\mathbf{y})^{-1}(\mathbf{y}^T\mathbf{x}).$$

The Schwarz inequality applies equally well to matrices. Let A and B be two matrices of dimension $m \times n$ and $m \times q$, respectively. Then, from the above expression, we have

$$B^T B \geqslant (A^T B)^T (A^T A)^{-1}(A^T B). \tag{A.16}$$

A.2.2 The Eigenvalue Problem

In many physical problems, one is interested in determining those values of a constant λ for which nontrivial solutions exist to a homogeneous set of equations of the form

$$
\begin{aligned}
a_{11}x_1 + a_{12}x_2 + \cdots + a_{1n}x_n &= \lambda x_1, \\
a_{21}x_1 + a_{22}x_2 + \cdots + a_{2n}x_n &= \lambda x_2, \\
&\vdots \\
a_{n1}x_1 + a_{n2}x_2 + \cdots + a_{nn}x_n &= \lambda x_n.
\end{aligned}
\tag{A.17}
$$

Such a problem is known as an *eigenvalue* problem (or *characteristic value* problem). The values of λ for which nontrivial solutions exist are called *eigenvalues* (also known as *characteristic values* or *latent roots*) of the problem or of the matrix A; corresponding vector solutions are known as *eigenvectors* (or *characteristic vectors*) of the matrix A. In matrix notation, Eq. (A.17) takes the form [3, 6]

$$
A\mathbf{x} = \lambda \mathbf{x} \quad \text{or} \quad (A - \lambda I)\mathbf{x} = \mathbf{0},
\tag{A.18}
$$

where I is the identity matrix of order n. This homogeneous problem possesses nontrivial solutions if and only if the determinant of the coefficient matrix $A - \lambda I$ vanishes. That is,

$$
|A - \lambda I| \equiv
\begin{vmatrix}
a_{11} - \lambda & a_{12} & \cdots & a_{1n} \\
a_{21} & a_{22} - \lambda & \cdots & a_{2n} \\
\vdots & \vdots & & \vdots \\
a_{n1} & a_{n2} & \cdots & a_{nn} - \lambda
\end{vmatrix} = 0.
\tag{A.19}
$$

The above condition requires that λ be a root of an algebraic equation of degree n, known as the characteristic equation. The n solutions $\lambda_1, \lambda_2, \ldots, \lambda_n$, which need not all be distinct, are the *characteristic numbers* or *latent roots* of the matrix A. Equation (A.19) when expanded yields an nth-order polynomial in λ as follows [1]:

$$
\lambda^n + a_{n-1}\lambda^{n-1} + \cdots + a_1\lambda + a_0 = 0.
\tag{A.20}
$$

The coefficients a_k which appear in Eq. (A.20) are given by complicated expressions involving various cofactors of the matrix A, with

$$
a_0 = |A| \quad \text{and} \quad a_1 = \text{tr}(A).
$$

Equation (A.20) is known to have n roots $\lambda_1, \lambda_2, \ldots, \lambda_n$, which are the solutions of the nth-order polynomial equation

$$(\lambda - \lambda_1)(\lambda - \lambda_2) \cdots (\lambda - \lambda_n) = 0. \tag{A.21}$$

These roots are the eigenvalues of A; they may be real or complex numbers:

$$\lambda_k = \alpha_k + i\beta_k,$$

where $i = \sqrt{-1}$. If all the eigenvalues are distinct, then there are n solutions for the eigenvectors obtained from

$$A\mathbf{x}_k = \lambda_k \mathbf{x}_k \qquad (k = 1, 2, \ldots, n), \tag{A.22}$$

where the eigenvector $\mathbf{x}_k$ is associated with the eigenvalue λ_k.

Let us now assume that we have an $n \times n$ diagonal matrix, Λ, of the form

$$\Lambda = \begin{bmatrix} \lambda_1 & 0 & \cdots & 0 \\ 0 & \lambda_2 & \cdots & 0 \\ \vdots & \vdots & \ddots & \vdots \\ 0 & 0 & \cdots & \lambda_n \end{bmatrix}$$

with distinct eigenvalues $\lambda_1, \lambda_2, \ldots, \lambda_n$. The matrix Λ is called the *matrix of eigenvalues*. Furthermore, if A is an $n \times n$ matrix and if we let $A_k(\lambda)$ denote the $k \times k$ matrix given by

$$A_k(\lambda) = \begin{bmatrix} \lambda & 1 & 0 & \cdots & 0 \\ 0 & \lambda & 1 & \cdots & 0 \\ \vdots & \vdots & \vdots & & \vdots \\ 0 & 0 & 0 & \cdots & 1 \\ 0 & 0 & 0 & \cdots & \lambda \end{bmatrix},$$

then it can be shown that if the eigenvalues of A are all real, then A is similar to a matrix of the form [1]

$$J(A) = \begin{bmatrix} A_{m_1}(\lambda_1) & 0 & \cdots & 0 \\ 0 & A_{m_2}(\lambda_2) & \cdots & 0 \\ \vdots & \vdots & & \vdots \\ 0 & 0 & \cdots & A_{m_p}(\lambda_p) \end{bmatrix}$$

where $m_1 + m_2 + \cdots + m_p = n$ and $\lambda_1, \lambda_2, \ldots, \lambda_p$ are the eigenvalues (not necessarily distinct) of A. The matrix $J(A)$ is called the *Jordan canonical form* of A (see also Section 4.4).

In the introduction to this section we noted that if two square matrices A and B are similar, then $B = P^{-1}AP$, where P is a nonsingular matrix. Consider now a square matrix A. The *diagonalization* of this square matrix can be carried out by finding a nonsingular square matrix P such that the matrix equation

$$P^{-1}AP = \Lambda$$

is fulfilled. In this equation, Λ is a square matrix diag $(\lambda_1, \ldots, \lambda_n)$, where $\lambda_1 \cdots \lambda_n$ are the n eigenvalues of the matrix A. Therefore,

$$AP = P\Lambda.$$

Finally, we have the Cayley–Hamilton theorem, which can be stated as follows [1, 3]:

Theorem: Let $f(\lambda) = 0$ be the characteristic equation of an arbitrary square matrix A; then A satisfies its own characteristic equation: $f(A) = 0$.

The Cayley–Hamilton theorem can be proved easily for the case of distinct eigenvalues. From the above discussion, we know that a nonsingular matrix P can be found such that $P^{-1}AP = \Lambda$ or $AP = P\Lambda$. This implies that by premultiplying in turn by $A, A^2, \ldots, A^n$ we have $A^2P = P\Lambda^2, \ldots, A^nP = P\Lambda^n$. Consequently, for any polynomial of the form

$$g(x) = \sum_{i=1}^{n} a_i x^i,$$

we have $g(A)P = P_g(\Lambda)$. Furthermore, this holds also for the characteristic polynomial. Thus,

$$f(A)P = Pf(\Lambda).$$

Next, we note that

$$f(\Lambda) = \text{diag} (f(\lambda_1), \ldots, f(\lambda_n)).$$

However, $f(\lambda_i) \triangleq 0$ for each i; since from the relation $f(A)P = Pf(\Lambda)$ the matrix P is nonsingular, then we have $f(A) = 0$.

A.2.3   Quadratic Forms

As the reader has already noticed, quadratic forms appear quite often in optimal control and estimation theory. For this reason, the quadratic form will be discussed briefly in this subsection. Let R^n denote the space of all column vectors $\mathbf{x} = [x_1 \ x_2 \ \cdots \ x_n]^T$ where the entries $x_1, x_2, \ldots, x_n$ are real numbers. A homogeneous polynomial of second degree, of the form

$$A \equiv a_{11}x_1^2 + a_{22}x_2^2 + \cdots + a_{nn}x_n^2$$

$$+ 2a_{12}x_1x_2 + 2a_{13}x_1x_2 + \cdots + 2a_{n-1,n}x_{n-1}x_n, \qquad (A.23)$$

is called a *quadratic form* in $x_1, x_2, \cdots, x_n$, where it is assumed that the elements a_{ij} and the variables x_i are real. Consider now the system of equations

$$a_{11}x_1 + a_{12}x_2 + \cdots + a_{1n}x_n = y_1,$$

$$a_{12}x_1 + a_{22}x_2 + \cdots + a_{2n}x_n = y_2, \qquad (A.24)$$

$$\vdots$$

$$a_{1n}x_1 + a_{2n}x_3 + \cdots + a_{nn}x_n = y_n,$$

which in matrix form can be written as

$$A\mathbf{x} = \mathbf{y}, \qquad (A.25)$$

where $A = [a_{ij}]$ is a symmetric matrix, that is, the elements satisfy the symmetry condition $a_{ji} = a_{ij}$. It is easily seen that Eq. (A.23) is equivalent to the relation $A \equiv (\mathbf{x}, \mathbf{y})$, so that we can write Eq. (A.23) in the form [4, 5]

$$A \equiv \mathbf{x}^T A \mathbf{x}. \qquad (A.26)$$

From this discussion we can state that every quadratic form can be written as $\mathbf{x}^T A \mathbf{x}$, where A is a uniquely determined symmetric matrix.

At this point we will discuss the concepts of positive definite, positive semidefinite, negative definite, and negative semidefinite matrices. Consider a symmetric real matrix P. The matrix P is called positive definite if

$$\mathbf{x}^T P \mathbf{x} > 0$$

for all $\mathbf{x} = 0$ in R^n. The positive definite property of P is also denoted by $P > 0$. A symmetric matrix S is called positive semidefinite if

$$\mathbf{x}^T S \mathbf{x} \geqslant 0$$

for all vectors $\mathbf{x}$. The notation for this case is $S \geqslant 0$. Now, if P is positive definite, then $N = -P$ is negative definite. Furthermore, if S is positive semidefinite, then $R = -S$ is negative semidefinite.

In Chapter 4, Section 4.1, the concepts of positive definiteness, and so on, of a quadratic form were defined in terms of the matrix eigenvalues. Another, perhaps easier way to understand the concepts of positive definiteness, is to consider the general real square matrix A. Then, the quadratic form has the following properties:

Positive definite $(A > 0)$:   if $x^T A x > 0$ for all nonzero x.
Positive semidefinite $(A \geqslant 0)$:   if $x^T A x \geqslant 0$ for all nonzero x.
Negative definite $(A < 0)$:   if $x^T A x < 0$ for all nonzero x.
Negative semidefinite $(A \leqslant 0)$:   if $x^T A x \leqslant 0$ for all nonzero x.

Finally, the above discussion can be summarized as follows; if x_i and x_j are of n-degree, the quadratic form can be expressed in vector–matrix form as*

$$q(\mathbf{x}) = q(x_1, \ldots, x_n) = \sum_{i=1}^{n} \sum_{j=1}^{n} Q_{ij} x_i x_j = \mathbf{x}^T Q \mathbf{x}.$$

In view of this discussion, a quadratic form is said to be positive (negative) definite if $q = 0$ for $\mathbf{x} = \mathbf{0}$ and $q > 0$ $(q < 0)$ for $\mathbf{x} \neq \mathbf{0}$. Furthermore, the quadratic form $q(\mathbf{x}) = \mathbf{x}^T Q \mathbf{x}$ is positive definite if and only if the n determinants $|Q_1|, \ldots, |Q_n|$ are all positive. The quadratic form is said to be positive semidefinite (negative semidefinite) if $q = 0$ for $\mathbf{x} = \mathbf{0}$ and $q \geqslant 0$ $(q \leqslant 0)$ for $\mathbf{x} \neq \mathbf{0}$.

A.2.4   The Matrix Inversion Lemma

The matrix inversion lemma is a useful identity that appears very often in sequential estimation algorithms. Consider the $n \times n$ matrix A partitioned as follows [2, 4, 7]:

$$A = \left[\begin{array}{c|c} A_{11} & A_{12} \\ \hline A_{21} & A_{22} \end{array} \right], \tag{A.27}$$

---

* If Q is a constant $r \times r$ matrix, then the quadratic form $\mathbf{x}^T Q \mathbf{x}$ can be expressed as

$$\mathbf{x}^T Q \mathbf{x} = \mathbf{x}^T [(Q + Q^T)/2] \mathbf{x},$$

and

$$Q_0 = (Q + Q^T)/2$$

is a symmetric matrix. That is, given any quadratic form $\mathbf{x}^T Q \mathbf{x}$, we can always replace it with a quadratic form $\mathbf{x}^T Q_0 \mathbf{x}$ where Q_0 is symmetric.

where $A_{11} = $ an $n_1 \times n_1$ nonsingular matrix,

$\quad A_{12} = $ an $n_1 \times n_2$ nonsingular matrix,

$\quad A_{21} = $ an $n_2 \times n_1$ nonsingular matrix,

$\quad A_{22} = $ an $n_2 \times n_2$ nonsingular matrix,

$\quad n = n_1 + n_2.$

Assuming that the necessary inverses of the above submatrices exist, then the matrix A can be partitioned as follows:

$$A^{-1} = \left[\begin{array}{c|c} A_{11}^{-1} + A_{11}^{-1} A_{12} \Delta^{-1} A_{21} A_{11}^{-1} & -A_{11}^{-1} A_{12} \Delta^{-1} \\ \hline -\Delta^{-1} A_{21} A_{11}^{-1} & \Delta^{-1} \end{array} \right], \quad \text{(A.28)}$$

where $\Delta = A_{22} - A_{21} A_{11}^{-1} A_{12}$.

From Eq. (A.28), the following identity holds:

$$(A_{11} - A_{12} A_{22}^{-1} A_{21})^{-1} = A_{11}^{-1} + A_{11}^{-1} A_{12} (A_{22} - A_{21} A_{11}^{-1} A_{12})^{-1} A_{21} A_{11}^{-1}. \quad \text{(A.29)}$$

The proof is by direct substitution. A special case of Eq. (A.29) is

$$(I + A)^{-1} = I - (I + A^{-1})^{-1}. \quad \text{(A.30)}$$

Another version of Eq. (A.29) is the following [7]:

$$(P^{-1} + H^T R^{-1} H)^{-1} = P - P H^T (H P H^T + R)^{-1} H P, \quad \text{(A.31)}$$

where P and R are nonsingular matrices of order $n \times n$ and $m \times m$, respectively, and H is an $m \times n$ matrix. Here it is noted that the matrix $H P H^T + R$ is often of lower order than $P^{-1} + H^T R^{-1} H$. The complete proof of this lemma is given in [4,7].

Some Useful Matrix Identities. Given the matrices A $(n \times n)$, $B(n \times r)$, and $C(r \times n)$ the following identities are true:

$$(A^{-1} + BC)^{-1} = A - AB(I + CAB)^{-1} CA,$$
$$(I + BCA)^{-1} = I - B(I + CAB)^{-1} CA,$$
$$(I + ABC)^{-1} = I - AB(I + CAB)^{-1} C,$$

where I is the identity or unit matrix.

REFERENCES

[1] Athans, M. and Falb, P. L.: *Optimal Control: An Introduction to the Theory and Its Applications*, McGraw-Hill Book Company, New York, 1966.

[2] Athans, M. and Schweppe, F. C.: *Gradient Matrices and Matrix Calculations*, M.I.T. Lincoln Laboratory, Technical Note 1965–53, 17 November 1965.

[3] Bellman, R.: *Introduction to Matrix Analysis*, McGraw-Hill Book Company, New York, 1960.

[4] Chui, C. K. and Chen, G.: *Kalman Filtering with Real-Time Applications*, 2nd Edition, Springer-Verlag, Berlin, Heidelberg, New York, 1991.

[5] Hildebrand, F. B.: *Methods of Applied Mathematics*, 2nd Edition, Prentice-Hall, Inc., Englewood Cliffs, New Jersey, 1965.

[6] Pipes, L. A.: *Matrix Methods for Engineering*, Prentice-Hall, Inc., Englewood Cliffs, New Jersey, 1963.

[7] Sage, A. P. and Melsa, J.L.: *Estimation Theory, with Applications to Communications and Control*, McGraw-Hill Book Company, New York, 1971.

[8] Schreier, O. and Sperner, E.: *Introduction to Modern Algebra and Matrix Theory*, 2nd Edition, Chelsea Publishing Company, New York, 1959.

APPENDIX B

MATRIX LIBRARIES

This appendix presents a brief discussion of applicable software tools that are available commercially. These tools are expertly written and are transportable to various computer systems. Furthermore, they support topics discussed in the text. Software can be categorized as (1) library packages, and (2) interactive packages. Two commonly used packages are:

1. EISPACK: includes functions for solving eigenvalue–eigenvector problems.
2. LINPACK: includes functions for solving and analyzing basic linear equations.

A package which carries most of the operations discussed in the text is the MATLAB package (see also Chapter 4, Section 4.3) available from Math Works, Inc. The above libraries form the foundation of MATLAB.

A derivative of MATLAB is the MATRIX$_x$ package, which incorporates many enhanced features of control, signal analysis, system identification, and nonlinear system analysis. Other commercially available subroutines which use many of the basic EISPACK and LINPACK libraries are those of (1) IMSL (International Mathematical and Statistical Libraries, Inc.) and (2) NAG (the Numerical Algorithms Group). *Canned* algorithms for most of the mathematical operations discussed in the text are available in the EISPACK and IMSL libraries.

Also presented in this appendix are a number of basic matrix operation programs coded in FORTRAN IV that often arise in estimation theory and aerospace software applications. The reader may use these subroutines as they are, or modify them to suit his need. These subroutines are presented here as a convenience to the reader and/or systems analyst and as a starting point for further research.

```
C
      SUBROUTINE MADD (R, A, B, NR, NC)
C
C     R = A + B
C
      DIMENSION A (NR, NC), B(NR, NC) , R(NR, NC)
C
      DO 10 I = 1, NR
      DO 20 J = 1, NC
      R (I, J) = A (I, J) + B(I, J)
20    CONTINUE
10    CONTINUE
      RETURN
      END
C
      SUBROUTINE MSUB (R, A, B, NR, NC)
C
C     R = A - B
C
      DIMENSION A(NR,NC) , B(NR,NC), R(NR,NC)
C
      DO 10 I = 1 , NR
      DO 20 J = 1 , NC
      R(I,J) = A(I, J) - B(I, J)
20    CONTINUE
10    CONTINUE
      RETURN
      END
C
      SUBROUTINE MMULT(R, A, B, NAR, NBR, NBC)
C
C     R = A*B
C
      DIMENSION R(NAR, NBC), A(NAR, NBR), B(NBR, NBC)
C
      DO 10 I = 1, NAR
      DO 20 J = 1, NBC
      R (I,J) = 0.
      DO 30 K = 1, NBR
      R(I, J) = R(I, J) + A(I, K) *B(K, J)
30    CONTINUE
20    CONTINUE
10    CONTINUE
      RETURN
      END
```

```
      SUBROUTINE SMULT (R, A, B, IA, JA, M, NCB, NRA)
C
      DIMENSION R (NRA, NCB), A(M), B(NCB, NCB), IA(M), JA(M)
C
C     SMULT IS A SUBROUTINE WHICH MULTIPLIES AN UPPER
      TRIANGULAR MATRIX BY A SPARSE MATRIX
C     R = THE RESULTANT MATRIX A * B
C     A = THE NONZERO ELEMENTS OF A SPARSE MATRIX READ IN
      VECTOR FORM.
C     B = A UPPER TRIANGULAR MATRIX.
C     IA = THE ROW INDICES OF THE NON ZERO ELEMENTS OF THE SPARE
C     MATRIX.
C     JA = THE CORRESPONDING COLUMN INDICES OF THE NON ZERO
      ELEMENTS OF THE SPARSE MATRIX.
C     NCB = THE NUMBER OF ROWS AND COLUMNS OF B.
C     M = THE NUMBER OF NON ZERO ELEMENTS IN THE SPARSE MATRIX.
C     NRA = THE NUMBER OF ROWS IN A
C
      DO 100 J = 1, NCB
C
C
      DO 10 I = 1. NRA
      R(I, J) = 0. 0
10    CONTINUE
C
      DO 30 K = 1, M
      IF ( JA(K) GT. J). GO TO 20
      R( IA(K), J) = R( IA(K), J) + A(K) * B (JA(K), J)
C
20    CONTINUE
30    CONTINUE
C
100   CONTINUE
C
      RETURN
      END


      SUBROUTINE MTRA (R,A,NR,NC)
C
C     R = TRANSPOSE (A)
C
      DIMENSION A(NR, NC), R(NC, NR)
C
      DO 10 I = 1, NR
      DO 20 J = 1, NC
      R(J, I) = A(I, J)
20    CONTINUE
10    CONTINUE
      RETURN
      END
```

```
      SUBROUTINE IDN(A, N)
C
C     A = IDENTITY MATRIX
C
      DIMENSION A(N, N)
C
      DO 20 I = 1, N
      DO 10 J = 1, N
      A(I, J) = 0. 0
10    CONTINUE
20    CONTINUE
      DO 30 I = 1, N
      A(I, I) = 1. 0
30    CONTINUE
      RETURN
      END
C
C
C
      SUBROUTINE MCON (R, A, C, NR, NC)
C
C     R = C * A,    C = CONSTANT
C
      DIMENSION R (NR, NC), A(NR, NC)
C
      DO 20 I = 1, NR
      DO 10 J = 1, NC
      R (I, J) = A (I, J) *C
10    CONTINUE
20    CONTINUE
      RETURN
      END
C
C
C
      SUBROUTINE ZER (R, NR, NC)
C
C     R = ZERO MATRIX
C
      DIMENSION R(NR, NC)
C
      DO 20 I = 1, NR
      DO 10 J = 1, NC
      R(I, J) = 0.
10    CONTINUE
20    CONTINUE
      RETURN
      END
```

```
      SUBROUTINE MEQU (R, A, NR, NC)
C
C     R = A
C
      DIMENSION R (NR, NC), A(NR, NC)

C
      DO 20 I = 1, NR
      DO 10 J = 1, NC
      R(I, J) = A (I, J)
10    CONTINUE
20    CONTINUE
      RETURN
      END

      SUBROUTINE MINV (RMI, RM, NR)
C
C     RMI = INVERSE (RM)
C
      COMMON /MICOM/ A(900), L(900), M(900)
C
      DIMENSION RMI (NR, NR), RM(NR, NR)
C
C     RM-REAL MATRIX
C     NR - NUMBER OF ROWS OF THIS SQUARE MATRIX
C
C     THE VECTORS A, L, AND M ARE WORK VECTORS WHICH MUST BE
C     DIMENSIONED AS THE SQUARE OF THE LARGEST MATRIX INVERSE
C     WHICH WILL BE COMPUTED. IF NR EXCEEDS THIS MAXIMUM THE
C     CALLING PROGRAM WILL STOP HERE.
C
      IF (NR. GT. 10) STOP
C
      DO 10 J = 1, NR
      IZ = NR * (J – 1)
      DO 20 I = 1, NR
      IJ = IZ + I
      A(IJ) = RM(I, J)
20    CONTINUE
10    CONTINUE
C
      CALL IMINV (A, NR, D, L, M)
C
      DO 30 J = 1, NR
      IZ = NR* (J – 1)
      DO 40 I = 1, NR
      IJ = IZ + I
      RMI (I, J) = A(I J)
40    CONTINUE
```

```
30    CONTINUE
C
      RETURN
      END
C
C
C
      SUBROUTINE IMINV (A, N, D, L, M)
C
      DIMENSION A(1), L(1), M(1)
C
C     A-INPUT AND OUTPUT SQUARE MATRIX
C     N-ORDER OF THIS SQUARE MATRIX
C     D-RESULTANT DETERMINANT
C     L-WORK VECTOR OF LENGTH N
C     M-WORK VECTOR OF LENGTH N
C        SEARCH FOR LARGEST ELEMENT
      D = 1. 0
      NK = - N
      DO 80 K = 1, N
      NK = NK + N
      L (K) = K
      M (K) = K
      KK = NK + K
      BIGA = A (KK)
      DO 20 J = K, N
      IZ = N* (J - 1)
      DO 20 I = K, N
      IJ = IZ + I
10    IF (ABS (BIGA) - ABS(A (IJ) ) ) 15, 20, 20
15    BIGA = A(IJ)
      L (K) = I
      M(K) = J
20    CONTINUE
C        INTERCHANGE ROWS
      J = L(K)
      IF (J - K) 35, 35, 25
25    KI = K - N
      DO 30 I = 1, N
      KI = KI + N
      HOLD = - A(KI)
      JI = KI - K + J
      A(KI) = A(JI)
30    A(JI) = HOLD
C            INTERCHANGE COLUMNS
```

```
   35  I = M(K)
       IF (I − K) 45, 45, 38
   38  JP = N*(I − 1)
       DO 40 J = 1, N
       JK = NK + J
       JI = JP + J
       HOLD = − A(JK)
       A(JK) = A(JI)
   40  A(JI) = HOLD
C      DIVIDE COLUMN BY MINUS PIVOT (VALUE OF PIVOT
C      ELEMENT IS CONTAINED IN BIGA)
   45  IF (ABS (BIGA) − 1. E − 20) 46, 46, 48
   46  D = 0. 0
       RETURN
   48  DO 55 I = 1, N
       IF (I − K) 50, 55, 50
   50  IK = NK + I
       A(IK) = A(IK)/(-BIGA)
   55  CONTINUE
C          REDUCE MATRIX
       DO 65 I = 1, N
       IK = NK + I
       HOLD = A(IK)
       IJ = I − N
       DO 65 J = 1, N
       IJ = IJ + N
       IF (I − K) 60, 65, 60
   60  IF (J − K) 62, 65, 62
   62  KJ = IJ − I + K
       A (IJ) = HOLD*A(KJ) + A(IJ)
   65  CONTINUE
C          DIVIDE ROW BY PIVOT
       KJ = K − N
       DO 75 J = 1, N
       KJ = KJ + N
       IF (J − K) 70, 75, 70
   70  A(KJ) = A (KJ) / BIGA
   75  CONTINUE
C          PRODUCT OF PIVOTS
       D = D* BIGA
C           REPLACE PIVOT BY RECIPROCAL
       A(KK) = 1.0 / BIGA
   80  CONTINUE
C          FINAL ROW AND COLUMN INTERCHANGE
       K = N
```

```
100 K = (K − I)
    IF (K) 150, 150, 105
105 I = L (K)
    IF (I − K) 120, 120, 108
108 JQ = N* (K − 1)
    JR = N* (I − 1)
    DO 110 J = 1, N
    JK = JQ + J
    HOLD = A (JK)
    JI = JR + J
    A(JK) = − A (JI)
110 A(JI) = HOLD
120 J = M (K)
    IF (J − K) 100, 100, 125
125 KI = K − N
    DO 130 I = 1, N
    KI = KI + N
    HOLD = A (KI)
    JI = KI − K + J
    A (KI) = − A (JI)
130 A (JI) = HOLD
    GO TO 100
150 RETURN
    END




    SUBROUTINE COVPRP (R, A, NR)
C
C   R = NORMALIZED A, A = COVARIANCE MATRIX
C
C   UPPER - T PART OF R CONTAINS CORRELATION
C   COEFFICIENTS DIAGONAL PART CONTAINS STANDARD DEVIATIONS
    LOWER-T PART OF R CONTAINS CROSS-COVARIANCES
C
    DIMENSION R (NR, NR), A(NR, NR)
C
    CALL MEQU (R, A, NR, NR)
    DO 10 I = 1, NR
    DO 20 J = 1, NR
    IF ( I. EQ. J ) GO TO 30
    TEST = A( I, I) * A( J, J)
    IF (TEST .NE. O.O ) GO TO 5
    R (I, J ) = 0.0
    GO TO 40
```

```
5          R (I, J) = A(I, J)/SQRT (A(I, I) *A(J, J))
           GO TO 40
30         R (I, J) = SQRT (A(I, J) )
40         CONTINUE
20         CONTINUE
10         CONTINUE
           RETURN
           END
C
           SUBROUTINE PRM(TITLE, P,NR,NC,LFN)
C
C          PRINT MATRIX P
C          TITLE = 6 CHARACTER LABEL
C          LFN = OUTPUT FILE
C
           DIMENSION P (NR, NC)
C
           WRITE (LFN ,100) TITLE
           DO 10 I = 1, NR
           WRITE (LFN, 200) I, (J,P(I, J), J = 1, NC)
      10   CONTINUE
     100 FORMAT (/, 1X, A6)
     220 FORMAT  (1X,  I3,  2X,  5 (I3,  1X,  G20.10),/(6X,  5  (I3,  1X,
           G20.10) ) )
           RETURN
           END
C
C
           SUBROUTINE PRMD (TITLE, P, N, LFN)
C
           DIMENSION P (N, N)
           WRITE (LFN, 100 ) TITLE
           WRITE (LFN, 200) (J, J, P (J,J), J = 1, N)
10         CONTINUE
100        FORMAT (/, 1X, A6)
200        FORMAT (4(3X, I3, 1X, I3, 1X, G20.10) )
           RETURN
           END


C
           FUNCTION ATANYX (Y, X)
C
C          4-QUADRANT ARC-TANGENT
C
           PI = 3.1415927
```

```
C
            IF (X .NE. 0.) GO TO 50
            IF (Y .LT. 0.) ATANYX = -PI/2.
            IF (Y .GT. 0.) ATANYX = PI/2.
            GO TO 100
C
50          Z = Y / X
            ATANYX = ATAN (Z)
            IF (Z .GT. 0. .AND. X .LT. 0. ) ATANYX = ATANYX - PI
            IF (Z .LE. 0. .AND. X .LT. 0.) ATANYX = ATANYX + PI
C
100         RETURN
            END
C
            SUBROUTINE FACTOR (P,U,D,N)
C
C           COMPUTE FACTORS U & D WHERE P = U * D * TRANS-
            POSE (U)
C
C           P = COVARIANCE MATRIX
C           U = UNIT UPPER - TRIANGULAR
C           D = DIAGONAL FACTOR STORED AS A VECTOR
C
            DIMENSION P (N,N), U (N,N), D (N)
C
C           ZERO THE LOWER TRIANGULAR MATRIX,
C           EXCLUDING THE DIAGONAL
C
            DO 20 I = 2,N
            DO 10 J = I, I - 1
            U (I, J) = 0.0
C
    10      CONTINUE
    20      CONTINUE
C
C           EPS : THRESHOLD AT WHICH AN ELEMENT OF D IS CONSIDERED
            TO BE ZERO. (P SINGULAR).
C
C
            EPS = 1.0E - 30
C
            J = N
C
            GO TO 150
C
```

```
      100    J = J − 1
C
      150    CONTINUE
C
             IF (J .LT. 2) GO TO 300
             U (J, J) = 1.
             D(J) = P (J, J)
C
             IF (D(J) .LT. EPS ) GO TO 160
             ALPHA = 1.0 / D(J)
             GO TO 170
C
      160    ALPHA = 0.0
             D (J) = 0.0
      170    J1 = J − 1
C
             DO 250 K = 1, J1
             BETA = P (K, J)
             U(K, J) = ALPHA*BETA
C
             DO 200 I = 1, K
             P (I, K) = P (I, K) − BETA*U(I, J)
      200    CONTINUE
C
      250    CONTINUE
C
             GO TO 100
C
      300    CONTINUE
C
             U(1, 1) = 1.
C
             D(1) = P(1, 1)
C
             IF (D(1) .LT. EPS) D(1) = 0.0
C
C
             RETURN
             END
C
C
             SUBROUTINE TRIINV (B, A, N)
C
C            B = INVERSE (A), WHERE A IS AN UPPER-
```

```
C              TRIANGULAR MATRIX
C
               DIMENSION B(N, N), A(N, N)
C
C              ZERO THE LOWER TRIANGULAR MATRIX,
C              EXCLUDING THE DIAGONAL
C
               DO 20 I = 2, N
               DO 10 J = 1, I − 1
               B(I, J) = 0.0
C
     10        CONTINUE
     20        CONTINUE
C
               B(1, 1) = 1./A(1, 1)
C
               DO 200 J = 2, N
               B (J, J) = 1./A(J, J)
               JM1 = J − 1
C
               DO 150 K = 1, JM1
               SUM = 0.
C
               DO 100 I = K, JM1
               SUM = SUM − B(K, I)*A(I, J)
     100       CONTINUE
C
               B (K, J) = SUM*B(J, J)
C
     150       CONTINUE
C
     200       CONTINUE
C
               RETURN
               END
C
```

REFERENCES

[1] Dongarra, J. J., Moler, C. B., Bunch, J. R., and Stewart, G. W.: *LINPACK User's Guide*, Society for Industrial and Applied Mathematics, Philadelphia, Pennsylvania, 1979.

[2] Garbow, B. S., Boyle, J. M., Dongarra, J. J., and Moler, C. B.: *Matrix Eigensystem Routines - EISPACK Guide Extension*, Lecture Notes in Computer Science, Vol. 51, Springer–Verlag, 1977.

[3] Golub, G. H. and Van Loan, C. F.: *Matrix Computations*, Johns Hopkins University Press, 1983.

[4] Moler, C., Shure, L., Little, J., and Bangert, S.: *Pro-MATLAB for Apollo Workstations*, The Math Works, Inc., Sherborn, Massachusetts, 1987.

[5] Smith, B. T., Boyle, J. M., Dongarra, J. J., Garbow, B. S., Ikebe, Y., Klema, C. B., and Moler, C. B.: *Matrix Eigensystem Routines-EISPACK Guide*, Lecture Notes in Computer Science, Vol. 6, 2nd Edition, Springer-Verlag, 1976.

INDEX

Accelerometer error model, 207
Adams–Moulton method, 311–313
Adjoint time, 236
Adjoint vector, 222
Admissible control, 226, 266–267, 279
Aided inertial system, 2, 347, 351
Alpha filter (α-filter), 329–331
Alpha-beta tracking filter (α-β filter), 329, 331–337
Alpha-beta-gamma tracking filter (α-β-γ filter), 329, 337–339
A priori information, 75
Augmented system, 201–204
Autocorrelation function (ACF), 25–26, 37–38
Average value, 11

Bandwidth, 35, 51, 236, 292
Bang-bang control, 285, 287–288
Bang-bang-off control, 230
Batch processing, 1, 67, 71, 121
Bayesian estimation, 3, 82–83
Bayes' rule, 22
Bellman's principle of optimality, 276
Beta distribution, 21
Bias, 48
Bierman U–D factorization algorithm, 127
Bivariate normal distribution, 10
Bolza problem, 219, 227
Borel field, 7
Boundary conditions, 222, 229
Brownian motion, 38, 48

Calculus of variations, 221–232
Canonical form, 206, 240, 266, 271
Carlson square root filter, 126–127, 171
Causal signal, 33
Cayley–Hamilton theorem, 385
Central limit theorem, 12–13
Centralized filter, 348–350
Central moments, 12
Characteristic function, 24
Chebychev inequality, 22, 57
Chi-square distribution, 16, 20
Cholesky factorization (or decomposition), 175–176, 183–185
Colored noise, 29, 201–204
Completely controllable, 159, 163
  observable, 160, 163
Conditional mean, 17
Constraints, 221, 227–228
  equality, 221
Continuous filter, 93
Control, 238, 240
Controllability, 159
Convergence, 154
Correlation:
  coefficient, 24, 85
  distance, 52–53, 138
  time, 35–36, 138
Costate, 238–239
Covariance, 23–24, 99
  analysis, 362–367
  matrix, 23–24, 97, 99

Covariance (*Continued*)
  propagation, 306
  update, 81, 122, 307
Cramér-Rao inequality, 77
  lower bound, 78, 90–91
Cramer's rule, 374
Cross-correlation, 26–27
Curve fitting, 65

Decentralized filter, 347–355
Density function, *see* Probability density
  function
Deterministic:
  control, 164–165
  system, 64
Diagonalization, 151–152
Dirac delta function, 96
Discrete filter, 111–125
Dispersion, *see* Variance
Divergence, 169–171
Doppler radar, 2, 319, 339, 351
Double precision, 171
Dual control, 162
Duality theorem, 162
Dynamic programming, 221, 274–285
Dynamics model, 198, 200

Efficient estimate, 11
Eigenvalue(s), 101, 149, 166–168, 383–385
Ergodic process, 8, 25–26
Error:
  analysis, 2, 305
  budget, 111, 325–326
  covariance, 97, 103, 121
  function, 14, 58
  initial, 136
  models, 104–105, 107, 136, 139
  system input, 47–49
Estimate, 96–97, 134
  efficient, 11
Estimation error, 68, 115, 303, 314
Estimator, 96
  maximum a posteriori (MAP), 74–75
  maximum likelihood, 73–75, 84
  minimum error variance, 69
Euclidean norm, 7, 158
Euler–Lagrange equation, 221, 224–226
Expected value, 11
Exponential distribution, 17, 20
Exponentially correlated noise, 36, 42
Exponentially time-correlated process,
  317–318
Extended Kalman filter, 190–195
Extrapolation, 115

Extrema, 222–223, 285
  local, 222
  sufficient conditions, 222, 267
  unconstrained, 233
Extremal, field of, 285

Fault detection and isolation (FDI), 355
Feedback, 166–168, 241
  control, 241
  matrix, 241, 243
Feedforward matrix, 166
Filter, 92–93, 140
  design, 305
  divergence, 169–171
  gain, 94
  model, 137, 139, 169, 302
  tuning, 117, 170
First-order Markov process, 36, 49–50
Fisher information matrix, 78
Fokker–Planck equation, 4
Forward Euler approximation, 112
Fourier transform pair, 32–33
Free terminal time problem, 266, 270
Fuel-optimal control, 220, 270, 295
Fundamental matrix, *see* State transition
  matrix
Fusion center, 351

Gain matrix, 94
Gamma distribution, 16, 17–18
Gaussian distribution, 8, 12–15
Gaussian noise, 42, 48, 50–51
Gaussian process, 9, 203
Gaussian white noise, 48
Global minimum, 267–268
Global Positioning System (GPS), 2, 319, 351
Gram–Schmidt orthogonalization, 183–185
Gyroscope drift, 209
Gyroscope error model, 209

Hamilton–Jacobi–Bellman equation, 284–285
Hamiltonian, 227, 266–267, 295
Hierarchical filtering, 384
Hilbert space, 4, 41, 258
Householder transformation, 130
Hypersurface, 285–286

Identity matrix, 374–375
Imbedding principle, 282
Inertial navigation system, 104–107, 208
Influence coefficients, 230
Information filter, 129–130
Information matrix, 129, 333, 355
Initial conditions, 95, 141–142

Inner product, 158, 382
Innovations, 109, 116, 169–171
Inverse of a matrix, 376–378
Inversion lemma, 387–388
Interactive multiple mode (IMM), 355

Jacobian, 193
Joint probability density, 24–25
Jordan canonical form, 150, 385
Joseph's form, 125

Kalman filter, 92
 continuous-time, 93–100
 discrete-time, 111–125
 duality theorem, 162
 gain, 97, 100, 109, 113
Kalman–Bucy filter, 96, 204
Kernel, 51
Kolmogorov, 3
Kronecker delta, 114, 375

Lagrange multipliers, 221–222, 227
Lagrangian, 221, 226
Laplace transform, 36, 146, 148–149, 151
Least squares curve fitting, 65
Least squares estimation, 3, 63–66
 weighted, 68
Likelihood function, 75
Linear filter, 32
Linear minimum-variance (LMV) estimation,
 69
Linear-quadratic Gaussian (LQG) problem,
 257–262
Linear-quadratic Gaussian loop transfer
 recovery (LQG/LTR) method, 263–264
Linear quadratic regulator (LQR), 232–246
 properties, 232
Linear regression, 63–64, 68
Linear system, 31–32, 94
Linearization, see Extended Kalman filter
Local extremum, 222
Lognormal distribution, 19
Lower triangular matrix, 375

Markov process, 36
Master filter, 348–349, 351–353
MATLAB, 120, 390
Matrix:
 adjoint, 158, 374
 algebra, 380–382
 cofactor, 372–373
 decomposition, 152–153
 determinant of, 372–374
 diagonal, 372, 374

diagonalization, 149, 151, 385
 Hermitian, 378
 idempotent, 377
 identity, 374–375
 inner product, 158, 382
 inverse, 376–378
 inversion lemma, 334, 387–388
 minor, 372
 negative definite, 100–101, 386–387
 negative semidefinite, 101, 386–387
 null, 246, 375
 orthogonal, 184–185
 positive definite, 24, 100–101, 233, 386–387
 positive semidefinite, 24, 100–101, 386–387
 product, 376–377, 380–381
 pseudoinverse, 67, 379
 quadratic form, 386–387
 rank, 160–161, 258, 378–379
 Riccati equation, 97, 103, 105, 306
 sensitivity, 95
 similar, 379
 skew-symmetric, 376
 state transition, 95, 115, 141–156
 symmetric, 378
 trace, 87, 137, 372, 381
 transpose, 376
 triangular, 375
 triangularization, 128
Mayer form, 227
Maximum a posteriori (MAP) estimator, 74,
 84, 357
Maximum likelihood estimate (MLE), 3, 63,
 73–80
Mean, 11–12
 conditional, 17
 square error, 40, 83–84, 356–357
Measurement:
 noise, 113–114, 139
 update, 121, 129, 172–179, 352
Minimum principle, see Pontryagin
 minimum principle
Minimum time, 268, 278, 287
Minimum variance estimate, 69
Moment, 10, 12–13
 central, 12
 joint, 24
Monte Carlo analysis, 305, 314–316
Multiple-mode adaptive estimation (MMAE),
 356
Multivariate normal distribution, 10

Noise, 128
 correlated, see Colored noise
 Gaussian, 48, 50–51

Noise (*Continued*)
  strength, 114–115
  white, 42–45
Nonlinear Kalman filter, 190–192
Norm, 154, 158
Numerical stability, 122, 130

Observability, 159
  complete, 160, 163
Observation model, 198
Observers, 96, 163–164, 167–168, 257, 259, 263–265
Omega, 2, 351
Optimal control, 241, 267, 293, 295
Optimal estimate, 117
Optimal filter, 96
Optimality principle, 276, 282–283

Parallel Kalman filter equations, 350–351
Parameter, 65, 82, 84
  estimation, 63, 82, 84–85
  sensitivity, 323
Peano–Baker formula, 144–145
Penalty cost, 233–234
Perfect measurements, 97
Performance analysis, 305
Performance index, 233, 295
Poisson distribution, 45–46, 58
Pole allocation, 168
Pontryagin minimum principle, 239, 265–268
Positive definite, 100–101
  property of P, 100
Positive semidefinite, 100
Power spectral density, 26–29, 35, 42–44, 49
Power spectrum, 29
Prediction, 115, 119
Probability, 6–8
  conditional, 17, 75, 356
  joint, 10
Probabilistic data association (PDA), 355
Probability fusion equations, 355–357
Probability density function (pdf), 8, 24–25
Probability distribution function (PDF), 6–9
Process noise, 96–97, 136, 139
Propagation, 99, 120, 123, 350
Proportional navigation (PN), 195–196, 234–235, 237
Pseudoinverse, 67

Quadratic forms, 386–387
Quadratic performance index, 233

Random bias, 48
Random constant, 48

Random variable, 6–7, 23–24, 74, 99
Random walk, 48–49, 317
Rank of a matrix, 160–161
Rayleigh distribution, 15–16, 21
Recursive least-squares estimation, 71–73
Recursive maximum-likelihood estimator, 80–82
Reference model, 321, 324
Regression, linear, 63–64, 68
Regulator, 219
  linear, 232
  continuous-time, 232–234
  discrete-time, 253–256
  linear-quadratic Gaussian, 256–262
Residuals, 65, 169
Riccati equation, 97
Robustness, 256–257, 263, 265
Root mean square (rms), 49, 137, 139–140, 302–303
Root sum square (rss), 319
Roundoff errors, 125, 170
Runge–Kutta method, 307–313
Runge–Kutta–Fehlberg routine, 195

Sample mean, 88
Schuler frequency, 109, 218
Schuler loop, 106, 217–218
Schwarz inequality, 22, 382
Second moment, 12, 57
Second-order Markov process, 48, 51, 147
Sensitivity analysis, 304, 321, 324–326
Separation theorem, 97, 167, 220
Series, infinite, 151
Signum, 287
Shaping filter, 42, 201–204
Smoothing, 1, 115
Spatial process, 52–53
Spectral factorization theorem, 202
Square root filter, 126–127
Square root information filter (SFIR), 129–130
Stability, 261–262
  asymptotic, 243, 259
  of filter, 130
  numerical, 122, 130
Standard deviation, 24, 137, 303, 330
State:
  reconstruction, 164
  terminal, 233
State space representation, 93, 205–206, 210
State transition matrix, 95, 115, 141–156
State variable, 93
  vector, 94, 113
Stationarity, 3, 38, 40, 92

wide sense, 3, 25
Stationary function, 223
Stochastic process, 8, 45
Stochastic difference equations, 111, 115
Strength of white noise, 114-115
Suboptimal filter, 320-324, 349
  reduced-order filter, 320, 349
Superposition principle, 32
Sweep method, 292
System error model, 47

TACAN, 2, 72
Taylor series, 145, 192, 194, 283
Temporal process, 52-53
Terminal state, 233-234
Terminal time, 232-233, 236
  free, 242, 266, 288
  unspecified, 233
Time-correlated noise, 201
Time optimal control, 220, 286, 295
Time update, 120, 180-190, 353
Trace of a matrix, 137, 372, 381
Transition matrix. *see* State transition matrix
Transversality condition, 226
Truncation error, 312-313
Truth model, 136, 302, 319
Tuning, filter, 117, 170, 201
Two-point boundary value problem
  (TPBVP), 228

U–D covariance factorization, 171-172

Unbiased estimate, 11, 68, 77-78, 116
Uniform distribution, 13, 18
Unity matrix, 374-375
Unmodeled dynamics, 132
Upper triangular matrix, 375
URAND, 315-316

Variable, random, 6-8
Variance, 12, 22, 135, 137
Variation, second, 222
Variational approach, 224
Vector:
  length, 7
  norm, 154, 158
Vertical deflection of gravity, 52-53

Weighted least squares, 68-69
White Gaussian noise process, 42, 47
White noise, 28-29, 42-45, 48
Wide-sense stationary, 25
Wiener filter, 38-41
  process, 48
Wiener–Hopf equation, 39-41
Wiener–Khintchine equations, 26, 37
Wiener–Kolmogorov procedure, 3,
  335
Wiener–Paley criterion, 33-34
Wordlength, 3, 170

Zero matrix. *see* Null matrix
Zero-mean noise, 42, 48, 96, 190

Printed in the United States
23808LVS00001B/108